THE ELEMENTS

THE ELEMENTS

.

Written and compiled by

JOHN EMSLEY

Reader in Inorganic Chemistry,
Department of Chemistry,
King's College, London

CLARENDON PRESS · OXFORD

Oxford University Press, Walton Street, Oxford OX2 6DP

Oxford New York Toronto
Delhi Bombay Calcutta Madras Karachi
Petaling Jaya Singapore Hong Kong Tokyo
Nairobi Dar es Salaam Cape Town
Melbourne Auckland

and associated companies in
Berlin Ibadan

Oxford is a trade mark of Oxford University Press

Published in the United States
by Oxford University Press, New York

First published 1989
Reprinted (with corrections) 1989

British Library Cataloguing in Publication Data
Emsley, John
The elements.
1. Chemical elements, chemical compounds—
Technical data—For schools
I. Title
540'.212
ISBN 0-19-855238-6
ISBN 0-19-855237-8 (pbk)

Library of Congress Cataloging in Publication Data
Emsley, J. (John)
The elements/written and compiled by John Emsley.
p.   cm.   Includes index.
1. Chemical elements—Handbooks, manuals, etc.   I. Title.
QD466.E48 1988   546—dc19   88-19011
ISBN 0-19-855238-6
ISBN 0-19-855237-8 (pbk)

Typeset by Cotswold Typesetting Ltd, Gloucester
Printed in Great Britain
by Bookcraft (Bath) Ltd
Midsomer Norton, Avon

Preface

Most people are aware that everything we see around us is composed of a limited number of chemical elements. Most scientists occasionally need to find information about individual elements, and it comes as a surprise to them to discover that such information is not always easily obtained. Even chemists may have difficulty in tracking down certain pieces of data. Chemists have at their disposal several excellent collections of numerical facts, but we tend to view our world through the eyes of analytical, organic, inorganic, or physical chemists, and compile our data accordingly.

Curiously, a handbook about the elements themselves may be seen as the province of all, and yet be the responsibility of no one. This may explain why it has remained neglected for so long, but hopefully the book you now hold fills this gap. It should not be seen as just a chemistry book, but as a general book to which all scientists could refer. That there is a widespread interest in the chemical elements was made clear to me a few years ago.

1984 was the 150th anniversary of the birth of Dimitri Mendeleyev, who devised the Periodic Table of the elements in 1869. To commemorate the event I wrote a short article for the popular science magazine *New Scientist*. In this, I also asked for readers' comments on the controversy over how the groups of the periodic table should be numbered. The response was remarkable—letters arrived by every post for the next six months. Most agreed that there was a need to change from the Roman numbers I–VIII (with the sub-divisions A and B) which were then in use, because of the conflicting ways in which these group numbers were used in different parts of the world.

However there was very little agreement on what should replace them. What agreement there was supported the use of Arabic numerals 1–18, and this was the system adopted for the *New Scientist* Periodic Table produced that year. Happily this was also the system favoured by the American Chemical Society and it had the support of the International Union of Pure and Applied Chemists. It was first proposed by the Swedish chemist Arne Ölander in 1958.

While the periodic table is universally recognized as the hallmark of chemistry, it generally conveys only the barest minimum of information about each element. The *New Scientist* version, slightly modified and reproduced on the inside front cover, contains only the chemical symbol of each element, its atomic number, and relative atomic mass. Other versions have been produced that crowded each box with up to ten or more pieces of data such as melting and boiling points, densities, crystal structures, electron arrangement, etc. Some companies produce special versions for physicists, crystallographers, spectroscopists, and even bird fanciers!

Many years ago I began my own collection of data about the elements. To start with I simply bought a large note book with over 100 pages and wrote the name of a different element at the top of each page. Over the years my collection of numerical information about each element grew until the book finally burst its binding as I stuck in more and more leaves. It is an edited version of this book that you now have in your hands. Not all the information about every element is here, nor could it be, since I have deliberately kept within the format of a double page for each element, although I have tried to include as much data as possible within this framework.

For each element I have grouped together properties under the headings chemical, physical, nuclear, and electronic, as well as giving the history and derivation of the element's name. I have also included something on its environmental and biological importance. All the data are given in SI units in the main tables, but there are full details of how to convert this information into other, commonly used, units in the Key chapter.

The final part of the book consists of a series of double tables of properties, arranged first in order of the elements and then in order of the property itself. Again the choice is partly my own but these tables were mainly determined by the need to include as many elements as possible for each property, and there are only a certain number of properties that are common to all, or most, of the elements.

Finally I make no apologies for including tables of such 'obvious' data as the elements in alphabetical order with their chemical symbols, and chemical symbols in alphabetical order with their full names (see p. 251 ff.). The need for these came home to me as I was preparing the *New Scientist* version of the periodic table. Opinion among other scientists is that chemists assume too much, and while we know without thinking that W is tungsten and Sn is tin, many of our fellow scientists need these symbols confirmed when they encounter them.

Hopefully this book will find its way on to the shelves of all scientists, and not just chemists. In particular I have in mind those scientists working in the media who have an uphill struggle to make chemistry interesting. With *The Elements* as a source book they have at their fingertips background information about an element with which they can back up their stories.

London   J.E.
December 1987

Contents

The key to *The Elements*

THE curious thing about numerical information is the way it varies slightly from book to book. Usually these variations are of the order of ±2 per cent, small but irritating. The need to standardize data has been recognized by several organizations, such as the International Union of Pure and Applied Chemists (IUPAC) and the National Bureau of Standards (NBS) in Washington DC. When such bodies set up committees to assess data and decide on the most reliable values, the job of an author like myself becomes much easier. For example, the table of thermodynamic data of the elements comes from an NBS book while the standard reduction potentials come from an IUPAC publication.

Rather than quote alternative values or a range of values for certain properties I have assumed that well established specialists in the collecting of data, such as those of the *CRC handbook of chemistry and physics* (reference 1) and *Lange's handbook of chemistry* (reference 2) provide reliable information, even though it may not be in SI units. Other compilations with extensive chemical data are references 3–6. In the sections which follow I give the sources of my information, explain the SI units used, and convert the data to other commonly used units.

Note: n.a. used in the tables means not available.

The elements: discovery, names, formulae, and relative atomic masses

IUPAC is the official body responsible for approving the names and formulae given to elements, and authenticating their relative atomic masses. The US names of aluminium (aluminum) and caesium (cesium) are so near to the recommended names that they present no problem, except that the latter name leads to a slightly different position in a list of elements arranged alphabetically. Changes are reported in *Pure and Applied Chemistry*, IUPAC's official journal. The relative atomic masses given here are taken from *Pure and Applied Chemistry*, 1986, **58**, 1677.

The most comprehensive volume on the discovery and history of the elements is M. E. Weeks and H. M. Leicester's, *Discovery of the elements*, published by the *Journal of Chemical Education*, Easton (PA), 1968. An outline of the discovery of each element is also given in reference 1. Individual articles in the *Journal of Chemical Education* from time to time give excellent histories of individual elements.

D. W. Ball, in the *Journal of Chemical Education*, 1985, **62**, 787, and J. G. Stark and H. G. Wallace, in *Education in Chemistry*, 1970, 152, explain how the names of the elements were chosen.

Chemical properties

Descriptions

The brief descriptions of the elements, their names, and their reactivity towards air, water, acids, and alkalis have been taken from references 7–9. The *Encyclopedia of the chemical elements*, edited by C. A. Hampel (Reinhold Book Corporation, New York, 1968) gives a comprehensive, if somewhat dated account, of the uses of each element.

1

Radii

To convert radii, which are given in picometres (pm), to metres divide by 10^{12}. To convert to nanometres (nm) divide by 1000; and to convert to angstroms (Å) divide by 100.

The radius of an atom depends upon several factors: oxidation state, degree of ionization, and coordination number (for metals this is generally 12). When it is part of a molecule two radii are defined: the covalent radius, which refers to the role it plays in forming bonds, and the van der Waals radius, which refers to the radius it presents to the world beyond the molecule. Many textbooks quote some of these radii, but the best sources appear to be references 2 and 4.

Electronegativity

This quantity is well understood but ill-defined. It refers to the potential of an atom to attract electrons to itself. The higher the electronegativity the stronger is this ability. Fluorine is the most electronegative of all elements. The units of electronegativity are rarely quoted.

Effective nuclear charge, Z_e

Like electronegativity, this quantity is easier to understand than calculate. Although Z_e can be defined for any electron within an atom, only the Z_e for the valence shell electrons is of interest to the chemist. Z_e is the charge due to the protons of the nucleus less a screening factor due to the other electrons of the atom. There are several slightly different ways of calculating this screening, and consequently there are several values for Z_e. Those quoted were computed by J. C. Slater (*Physical Reviews*, 1930, **36**), E. Clementi and D. L. Raimondi, (*Journal of Chemical Physics*, 1963, **38**, 2686), E. Clementi, D. L. Raimondi, and W. P. Reinhardt (*Journal of Chemical Physics*, 1967, **47**, 1300), and C. Froese-Fischer (*Atomic Data*, 1972, **4**, 301, and *Atomic Data and Nuclear Data Tables*, 1973, **12**, 87).

Standard reduction potentials

In these diagrams the element is arranged with the highest oxidation state on the left. Potentials are given in volts. The higher the value of $E^\ominus$, the stronger is the oxidant as an oxidizing agent; the lower $E^\ominus$, the stronger is the reductant as a reducing agent. These diagrams are taken from *Standard potentials in aqueous solutions*, edited by A. J. Bard, R. Parsons, and J. Jordan. (Marcel Dekker (for IUPAC), New York, 1985). Other compilations of $E^\ominus$ data are to be found in references 1–6.

Oxidation states

Although it is merely a formalism, the concept of oxidation number is much used in describing the changes that happen to an element in its chemical reactions. Consequently it is a useful way of classifying and explaining the compounds of that element. From the multitude of compounds known for elements in their various oxidation states I have chosen where possible to give the oxides, hydroxides or acids, hydrides, fluorides, and chlorides (after which 'etc.' means the corresponding bromides and iodides), and the species present in aqueous solutions of simple salts of the element (denoted 'aq'). Salts, complexes, and organometallic compounds are also given if these are special; otherwise I have merely indicated that such substances exist, and references 7–9 should be consulted for further details.

Covalent bonds

To convert bond lengths, r, given in picometres, to metres, divide by 10^{12}; to convert them to nanometres (nm) divide by 1000. To convert them to angstroms (Å) divide by 100. To convert bond energies, i.e. bond enthalpies, E, given in kJ mol^{-1} to kcal mol^{-1}, divide by 4.184.

The values are taken from various sources, notably bond lengths from references 1 and 4, and bond enthalpies from references 1, 3, and 4, and from *Bond energies, ionization potentials and electron affinities*, by V. I. Vedeneyev, L. V. Gurvich, V. N. Kondrat'yev, V. A. Mededev, and Ye. L. Frankevich (Edward Arnold, London, 1966). *SI chemical data* by G. H. Aylward and T. J. V. Findley (Wiley, Sydney, 1971) also gives r and E for many bonds.

Physical properties

Melting points and boiling points

These are given in Kelvin (K). They can be converted to degrees Celsius (C) by subtracting 273.15. The values quoted are based on reference 1, but are also given with slight variations in all major data books (references 2–9).

Enthalpy of fusion, ΔH_{fusion}

These, given in kJ mol^{-1}, can be converted to kcal mol^{-1} by dividing by 4.184.

The values are taken mainly from R. Loebel's compilation in reference 1 (where they are given in c.g.s. units), supplemented by references 4 and 6, where they are in SI units.

Thermodynamic properties

To convert kJ mol^{-1} to kcal mol^{-1} divide by 4.184. To convert entropies in J K^{-1} mol^{-1} to eu (entropy units, i.e. cal K^{-1} mol^{-1}) divide by 4.184. To convert specific heats, C_p, to cal g^{-1} K^{-1} divide first by 4.184 and then by the relative atomic mass of the element concerned.

The thermodynamic properties are taken from *The NBS tables of chemical thermodynamic properties*, by D. D. Wagman, W. H. Evans, and V. B. Parker, which was published jointly by the American Chemical Society and the American Institute of Physics for the National Bureau of Standards, Washington DC, in 1982. Although thermodynamic data for the elements are to be found in references 1–5, the NBS compilation is preferred and is in SI units.

Density

Since the basic SI unit of weight is the kilogramme (kg) and of length the metre (m), the preferred unit of density is kg m^{-3}. The more common unit however is g cm^{-3} and to convert from kg m^{-3} to g cm^{-3} divide by 1000.

Many sources list densities at only one temperature, such as reference 2 (293 K), others at various temperatures, e.g. reference 6, which reports them in SI units. The densities used here are based mainly on reference 1 for the solid elements. In the same work can be found the densities of the liquid elements at their melting points, compiled by G. Lang.

Thermal conductivity

The SI unit for this property is watts per metre per Kelvin (W m^{-1} K^{-1}). To convert to W cm^{-1} K^{-1} divide by 100.

The values are taken from C. Y. Ho, R. W. Powell, and P. E. Liley (*Journal of Physical Chemistry Reference Data*, 1974, **3**, suppl. 1), with the values for carbon being taken from reference 1 for the directions perpendicular and parallel to the graphite axis. Reference 2 also reports the thermal conductivity of the elements.

Electrical resistivity

The SI units are ohm metres (Ω m) and the electrical resistivity of metals are of the order of 10^{-8} Ω m. To convert to the more common units of $\mu\Omega$ cm multiply the Ω m values by 10^8.

The data are taken from reference 1. Reference 6, pp. 102–3, also gives the electrical resistivity in Ω m at temperatures of 78, 273, 373, 573, and 1473 K. Values in non-SI units are given in reference 3, pp. 580–684.

Mass magnetic susceptibility, χ

This is obtained from the volume magnetic susceptibility, κ, which is unitless, by dividing by the density (kg m^{-3} in SI units). To convert mass magnetic susceptibilities from the SI units of kg^{-1} m^3 to c.g.s. units of g^{-1} cm^3, multiply by $1000/4\pi$, i.e. 79.6. To convert to molar magnetic susceptibility multiply first by 79.6 then by the relative atomic mass of the element. The use of SI units for magnetic properties is discussed by T. I. Quickenden and R. C. Marshall, in the *Journal of Chemical Education*, 1972, **49**, 114.

The magnetic susceptibility data are taken from *Constantes sélectionnées. Diamagnétisme et paramagnétisme*, by G. Foëx (Masson et Cie, Paris, 1957), and *Modern magnetism* (4th edition), by L. F. Bates (Cambridge University Press, Cambridge, 1963). Values for certain of the lanthanides were taken from J. M. Lock (*Proceedings of the Physical Society*, 1957, **B70**, 476 and 566). Reference 1 reports molar magnetic susceptibilities in c.g.s. units.

Coefficient of linear thermal expansion, α

This is the same in SI and c.g.s. units, i.e. K^{-1}. It is often reported as $10^6\alpha$. The data are taken from reference 3 (pp. 580–684).

Molar volume (or atomic volume)

This represents a slight problem for SI whose unit of volume is m^3. Traditionally atomic volume is obtained by dividing the relative atomic mass of an element in g by the density in g cm^{-3}, and is consequently traditionally expressed in cm^3. To express the molar volume in m^3 divide the values given by 10^6. Expressing it in this way, however, implies a density measured in g m^{-3}, which is not a recognized way of expressing this quantity.

Since atomic volume for an element depends upon the density, it also depends upon the phase, the allotrope, and the temperature. The values reported here are based where possible on the solid state at room temperature (298 K or as indicated), and are taken mainly from C. N. Singman (*Journal of Chemical Education*, 1984, **61**, 137).

Lattice structure

To convert the cell parameters to angstroms (Å) divide by 100. To convert them to nm divide by 1000.

The abbreviation b.c.c. means body-centred cubic, f.c.c. means face-centred cubic, and h.c.p. means hexagonal close packed. The data are taken from *Landolt–Bornstein*, New Series, Group III, vol. 6, edited by K. H. Hellwege and A. M. Hellwege. Crystal structure data are also to be found in references 3, 5, and 9; and *The Structures of the Elements* by J. Donohue (John Wiley & Sons, 1974).

X-ray diffraction mass absorption coefficients

These are taken from *International tables for X-ray crystallography*, Vol. III,

Physical and Chemical Tables, section 3.2, published for the International Union of Crystallographers by The Kynock Press (Birmingham UK, 1962).

Nuclear properties

Thermal neutron capture cross-section

The barn is defined as 10^{-24} cm^2, which is 10^{-28} m^2. The values are taken from reference 1.

Key isotopes

This lists all stable isotopes and the longest lived radioactive nuclides, plus those used in research. The per cent natural abundance is given as 'trace' for certain short-lived nuclides that are part of a natural decay series. The half-life, $T_{1/2}$, is expressed in seconds (s), minutes (m), hours (h), days (d), or years (y). The decay mode is shown as β^- for negative emission, β^+ for positron emission, α for alpha decay, EC for electron capture, IT for isomeric transition, and SF for spontaneous fission. Some nuclei decay by two routes. The energy (in mega electronvolts, MeV) of the radiation is given in parentheses. γ indicates the emission of gamma radiation.

The nuclide data is taken from reference 1. Similar tables are given in references 2 and 3. The most comprehensive listing is to be found in the *Table of isotopes* (7th edition) by C. M. Lederer and V. S. Shirley (John Wiley & Sons, New York, 1978). This work lists the full data on 2600 known isotopes, and its compilation was supported by the US NBS Office of Standard Reference Data.

Nuclear spin (I) is reported in units of $h/2\pi$. The nuclear magnetic moment (μ) is reported in nuclear magnetrons with diamagnetic correction. Both I and μ are taken from *Nuclear spins and moments* by G. H. Fuller (*Journal of Physical Chemistry Reference Data*, 1976, **5**, 835).

Use is made of isotopes in various areas of research, denoted by 'NMR', 'tracer', or 'med' short for medical. Nuclear magnetic resonance spectroscopy (NMR) details are given directly below the Key Isotopes table. The radioactive isotopes used as tracers and in medicine are taken from the *Merck index* (10th edition), edited by M. Windholz (Merck & Co., Inc., Rahway (NY), 1983); *Isotopes, products and services catalog* (Oak Ridge National Laboratory, P.O. Box X, Oak Ridge, Tennessee 37831, USA); and *Biochemicals* (The Radiochemical Centre, P.O. Box 16, Amersham, Bucks, HP7 9LL, UK). *Radionuclide tracers* by M. F. L'Annunziata (Academic Press, New York, London, 1987), also contains appendices of available isotopes and their radiation characteristics.

Nuclear magnetic resonance

Where data for two nuclei are given, the one enclosed in square brackets is less frequently used for NMR studies. Nuclear spins are given in the Key Isotopes table.

Relative sensitivity is at constant field for equal numbers of nuclei. Absolute sensitivity is the relative sensitivity multiplied by the natural abundance (given in the Key Isotopes table as per cent, but here used as a fraction of unity). Also known as receptivity, it is commonly quoted relative to ^{13}C = 1.00. The ratio of ^{1}H to ^{13}C receptivities is 1 to 5680.

The magnetogyric ratio is given in rad T^{-1} s^{-1}, but is often quoted as γ values which are in units of rad T^{-1} s^{-1} multiplied by 10^7. Thus for ^{1}H $\gamma = 26.75$, as opposed to 26.75×10^{-7} rad T^{-1} s^{-1}. γ is a constant of proportionality between frequency and field strength, and so the units are frequency (rad s^{-1})/field (T), i.e. rad T^{-1} s^{-1}.

Quadrupole moments are given in units of m^2. They can be converted to cm^2 by multiplying by 10^4, or to barns by multiplying by 10^{28}.

Frequency is quoted relative to the ^1H signal of $Si(CH_3)_4$, which is exactly 100 MHz in a magnetic field of 2.3488 T. For NMR spectrometers with ^1H at 60, 90, 200, 250, 360, or 400 MHz the frequency and field vary in direct proportion; in other words multiply the frequencies given in the tables by 0.6, 0.9, 2, 2.5, 3.6, or 4 respectively.

The data for the NMR tables were compiled from the *Handbook of high resolution multinuclear NMR*, by C. Brevard and P. Granger (John Wiley & Sons, New York, 1981), *NMR and the periodic table*, by R. K. Harris and B. E. Mann (Academic Press, London, 1978), *Multinuclear NMR* by J. Mason (Plenum Press, New York and London, 1987) and from Bruker Scientific Instruments publications.

Electron shell properties

Ground state electron configuration and term symbol

These are given in most inorganic textbooks, such as references 7 and 8. The data given here were taken from the *Handbook of atomic data*, by S. Fraga, J. Karwowski, and K. M. S. Saxena (Elsevier, Amsterdam, 1976).

Electron affinity

To convert the values given in kJ mol^{-1} to electron volts (eV) divide by 96.486; to convert to MJ mol^{-1} divide by 1000.

Electron affinity is conventionally reported as positive if the addition of an electron to an atom releases energy, as it almost invariably does for the step $M \rightarrow M^-$, and negative if the process is energy absorbing, as it is for the addition of a second electron, i.e. $M^- \rightarrow M^{2-}$. This energy convention is the opposite of that used in reporting ionization energies and most other energy changes.

Although many chemistry textbooks report some electron affinities, and the major compilations, references 1 and 2, give extensive lists (in eV), the data reported here come from H. Hotop and W. C. Lineberger (*Journal of Physical Chemistry Reference Data*, 1975, **4**, 539), also reported in eV. An earlier compilation by R. J. Zollweg (*Journal of Chemical Physics*, 1969, **50**) also gives some values for the heavier elements. Their variation from element to element is discussed by E. C. M. Chen and W. E. Wentworth (*Journal of Chemical Education*, 1975, **52**, 486).

Ionization energies (ionization potentials)

To convert from kJ mol^{-1} to electron volts (eV) divide the values given by 96.486; to convert to MJ mol^{-1} divide by 1000.

Ionization energies are known to a high degree of accuracy for removal of the first, second, third, fourth, and fifth electrons for most elements, and for subsequent electron removal from the lighter elements. Values given in parentheses are less reliable. For some elements ionization energies beyond the tenth are available.

Many textbooks and compilations list ionization energies, some in SI units such as reference 4. The NBS has at different times published ionization data, beginning with E. C. Moore's *Atomic energy levels* Volume III (NBS Circular 467, 1958). Reference 1 quotes *Analyses of optical spectra*, NSRDS-NBS 34 (Office of Standard Reference Data, NBS, Washington DC, 1970). The values for the lanthanides and actinides were taken from W. C. Martin, L. Hagan, J. Reader, and J. Sugar (*Journal of Physical Chemistry Reference Data*, 1974, **3**, 771).

Principal lines in atomic spectrum

The stronger lines in the spectrum are listed with the strongest shown in bold. Lines arising from the neutral atom are indicated by Species I, and those arising from the singly charged ion M^+ are Species II. The most sensitive line in the neutral atom spectrum is denoted by U1, the next most sensitive by U2, etc. The corresponding order of sensitivity for the M^+ ion spectrum is shown as V1, V2, etc. Application in atomic absorption spectrometry is indicated by AA and in atomic emission spectrometry by AE.

For some transuranium elements with many intense lines only the first seven lines identified as arising from the neutral atom are given.

Data are derived from *Tables of spectral line intensities* (2nd edition) by W. F. Meggers, C. H. Corliss, and B. F. Scribner (NBS Monograph 145, Part I, 1975); *Line spectra of the elements* by J. Reader and C. H. Corliss in reference 1; *Atomic energy levels and Grotrian diagrams* Vols I, II, and III, by S. Bashkin and J. O. Stoner (Elsevier, Amsterdam, 1976); *CRC handbook of spectroscopy*, Vol. 1, by J. W. Robinson (CRC Press, Boca Raton, FL, 1974); *Pye Unicam atomic absorption data book* (3rd edition) by P. J. Whiteside (Pye Unicam, Cambridge, UK, 1979); and the *Handbook of the physicochemical properties of the elements*, edited by G. V. Samsonov, pp. 30–67 (IFI-Plenum, New York, 1968).

Environmental and biological properties

Abundances

For atmosphere, Earth's crust, and seawater the units are parts per million (p.p.m.), defined in the case of the atmosphere as cubic centimetres per cubic metre, in the case of the Earth's crust as grammes per metric tonne (1000 kg), which is the same as milligrammes per kg, and for seawater as milligrammes per litre.

The atmospheric abundances are taken from reference 6; the crustal abundances from *Geochemistry* by W. S. Fyfe (Oxford University Press, Oxford, 1974), as reproduced in reference 7; the concentrations in seawater come from references 3 and 6 and the *McGraw-Hill encyclopedia of science and technology*, p. 627 (McGraw-Hill, New York, 1970).

Biological role

Information about essential elements is from the *Handbook of vitamins, minerals and hormones* (2nd edition) by R. J. Kutsky (Van Nostrand Reinhold, New York, 1981). Toxicological information about the elements depends on individual compounds, as well as the elements themselves. Useful sources of information are *Metal toxicology in mammals* (2 volumes) by T. D. Luckey and B. Venugopal (Plenum Press, New York, 1977), and the *Handbook of toxicity of inorganic compounds*, edited by H. G. Seiler and H. Sigel (Marcel Dekker, New York, 1988), which covers a majority of elements in alphabetical order.

General references

1. Weast, R. C. (ed.) *CRC handbook of chemistry and physics*, (63rd edn), CRC Press, Boca Raton, Fl 1982.
2. Dean, J. A. (ed.) *Lange's handbook of chemistry*, (13th edn). McGraw-Hill, New York, 1985.
3. Moses, A. J. *The practising scientist's handbook*. Van Nostrand Reinhold, New York, 1978.
4. Ball, M. C. and Norbury, A. H. *Physical data for inorganic chemists*. Longman, London, 1974.

5. Samsonov, G. V. (ed.), *Handbook of the physicochemical properties of the elements*. IFI-Plenum, New York, 1968.
6. Kaye, G. W. C. and Laby, T. H. *Tables of physical and chemical constants* (14th edn), Longman, London, 1973.
7. Greenwood, N. N. and Earnshaw, A. *Chemistry of the elements*. Pergamon Press, Oxford, 1984.
8. Cotton, F. A. and Wilkinson, G. *Advanced inorganic chemistry* (4th edn), John Wiley & Sons, New York, 1980.
9. Bailar, J. C., Emeleus, H. J., Nyholm, R., and Trotman-Dickenson, A. F. (eds). *Comprehensive inorganic chemistry*, (5 vols) Pergamon Press, Oxford, 1973.

THE
ELEMENTS

Ac

Atomic number: 89

Relative atomic mass ($^{12}C = 12.0000$): **(227)**

Chemical properties

Occurs as part of ^{235}U decay series but made by neutron bombardment of ^{226}Ra. Soft, silvery-white metal which glows in the dark. Reacts with water to evolve H_2.

Radii/pm: Ac^{3+} 118; atomic 187.8

Electronegativity: 1.1 (Pauling); 1.00 (Allred-Rochow)

Effective nuclear charge: 1.80 (Slater)

Standard reduction potentials $E^{\ominus}/V$

	III		0
Acidic solution	Ac^{3+}	$\xrightarrow{-2.13}$	Ac
Basic solution	$Ac(OH)_3$	$\xrightarrow{-2.6}$	Ac

Oxidation states

Ac^0 ($d^1 s^2$)
Ac^{III} ([Rn]) Ac_2O_3, $Ac(OH)_3$ insoluble
AcH_2 and AcH_3 are probably Ac^{III} compounds

Physical properties

Melting point/K: 1320 ± 50

Boiling point/K: 3470 ± 300

ΔH_{fusion}/kJ mol^{-1}: 14.2

ΔH_{vap}/kJ mol^{-1}: 293

Thermodynamic properties (298.15 K, 0.1 MPa)

State	$\Delta_f H^{\ominus}$/kJ mol^{-1}	$\Delta_f G^{\ominus}$/kJ mol^{-1}	$S^{\ominus}$/J K^{-1} mol^{-1}	C_p/J K^{-1} mol^{-1}
Solid	0	0	56.5	27.2
Gas	406	366	188.1	20.84

Density/kg m^{-3}: 10 060 [293 K]

Thermal conductivity/W m^{-1} K^{-1}: 12 [300 K]

Electrical resistivity/Ω m: n.a.

Mass magnetic susceptibility/kg^{-1} m^3: n.a.

Molar volume/cm^3: 22.6

Coefficient of linear thermal expansion/K^{-1}: 14.9×10^{-6}

Lattice structure (cell dimensions/pm), space group

f.c.c. ($a = 531.1$), Fm3m

X-ray diffraction: mass absorption coefficients (μ/ρ)/cm^2 g^{-1}:
CuK_{α} n.a. MoK_{α} n.a.

Actinium

Atomic number: 89
Thermal neutron capture cross-section/barns: 810 ± 20 (^{227}Ac)
Number of isotopes (including nuclear isomers): 11
Isotope mass range: $221 \rightarrow 231$

Key isotopes

Nuclide	Atomic mass	Natural abundance (%)	Half life $T_{1/2}$	Decay mode and energy (MeV)	Nuclear spin I	Nuclear magnetic moment μ	Uses
^{225}Ac		0	10.0d	α(5.829)			tracer
^{227}Ac	227.0278	trace	21.6y	β(0.043 99%; α; γ	3/2	+1.1	NMR
^{228}Ac		trace	6.13h	β^-(2.14); γ			

NMR

^{227}Ac

Relative sensitivity (^{1}H = 1.00)   —
Absolute sensitivity (^{1}H = 1.00)   —
Receptivity (^{13}C = 1.00)   —
Magnetogyric ratio/rad T^{-1} s^{-1} 3.5×10^7
Quadrupole moment/m^2 1.7×10^{-28}
Frequency (^{1}H = 100 MHz; 2.3488 T)/MHz   13.1

Ground state electron configuration: [Rn]$6d^1 7s^2$
Term symbol: $^2D_{3/2}$
Electron affinity (M$\rightarrow$M$^-$)/kJ mol^{-1}: n.a.

Ionization energies/kJ mol^{-1}

1. M $\rightarrow$ M$^+$	499		6. M$^{5+} \rightarrow$ M^{6+}	(7 300)
2. M$^+ \rightarrow$ M^{2+}	1170		7. M$^{6+} \rightarrow$ M^{7+}	(9 200)
3. M$^{2+} \rightarrow$ M^{3+}	1900		8. M$^{7+} \rightarrow$ M^{8+}	(10 500)
4. M$^{3+} \rightarrow$ M^{4+}	(4700)		9. M$^{8+} \rightarrow$ M^{9+}	(11 900)
5. M$^{4+} \rightarrow$ M^{5+}	(6000)		10. M$^{9+} \rightarrow$ M^{10+}	(15 800)

Principal lines in atomic spectrum

Wavelength/nm	Species	Sensitivity	Application
386.312	II		
408.844	II		
416.840	II		
417.998	I		
729.040*	I		

* Lowest energy transition from ground state to nearest empty orbital

Abundance: Only traces in uranium ores
Biological role: None; toxic due to radioactivity

AI

Atomic number: 13
Relative atomic mass ($^{12}C = 12.0000$): 26.981 54

Chemical properties

Hard, strong, silvery-white metal, protected by oxide film from reacting with air and water. Soluble in hot concentrated HCl and NaOH solution. Hundreds of uses as metal and alloys.

Radii/pm: Al^{3+} 57; covalent 125; atomic 143.1: van der Waals 205

Electronegativity: 1.61 (Pauling); 1.47 (Allred-Rochow)

Effective nuclear charge: 3.50 (Slater); 4.07 (Clementi); 3.64 (Froese–Fischer)

Standard reduction potentials $E^{\ominus}$/V

	III		0
Acid solution	Al^{3+}	-1.676	Al
	AlF_6^{3-}	-2.067	Al
Alkaline solution	$Al(OH)_3$	-2.300	Al
	$Al(OH)_4^-$	-2.310	Al

Covalent bonds

Covalent bonds	r/pm	E/kJ mol^{-1}
Al–H	c. 170	285
Al–C	224	225
Al–O	162	585
Al–F	163	665
Al–Cl	206	498
Al–Al	286	c. 200

Oxidation states

Al^0 (s^2p^1)
Al^I (s^2) AlCl in gas phase
Al^{III} ([Ne]) Al_2O_3 (amphoteric)
  $AlO(OH)$, $Al(OH)_3$
  $Al(H_2O)_6^{3+}$ (aq)
  Al^{3+} salts, AlH_3
  $LiAlH_4$, AlF_3,
  Na_3AlF_6, Al_2Cl_6

Physical properties

Melting point/K: 933.52
Boiling point/K: 2740
ΔH_{fusion}/kJ mol^{-1}: 10.67
ΔH_{vap}/kJ mol^{-1}: 290.8

Thermodynamic properties (298.15 K, 0.1 MPa)

State	$\Delta_f H^{\ominus}$/kJ mol^{-1}	$\Delta_f G^{\ominus}$/kJ mol^{-1}	$S^{\ominus}$/J K^{-1} mol^{-1}	C_p/J K^{-1} mol^{-1}
Solid	0	0	28.33	24.35
Gas	326.4	285.7	165.54	21.38

Density/kg m^{-3}: 2698 [293 K]; 2390 [liquid at m.p.]
Thermal conductivity/W m^{-1} K^{-1}: 237 [300 K]
Electrical resistivity/Ω m: 2.6548×10^{-8} [293 K]
Mass magnetic susceptibility/kg^{-1} m^3: $+7.7 \times 10^{-9}$ (s)
Molar volume/cm^3: 10.00
Coefficient of linear thermal expansion/K^{-1}: 23.03×10^{-6}

Lattice structure (cell dimensions/pm), space group

f.c.c. ($a = 404.959$), Fm3m

X-ray diffraction: mass absorption coefficients (μ/ρ)/cm^2 g^{-1}:
CuK_α 48.6 MoK_α 5.16

Aluminium (Aluminum)

Atomic number: 13
Thermal neutron capture cross-section/barns: 0.232
Number of isotopes (including nuclear isomers): 8
Isotope mass range: 24→30

Key isotopes

Nuclide	Atomic mass	Natural abundance (%)	Half life $T_{1/2}$	Decay mode and energy (MeV)	Nuclear spin I	Nuclear magnetic moment μ	Uses
^{26}Al		0	7.4×10^5y	β^+(4.003); EC; γ	5		tracer
^{27}Al	26.981 54	100	stable		5/2	+3.6413	NMR

NMR

	^{27}Al
Relative sensitivity (^{1}H = 1.00)	0.21
Absolute sensitivity (^{1}H = 1.00)	0.21
Receptivity (^{13}C = 1.00)	1.17×10^3
Magnetogyric ratio/rad T^{-1} s^{-1}	6.9704×10^7
Quadrupole moment/m^2	0.4193×10^{-28}
Frequency (^{1}H = 100 MHz; 2.3488 T)/MHz	26.057

Reference: $Al(H_2O)_6^{3+}$

Ground state electron configuration: [Ne] $3s^2 3p^1$
Term symbol: $^2P_{1/2}$
Electron affinity (M→M$^-$)/kJ mol^{-1}: 44

Ionization energies/kJ mol^{-1}

1. M →M$^+$	577.4		6. M^{5+}→M^{6+}	18 376
2. M$^+$→M^{2+}	1816.6		7. M^{6+}→M^{7+}	23 293
3. M^{2+}→M^{3+}	2744.6		8. M^{7+}→M^{8+}	27 457
4. M^{3+}→M^{4+}	11 575		9. M^{8+}→M^{9+}	31 857
5. M^{4+}→M^{5+}	14 839		10. M^{9+}→M^{10+}	38 459

Principal lines in atomic spectrum

Wavelength/nm	Species	Sensitivity	Application
308.215	I	U4	AA AE
309.271	I	U3	AA AE
309.284	I		
349.401*	I	U2	AA AE
396.152	I	U1	AA AE

* Lowest energy transition from ground state to nearest empty orbital

Abundance: Earth's crust 83 000 p.p.m.; seawater 0.01 p.p.m.
Biological role: None; accumulates in body from daily intake (*c.* 20 mg per day); implicated in Alzheimer's disease (senile dementia)

Atomic number:	**95**
Relative atomic mass ($^{12}C = 12.0000$):	**(243)**

Chemical properties

Radioactive, silvery metal produced in 100 g quantities as ^{243}Am from neutron bombardment of ^{239}Pu. Attacked by air, steam, and acids, but not alkalis.

Radii/pm: Am^{6+} 80; Am^{5+} 86; Am^{4+} 92; Am^{3+} 107; atomic 184
Electronegativity: 1.3 (Pauling); n.a. (Allred-Rochow)
Effective nuclear charge: 4.65 (Slater)

Standard reduction potentials $E^{\ominus}/V$

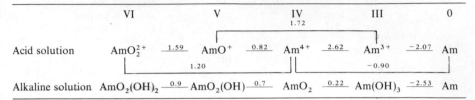

	VI		V		IV		III		0

Acid solution
AmO_2^{2+} —1.59— AmO^+ —0.82— Am^{4+} —2.62— Am^{3+} —−2.07— Am
(IV to III: 1.72)
(VI to IV: 1.20)
(IV to III overall: −0.90)

Alkaline solution $AmO_2(OH)_2$ —0.9— $AmO_2(OH)$ —0.7— AmO_2 —0.22— $Am(OH)_3$ —−2.53— Am

Oxidation states

Am^{II} (f^7)	AmO, $AmCl_2$, etc.	
Am^{III} (f^6)	Am_2O_3, AmF_3, $AmCl_3$, etc. $[AmCl_6]^{3-}$, Am^{3+}(aq)	
Am^{IV} (f^5)	AmO_2, AmF_4	
Am^{V} (f^4)	AmO_2^+ (aq)	} unstable due to reduction by radioactive
Am^{VI} (f^3)	AmO_2^{2+} (aq)	} decay products

Physical properties

Melting point/K: 1267
Boiling point/K: 2880
ΔH_{fusion}/kJ mol^{-1}: 14.4
ΔH_{vap}/kJ mol^{-1}: 238.5

Thermodynamic properties (298.15 K, 0.1 MPa)

State	$\Delta_f H^{\ominus}$/kJ mol^{-1}	$\Delta_f G^{\ominus}$/kJ mol^{-1}	$S^{\ominus}$/J K^{-1} mol^{-1}	C_p/J K^{-1} mol^{-1}
Solid	0	0	n.a.	n.a.
Gas	n.a.	n.a.	n.a.	n.a.

Density/kg m^{-3}: 13 670 [293 K]
Thermal conductivity/W m^{-1} K^{-1}: 10 est. [300 K]
Electrical resistivity/Ω m: 68×10^{-8}
Mass magnetic susceptibility/kg^{-1} m^3: $+5 \times 10^{-8}$ (s)
Molar volume/cm^3: 17.78
Coefficient of linear thermal expansion/K^{-1}: n.a.

Lattice structure (cell dimensions/pm), space group

α-Am h.c.p. ($a = 346.80$; $c = 1124.0$), P6$_3$/mmc
β-Am f.c.c. ($a = 489.4$), Fm3m
$T(\alpha \rightarrow \beta) = 1347$ K

X-ray diffraction: mass absorption coefficients (μ/ρ)/cm^2 g^{-1}:
n.a.

Discovered in 1944 by G. T. Seaborg, R. A. James, L. O. Morgan and A. Ghiorso at Chicago, USA
[English, *America*]

Americium

Nuclear properties

Atomic number: 95

Thermal neutron capture cross-section/barns: 180 ± 20 (^{243}Am)

Number of isotopes (including nuclear isomers): 13

Isotope mass range: $237 \rightarrow 247$

Key isotopes

Nuclide	Atomic mass	Natural abundance (%)	Half life $T_{1/2}$	Decay mode and energy (MeV)	Nuclear spin I	Nuclear magnetic moment μ	Uses
^{241}Am	241.0567	0	458y	α(5.640); γ	5/2	+1.59	tracer; medical
^{243}Am	243.0614	0	7.4×10^3y	α(5.439); γ	5/2	+1.4	NMR

NMR

^{243}Am

Relative sensitivity (^{1}H $= 1.00$)	—
Absolute sensitivity (^{1}H $= 1.00$)	—
Receptivity (^{13}C $= 1.00$)	—
Magnetogyric ratio/rad T^{-1} s^{-1}	1.54×10^7
Quadrupole moment/m^2	4.9×10^{-28}
Frequency (^{1}H $= 100$ MHz; 2.3488 T)/MHz	5.76

Electron shell properties

Ground state electron configuration: [Rn]$5f^7 7s^2$

Term symbol: $^8S_{7/2}$

Electron affinity (M$\rightarrow$M$^-$)/kJ mol^{-1}: n.a.

Ionization energies/kJ mol^{-1}

1. M $\rightarrow$M$^+$ 578.2	6. M$^{5+}\rightarrow$M^{6+}
2. M $^+\rightarrow$M^{2+}	7. M$^{6+}\rightarrow$M^{7+}
3. M$^{2+}\rightarrow$M^{3+}	8. M$^{7+}\rightarrow$M^{8+}
4. M$^{3+}\rightarrow$M^{4+}	9. M$^{8+}\rightarrow$M^{9+}
5. M$^{4+}\rightarrow$M^{5+}	10. M$^{9+}\rightarrow$M^{10+}

Principal lines in atomic spectrum

Wavelength/nm	Species	Sensitivity	Application
351.013	I		
356.919	I		
367.312	I		
428.926	I		
605.464	I		

Abundance: 0

Biological role: Toxic due to radioactivity

Sb

Atomic number: 51
Relative atomic mass ($^{12}C = 12.0000$): 121.75

Chemical properties

Metalloid element, various allotropes, of which metal is bright, silvery, hard, and brittle. Occurs in nature as Sb_2S_3. Stable in dry air and not attacked by dilute acids or alkalis. Used to harden other metals as alloys.

Radii/pm: Sb^{5+} 62; Sb^{3+} 89; covalent 141; atomic 182; van der Waals 220; Sb^{2-} 245

Electronegativity: 2.05 (Pauling); 1.82 (Allred-Rochow)

Effective nuclear charge: 6.30 (Slater); 9.99 (Clementi); 12.37 (Froese–Fischer)

Standard reduction potentials $E^{\ominus}/V$

	V	IV	III	0	−III
Acid solution	Sb_2O_5	——— 0.605 ———	SbO^+	—0.204— Sb	
Neutral solution	Sb_2O_5	—1.055— Sb_2O_4 —0.342—	Sb_4O_6	—0.150— Sb	—0.510— SbH_3
		——— 0.699 ———			
Alkaline solution	$Sb(OH)_6^-$	——— −0.465 ———	$Sb(OH)_4^-$	—0.639— Sb —1.338—	SbH_3

Covalent bonds r/pm E/kJ mol^{-1}

	r/pm	E/kJ mol^{-1}
Sb–H	170.7	257
Sb–C	220	215
Sb–O	200	314
Sb–F	203	389
Sb–Cl	233	313
Sb–Sb	290	299

Oxidation states

Sb^{-III}	SbH_3
Sb^{III}	Sb_4O_6, SbO_3^{3-} (aq)
	SbF_3, $SbCl_3$, etc.
	$[SbF_5]^{2-}$, Sb_2S_3
Sb^V	Sb_4O_{10}, $[Sb(OH)_6]^-$ (aq)
	SbF_5, $SbCl_5$, $[SbCl_6]^-$
	$[SbBr_6]^-$

Physical properties

Melting point/K: 903.89
Boiling point/K: 1908
ΔH_{fusion}/kJ mol^{-1}: 20.9   **ΔH_{vap}/kJ mol^{-1}:** 165.8

Thermodynamic properties (298.15 K, 0.1 MPa)

State	$\Delta_f H^{\ominus}$/kJ mol^{-1}	$\Delta_f G^{\ominus}$/kJ mol^{-1}	$S^{\ominus}$/J K^{-1} mol^{-1}	C_p/J K^{-1} mol^{-1}
Solid	0	0	45.69	25.23
Gas	262.3	222.1	180.27	20.79

Density/kg m^{-3}: 6691 [293 K]; 6483 [liquid at m.p.]
Thermal conductivity/W m^{-1} K^{-1}: 243 [300 K]
Electrical resistivity/Ω m: 39.0×10^{-8} [273 K]
Mass magnetic susceptibility/kg^{-1} m^3: -1.0×10^{-8} (s)
Molar volume/cm^3: 18.20
Coefficient of linear thermal expansion/K^{-1}: 8.5×10^{-6}

Lattice structure (cell dimensions/pm), space group

Grey rhombohedral ($a = 430.84$; $c = 1124.7$), R$\bar{3}$m
(Grey) cubic ($a = 298.6$), Pm3m
Metal h.c.p. ($a = 336.9$, $c = 533$), P6$_3$/mmc

X-ray diffraction: mass absorption coefficients (μ/ρ)/cm^2 g^{-1}:
CuK$_\alpha$ 270   MoK$_\alpha$ 33.1

Antimony

[Greek, *anti + monos* = not alone; Latin, *stibium*]

Atomic number: 51

Thermal neutron capture cross-section/barns: 5 ± 1

Number of isotopes (including nuclear isomers): 29

Isotope mass range: $112 \rightarrow 133$

Key isotopes

Nuclide	Atomic mass	Natural abundance (%)	Half life $T_{1/2}$	Decay mode and energy (MeV)	Nuclear spin I	Nuclear magnetic moment μ	Uses
^{121}Sb	120.9038	57.3	stable		5/2	$+3.3592$	NMR
^{122}Sb		0	2.80d	β^- (1.972); β^+; EC; γ	2	-1.90	tracer
^{123}Sb	122.9041	42.7	stable		7/2	$+2.5466$	NMR
^{124}Sb		0	60.4d	β^- (2.916); γ	3	± 1.3	tracer
^{125}Sb		0	2.71y	β^- (0.764); γ	7/2	± 2.61	tracer

NMR

	^{121}Sb	[^{123}Sb]
Relative sensitivity (^{1}H = 1.00)	0.16	4.57×10^{-2}
Absolute sensitivity (^{1}H = 1.00)	9.16×10^{-2}	1.95×10^{-2}
Receptivity (^{13}C = 1.00)	520	111
Magnetogyric ratio/rad T^{-1} s^{-1}	6.4016×10^7	3.4668×10^7
Quadrupole moment/m^2	-0.53×10^{-28}	-0.68×10^{-28}
Frequency (^{1}H = 100 MHz; 2.3488 T)/MHz	23.930	12.959

Reference: $Et_4N^+SbCl_6^-$

Ground state electron configuration: [Kr] $4d^{10}5s^25p^3$

Term symbol: $^4S_{3/2}$

Electron affinity (M→M$^-$)/kJ mol^{-1}: 101

Ionization energies/kJ mol^{-1}

1. M → M$^+$	833.7	6. M^{5+}→M^{6+}	10 400
2. M$^+$→M^{2+}	1794	7. M^{6+}→M^{7+}	(12 700)
3. M^{2+}→M^{3+}	2443	8. M^{7+}→M^{8+}	(15 200)
4. M^{3+}→M^{4+}	4260	9. M^{8+}→M^{9+}	(17 800)
5. M^{4+}→M^{5+}	5400	10. M^{9+}→M^{10+}	(20 400)

Principal lines in atomic spectrum

Wavelength/nm	Species	Sensitivity	Application
206.833	I	U1	AA AE
217.581	I	U2	AA AE
231.147*	I		AA AE
252.852	I		AE
259.805	I		AE

* Lowest energy transition from ground state to nearest empty orbital

Abundance: Earth's crust 0.2 p.p.m.; seawater 0.0005 p.p.m.

Biological role: Toxic; stimulatory

Atomic number: **18**

Relative atomic mass ($^{12}C = 12.0000$): **39.948**

Chemical properties

Colourless, odourless gas comprising 1 per cent of the atmosphere. Obtained from liquid air. Used as inert atmosphere in lamps and high temperature metallurgy.

Radii/pm: atomic 174; van der Waals 191

Electronegativity: n.a. (Pauling); n.a. (Allred-Rochow)

Effective nuclear charge: 6.75 (Slater); 6.76 (Clementi); 7.52 (Froese–Fischer)

Oxidation states

Ar^0	$Ar_8(H_2O)_{46}$ and $Ar(quinol)_3$. These are not true compounds but clathrates in which argon atoms are trapped inside lattice of other molecules

Physical properties

Melting point/K: 83.78
Boiling point/K: 87.29
ΔH_{fusion}/kJ mol^{-1}:   1.21
ΔH_{vap}/kJ mol^{-1}:   6.53

Thermodynamic properties   (298.15 K, 0.1 MPa)

State	$\Delta_f H^{\ominus}$/kJ mol^{-1}	$\Delta_f G^{\ominus}$/kJ mol^{-1}	$S^{\ominus}$/J K^{-1} mol^{-1}	C_p/J K^{-1} mol^{-1}
Gas	0	0	154.843	20.786

Density/kg m^{-3}:   1656 [40 K]; 13806 [liquid b.p.], 1.784 [273 K]
Thermal conductivity/W m^{-1} K^{-1}:   0.0177 [300 K]$_g$
Mass magnetic susceptibility/kg^{-1} m^3: -6.16×10^{-9} (g)
Molar volume/cm^3:   24.12 [40 K]

Lattice structure (cell dimensions/pm), space group

f.c.c. (40 K) ($a = 531.088$), Fm3m

X-ray diffraction: mass absorption coefficients (μ/ρ)/cm^2 g^{-1}:
CuK$_\alpha$ 123   MoK$_\alpha$ 13.5

Argon

Atomic number: 18

Thermal neutron capture cross-section/barns: 0.650 ± 0.030

Number of isotopes (including nuclear isomers): 8

Isotope mass range: $35 \rightarrow 42$

Nuclear properties

Key isotopes

Nuclide	Atomic mass	Natural abundance (%)	Half life $T_{1/2}$	Decay mode and energy (MeV)	Nuclear spin I	Nuclear magnetic moment μ	Uses
^{36}Ar	35.967 55	0.337	stable		0		
^{37}Ar	36.9667	0	35d	EC (0.814); no γ	3/2	+0.95	
^{38}Ar	37.962 72	0.063	stable		0		
^{39}Ar	38.964	0	269y	β^- (3.44); no γ	7/2	−1.3	
^{40}Ar	39.9624	99.600	stable		0		

Ground state electron configuration: [Ne] $3s^2 3p^6$

Term symbol: $^1S_0$

Electron affinity $(M \rightarrow M^-)$/kJ mol^{-1}: −35 (calc.)

Electron shell properties

Ionization energies/kJ mol^{-1}

1.	M $\rightarrow$ M$^+$	1520.4	6. M$^{5+} \rightarrow$ M^{6+}	8 811
2.	M$^+ \rightarrow$ M^{2+}	2665.2	7. M$^{6+} \rightarrow$ M^{7+}	12 021
3.	M$^{2+} \rightarrow$ M^{3+}	3928	8. M$^{7+} \rightarrow$ M^{8+}	13 844
4.	M$^{3+} \rightarrow$ M^{4+}	5770	9. M$^{8+} \rightarrow$ M^{9+}	40 759
5.	M$^{4+} \rightarrow$ M^{5+}	7238	10. M$^{9+} \rightarrow$ M^{10+}	46 186

Principal lines in atomic spectrum

Wavelength/nm	Species	Sensitivity	Application
106.660* (vac.)	I		
696.543	I	U3	AE
706.722	I	U3	AE
750.387	I	U4	AE
811.531	I	U2	AE

* Lowest energy transition from ground state to nearest empty orbital

Abundance: Atmosphere 9300 p.p.m. (volume); earth's crust 0.04 p.p.m.; sea-water 0.6 p.p.m.

Biological role: None

Atomic number: 33
Relative atomic mass ($^{12}C = 12.0000$): **74.9216**

Chemical properties

Metalloid, several allotropes. Grey arsenic is metallic—soft and brittle, tarnishes, burns in O_2, resists attack by water, acids, and alkalis. Attacked by hot acids and molten NaOH. Uses: alloys, semiconductors.

Radii/pm: As^{5+} 46; As^{3+} 69; covalent 121; atomic 125; van der Waals 200

Electronegativity: 2.18 (Pauling); 2.20 (Allred-Rochow)

Effective nuclear charge: 6.30 (Slater); 7.45 (Clementi); 8.98 (Froese–Fischer)

Standard reduction potentials $E^{\ominus}/V$

	V		III		0		−III
Acid solution	H_3AsO_4	$\xrightarrow{0.560}$	$HAsO_2$	$\xrightarrow{0.240}$	As	$\xrightarrow{-0.225}$	AsH_3
Alkaline solution	AsO_4^{3-}	$\xrightarrow{-0.67}$	AsO_2^-	$\xrightarrow{-0.68}$	As	$\xrightarrow{1.37}$	AsH_3

Covalent bonds	r/pm	E/kJ mol^{-1}
As–H	151.9	c. 245
As–C	198	200
As–O	178	477
As–F	171	464
As–Cl	216	293
As–As	244	348

Oxidation states

As^{-III}	AsH_3
As^{III}	As_4O_6, H_3AsO_3, $H_2AsO_3^-$ (aq), AsF_3, $AsCl_3$, etc.
As^V	As_4O_{10}, H_3AsO_4, $H_2AsO_4^-$, etc. (aq), $NaAsO_3$, AsF_5

Physical properties

Melting point/K: 1090 (α) under pressure
Boiling point/K: 889 (sublimes)
ΔH_{fusion}/kJ mol^{-1}: 27.7
ΔH_{vap}/kJ mol^{-1}: 31.9

Thermodynamic properties (298.15 K, 0.1 MPa)

State	$\Delta_f H^{\ominus}$/kJ mol^{-1}	$\Delta_f G^{\ominus}$/kJ mol^{-1}	$S^{\ominus}$/J K^{-1} mol^{-1}	C_p/J K^{-1} mol^{-1}
Solid (α)	0	0	35.1	24.64
Gas	302.5	261.0	174.21	20.786

Density/kg m^{-3}: 5780 (α); 4700 (β) [293 K]
Thermal conductivity/W m^{-1} K^{-1}: 50.0 (α) [300 K]
Electrical resistivity/Ω m: 26×10^{-8} [273 K]
Mass magnetic susceptibility/kg^{-1} m^3: -9.17×10^{-10} (α); -3.97×10^{-9} (β)
Molar volume/cm^3: 12.95 (α); 15.9 (β)
Coefficient of linear thermal expansion/K^{-1}: 4.7×10^{-6}

Lattice structure (cell dimensions/pm), space group

α-As rhombohedral ($a = 413.18$; $\alpha = 54°10'$), R$\bar{3}$m, metallic form
β-As hexagonal ($a = 376.0$, $c = 10.548$), yellow
grey amorphous
$T(\alpha \rightarrow \beta) = 501$ K
$T(\beta \rightarrow \text{grey}) = $ room temperature

X-ray diffraction: mass absorption coefficients (μ/ρ)/cm^2 g^{-1}:
CuK$_\alpha$ 83.4   MoK$_\alpha$ 69.7

Probably first isolated by Albertus Magnus (1193–1280)

[.Greek, *arsenikon* = yellow orpiment]

Arsenic

Atomic number: 33
Thermal neutron capture cross-section/barns: 4.30 ± 0.10
Number of isotopes (including nuclear isomers): 14
Isotope mass range: $69 \rightarrow 81$

Nuclear properties

Key isotopes

Nuclide	Atomic mass	Natural abundance (%)	Half life $T_{1/2}$	Decay mode and energy (MeV)	Nuclear spin I	Nuclear magnetic moment μ	Uses
^{73}As			80.3d	EC(0.37); γ			tracer
^{74}As	73.924		17.9d	β^-(1.36); β^+; EC; γ			tracer
^{75}As	74.9216	100	stable		3/2	+1.439	NMR
^{76}As	75.922		26.5h	β^-(2.97); γ	2	−0.905	tracer

NMR
^{75}As

Relative sensitivity (^{1}H = 1.00) 2.51×10^{-2}
Absolute sensitivity (^{1}H = 1.00) 2.51×10^{-2}
Receptivity (^{13}C = 1.00)   143
Magnetogyric ratio/rad T^{-1} s^{-1} 4.5804×10^7
Quadrupole moment/m^2 0.3×10^{-28}
Frequency (^{1}H = 100 MHz; 2.3488 T)/MHz   17.126
Reference: KAsF$_6$

Ground state electron configuration: [Ar] $3d^{10}4s^24p^3$
Term symbol: $^4S_{3/2}$
Electron affinity (M→M$^-$)/kJ mol^{-1}: 77

Electron shell properties

Ionization energies/kJ mol^{-1}

1. M $\rightarrow$ M$^+$	947.0	6. M$^{5+} \rightarrow$ M^{6+}	12 305
2. M$^+ \rightarrow$ M^{2+}	1798	7. M$^{6+} \rightarrow$ M^{7+}	(15 400)
3. M$^{2+} \rightarrow$ M^{3+}	2735	8. M$^{7+} \rightarrow$ M^{8+}	(18 900)
4. M$^{3+} \rightarrow$ M^{4+}	4837	9. M$^{8+} \rightarrow$ M^{9+}	(22 600)
5. M$^{4+} \rightarrow$ M^{5+}	6042	10. M$^{9+} \rightarrow$ M^{10+}	(26 400)

Principal lines in atomic spectrum

Wavelength/nm	Species	Sensitivity	Application
193.696	I		AA
197.197*	I		AA
228.812	I	U2	AE
234.984	I	U3	AE
245.653	I	U4	AE

* Lowest energy transition from ground state to nearest empty orbital

Abundance: Earth's crust 1.8 p.p.m.; seawater 0.003 p.p.m.
Biological role: May be essential, but toxic in small doses; stimulatory; suspected carcinogen

<table>
<tr><td>## At</td><td>**Atomic number:** 85
Relative atomic mass ($^{12}C = 12.0000$): **(210)**</td></tr>
</table>

Chemical properties

Radioactive element obtainable in various ways, e.g. neutron bombardment of ^{200}Bi produces ^{211}At, but not in weighable amounts.

Radii/pm: At^{5+} 57; At^- 227

Electronegativity: 2.2 (Pauling); 1.96 (Allred-Rochow)

Effective nuclear charge: 7.60 (Slater); 15.16 (Clementi); 19.61 (Froese–Fischer)

Standard reduction potentials $E^{\ominus}$/V

	V		I		0		−I
Acid solution	$HAtO_3$	—1.4—	HAtO	—0.7—	At_2	—0.2—	At^-
Alkaline solution	AtO_3^-	—0.5—	AtO^-	—0.0—	At_2	—0.2—	At^-

Covalent bonds

	r/pm	E/kJ mol^{-1}
At–At	$c.$ 290 (estimated)	110

Oxidation states

At^{-1} ([Rn]) At^- (aq)
At^I (s^2p^4) $AtBr_2^-$
At^{III} (s^2p^2) AtO_3^- (aq)

Physical properties

Melting point/K: 575 (est.)
Boiling point/K: 610 (est.)
ΔH_{fusion}/kJ mol^{-1}: 23.8
ΔH_{vap}/kJ mol^{-1}: n.a.

Thermodynamic properties (298.15 K, 0.1 MPa)

State	$\Delta_f H^{\ominus}$/kJ mol^{-1}	$\Delta_f G^{\ominus}$/kJ mol^{-1}	$S^{\ominus}$/J K^{-1} mol^{-1}	C_p/J K^{-1} mol^{-1}
Solid (α)	0	0	n.a.	n.a.
Gas	n.a.	n.a.	n.a.	n.a.

Density/kg m^{-3}: n.a.
Thermal conductivity/W m^{-1} K^{-1}: 1.7 [300 K]
Mass magnetic susceptibility/kg^{-1} m^3: n.a.
Molar volume/cm^3: n.a.

Lattice structure (cell dimensions/pm), space group

n.a.

X-ray diffraction: mass absorption coefficients (μ/ρ)/cm^2 g^{-1}:
n.a.

Produced by D. R. Corson, K. R. Mackenzie, and E. Segré in
1940 at the University of California, USA

[Greek, *astatos* = unstable]

Astatine

Atomic number: 85

Thermal neutron capture cross-section/barns: n.a.

Number of isotopes (including nuclear isomers): 21

Isotope mass range: $200 \rightarrow 219$

Key isotopes

Nuclide	Atomic mass	Natural abundance (%)	Half life $T_{1/2}$	Decay mode and energy (MeV)	Nuclear spin I	Nuclear magnetic moment μ	Uses
^{210}At		0	8.3h	EC(3.87 calc.); α(5.63)<1%			
^{211}At	210.9875	0	7.21h	EC(0.97 calc.) 59%; α(5.981) 41%	9/2		

Ground state electron configuration: $[Xe] 4f^{14}5d^{10}6s^26p^5$

Term symbol: $^3P_{3/2}$

Electron affinity $(M \rightarrow M^-)$/kJ mol^{-1}: 270

Ionization energies/kJ mol^{-1}

1.	$M \rightarrow M^+$	930	6.	$M^{5+} \rightarrow M^{6+}$	(7 500)
2.	$M^+ \rightarrow M^{2+}$	1600	7.	$M^{6+} \rightarrow M^{7+}$	(8 800)
3.	$M^{2+} \rightarrow M^{3+}$	(2900)	8.	$M^{7+} \rightarrow M^{8+}$	(13 300)
4.	$M^{3+} \rightarrow M^{4+}$	(4000)	9.	$M^{8+} \rightarrow M^{9+}$	(15 400)
5.	$M^{4+} \rightarrow M^{5+}$	(4900)	10.	$M^{9+} \rightarrow M^{10+}$	(17 700)

Principal lines in atomic spectrum

Wavelength/nm	Species	Sensitivity	Application
216.225	I		
224.401	I		

Abundance: Trace amount in some minerals

Biological role: None; toxic due to radioactivity

Ba	**Atomic number: 56**
	Relative atomic mass ($^{12}C = 12.0000$): 137.327

Chemical properties

Relatively soft, silvery-white metal. Obtained from BaO on heating with aluminium. Attacked by air and water. Principal ore is barytes, $BaSO_4$.

Radii/pm: Ba^{2+} 143; atomic 217.3; covalent 198

Electronegativity: 0.89 (Pauling); 0.97 (Allred-Rochow)

Effective nuclear charge: 2.85 (Slater); 7.58 (Clementi); 10.27 (Froese–Fischer)

Standard reduction potentials $E^{\ominus}/V$

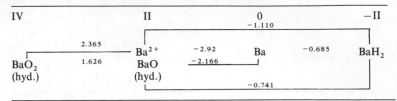

Oxidation states

Ba ^{II} ([Xe]) BaH_2, BaO, $Ba(OH)_2$ basic, BaO_2 (peroxide), Ba^{2+}(aq), BaF_2, $BaCl_2$, etc., $BaCO_3$, $BaSO_4$ insoluble, many other salts

Physical properties

Melting point/K: 1002
Boiling point/K: 1910
ΔH_{fusion}/kJ mol^{-1}: 7.66
ΔH_{vap}/kJ mol^{-1}: 150.9

Thermodynamic properties (298.15 K, 0.1 MPa)

State	$\Delta_f H^{\ominus}$/kJ mol^{-1}	$\Delta_f G^{\ominus}$/kJ mol^{-1}	$S^{\ominus}$/J K^{-1} mol^{-1}	C_p/J K^{-1} mol^{-1}
Solid	0	0	62.8	28.07
Gas	180	146	170.243	20.786

Density/kg m^{-3}: 3594 [293 K]; 3325 [liquid at m.p.]
Thermal conductivity/W m^{-1} K^{-1}: 18.4 [300 K]
Electrical resistivity/Ω m: 50×10^{-8} [273 K]
Mass magnetic susceptibility/kg^{-1} m^3: $+1.9 \times 10^{-9}$ (s)
Molar volume/cm^3: 38.21
Coefficient of linear thermal expansion/K^{-1}: 18.1–21.0×10^{-6}

Lattice structure (cell dimensions/pm), space group

b.c.c. ($a = 502.5$), Im3m

High pressure form: ($a = 390.1$, $c = 615.5$), P6$_3$mmc

X-ray diffraction: mass absorption coefficients (μ/ρ)/cm^2 g^{-1}:
CuK$_\alpha$ 330   MoK$_\alpha$ 43.5

Barium

Atomic number: 56
Thermal neutron capture cross-section/barns: 1.2 ± 0.1
Number of isotopes (including nuclear isomers): 25
Isotope mass range: $123 \rightarrow 143$

Nuclear properties

Key isotopes

Nuclide	Atomic mass	Natural abundance (%)	Half life $T_{1/2}$	Decay mode and energy (MeV)	Nuclear spin I	Nuclear magnetic moment μ	Uses
^{130}Ba	129.9062	0.106	stable				
^{132}Ba	131.9057	0.101	stable				
^{133}Ba		0	7.2y	EC(0.488); γ			tracer
^{134}Ba	133.9043	2.417	stable		0		
^{135}Ba	134.9056	6.592	stable		3/2	+0.8365	NMR
^{135m}Ba		0	28.7h	IT(0.268); γ			tracer
^{136}Ba	135.9044	7.854	stable		0		
^{137}Ba		11.32	stable		3/2	+0.9357	NMR
^{138}Ba	137.9050	71.70	stable		0		
^{140}Ba		0	12.80d	β^-(1.05); γ			tracer

NMR

	$[^{135}\text{Ba}]$	^{137}Ba
Relative sensitivity (^{1}H = 1.00)	4.90×10^{-4}	7.76×10^{-4}
Absolute sensitivity (^{1}H = 1.00)	3.22×10^{-4}	7.76×10^{-4}
Receptivity (^{13}C = 1.00)	1.83	4.41
Magnetogyric ratio/rad T^{-1} s^{-1}	2.6575×10^7	2.9728×10^7
Quadrupole moment/m^2	0.18×10^{-28}	0.28
Frequency (^{1}H = 100 MHz; 2.3488 T)/MHz	9.934	11.113

Reference: $BaCl_2$ (aq)

Ground state electron configuration: $[\text{Xe}]6s^2$
Term symbol: $^1S_0$
Electron affinity $(M \rightarrow M^-)$/kJ mol^{-1}: -46

Electron shell properties

Ionization energies/kJ mol^{-1}

1. $M \rightarrow M^+$	502.8	6. $M^{5+} \rightarrow M^{6+}$	(7 700)
2. $M^+ \rightarrow M^{2+}$	965.1	7. $M^{6+} \rightarrow M^{7+}$	(9 000)
3. $M^{2+} \rightarrow M^{3+}$	(3600)	8. $M^{7+} \rightarrow M^{8+}$	(10 200)
4. $M^{3+} \rightarrow M^{4+}$	(4700)	9. $M^{8+} \rightarrow M^{9+}$	(13 500)
5. $M^{4+} \rightarrow M^{5+}$	(6000)	10. $M^{9+} \rightarrow M^{10+}$	(15 100)

Principal lines in atomic spectrum

Wavelength/nm	Species	Sensitivity	Application
350.111	I		AA
455.403	II	V1	AA AE
493.409	II	V2	AA AE
553.548	I	U1	AA AE
577.767	I	U2	AA AE
791.134*	I		

* Lowest energy transition from ground state to nearest empty orbital

Abundance: Earth's crust 390 p.p.m.; seawater 0.03 p.p.m.
Biological role: Toxic; stimulatory

Bk	Atomic number: 97
	Relative atomic mass ($^{12}C = 12.0000$): (247)

Chemical properties

Radioactive silvery metal, made in mg quantities as ^{294}Bk by neutron bombardment of ^{239}Pu. Attacked by oxygen, steam, and acids, but not by alkalis

Radii/pm: Bk^{2+} 118; Bk^{3+} 98; Bk^{4+} 87
Electronegativity: 1.3 (Pauling); n.a. (Allred-Rochow)
Effective nuclear charge: 1.65 (Slater)

Standard reduction potentials $E^{\ominus}$/V

	III	IV	0
		−1.05	
Acid solution	Bk^{4+}	$\xrightarrow{1.67}$ Bk^{3+} $\xrightarrow{-2.01}$	Bk

Oxidation states

Bk (f^9)	BkO
Bk^{III} (f^8)	Bk_2O_3, BkF_3, $BkCl_3$ etc.,
	$[BkCl_6]^{3-}$, Bk^{3+} (aq)
Bk^{IV} (f^7)	BkO_2, BkF_4

Physical properties

Melting point/K: n.a.
Boiling point/K: n.a.
ΔH_{fusion}/kJ mol^{-1}: n.a.
ΔH_{vap}/kJ mol^{-1}: n.a.

Thermodynamic properties (298.15 K, 0.1 MPa)

State	$\Delta_f H^{\ominus}$/kJ mol^{-1}	$\Delta_f G^{\ominus}$/kJ mol^{-1}	$S^{\ominus}$/J K^{-1} mol^{-1}	C_p/J K^{-1} mol^{-1}
Solid	0	0	n.a.	n.a.
Gas	n.a.	n.a.	n.a.	n.a.

Density/kg m^{-3}: 14 790 [293 K]
Thermal conductivity/W m^{-1} K^{-1}: 10 (est.) [300 K]
Electrical resistivity/Ω m: n.a.
Mass magnetic susceptibility/kg^{-1} m^3: n.a.
Molar volume/cm^3: 16.70
Coefficient of linear thermal expansion/K^{-1}: n.a.

Lattice structure (cell dimensions/pm), space group

n.a.

X-ray diffraction: mass absorption coefficients (μ/ρ)/cm^2 g^{-1}:
n.a.

Produced in December 1949 by S. G. Thompson, A. Ghiorso, and
G. T. Seaborg at Berkeley, California, USA

[English, *Berkeley*]

Berkelium

Atomic number: 97

**Thermal neutron capture
cross-section**/barns: 1000 ± 500 (^{249}Bk)

Number of isotopes (including nuclear isomers): 8

Isotope mass range: $243 \rightarrow 250$

**Nuclear
properties**

Key isotopes

Nuclide	Atomic mass	Natural abundance (%)	Half life $T_{1/2}$	Decay mode and energy (MeV)	Nuclear spin I	Nuclear magnetic moment μ	Uses
^{247}Bk	247.0702	0	1.4×10^3y	$\alpha(5.86)$; γ			
^{249}Bk		0	314d	β^-	7/2		

Ground state electron configuration: $[Rn]5f^9 7s^2$

Term symbol: $^6H_{15/2}$

Electron affinity $(M \rightarrow M^-)/kJ\ mol^{-1}$: n.a.

**Electron
shell
properties**

Ionization energies/kJ mol^{-1}

1. $M \rightarrow M^+$	601	6. $M^{5+} \rightarrow M^{6+}$	
2. $M^+ \rightarrow M^{2+}$		7. $M^{6+} \rightarrow M^{7+}$	
3. $M^{2+} \rightarrow M^{3+}$		8. $M^{7+} \rightarrow M^{8+}$	
4. $M^{3+} \rightarrow M^{4+}$		9. $M^{8+} \rightarrow M^{9+}$	
5. $M^{4+} \rightarrow M^{5+}$		10. $M^{9+} \rightarrow M^{10+}$	

Selected lines in atomic spectrum

Wavelength/nm	Species	Sensitivity	Application
329.972	I		
325.219	I		
328.875	I		
329.935	I		
333.526	I		
340.828	I		
342.695	I		

Abundance: Nil

Biological role: Toxic due to radioactivity

Be

Atomic number: 4

Relative atomic mass ($^{12}C=12.0000$): **9.012 18**

Chemical properties

Silvery-white, lustrous, relatively soft metal, obtained e.g. by the electrolysis of fused $BeCl_2$. Unaffected by air or water. Used in alloys with copper and nickel. Occurs as beryl $Be_3Al_2Si_6O_{18}$.

Radii/pm: Be^{2+} 34; atomic 113.3; covalent 89

Electronegativity: 1.57 (Pauling); 1.47 (Allred-Rochow)

Effective nuclear charge: 1.95 (Slater); 1.91 (Clementi); 2.27 (Froese–Fischer)

Standard reduction potentials $E^\ominus$/V

II		0
Be^{2+}	$\xrightarrow{-1.97}$	Be

Covalent bonds r/pm E/kJ mol^{-1}

	r/pm	E/kJ mol^{-1}
Be–H	163	226
Be–O	133	523
Be–F	143	615
Be–Cl	177	293
Be–C	193	
Be–Be	222.6	

Oxidation states

Be^{II}	BeO, $Be(OH)_2$, BeH_2, $[Be(H_2O)_4]^{2+}$aq BeF_2, $BeCl_2$ etc., $BeCO_3$, salts

Physical properties

Melting point/K: 1551 ± 5

Boiling point/K: 3243 (under pressure)

ΔH_{fusion}/kJ mol^{-1}: 9.80

ΔH_{vap}/kJ mol^{-1}: 308.8

Thermodynamic properties (298.15 K, 0.1 MPa)

State	$\Delta_f H^\ominus$/kJ mol^{-1}	$\Delta_f G^\ominus$/kJ mol^{-1}	$S^\ominus$/J K^{-1} mol^{-1}	C_p/J K^{-1} mol^{-1}
Solid	0	0	9.50	16.44
Gas	324.6	286.6	136.269	20.786

Density/kg m^{-3}: 1847.7 [293 K]

Thermal conductivity/W m^{-1} K^{-1}: 200 [300 K]

Electrical resistivity/Ω m: 4.0×10^{-8} [293 K]

Mass magnetic susceptibility/kg^{-1} m^3: -1.3×10^{-8} (s)

Molar volume/cm^3: 4.88

Coefficient of linear thermal expansion/K^{-1}: 11.5×10^{-6}

Lattice structure (cell dimensions/pm), space group

α-Be   h.c.p. ($a=228.55$, $c=358.32$), P6$_3$/mmc

β-Be   b.c.c. ($a=255.15$), Im3m

$T(\alpha \rightarrow \beta)=1523$ K

X-ray diffraction: mass absorption coefficients (μ/ρ)/cm^2 g^{-1}: CuK$_\alpha$ 1.50   MoK$_\alpha$ 0.298

Discovered in 1797 by N.-L. Vauquelin. Isolated in 1828 by F. Wöhler
at Berlin, Germany and independently by A. A. B. Bussy at Paris, France
[Greek, *beryllos* = beryl]

Beryllium

Atomic number: 4

Thermal neutron capture cross-section/barns: 0.0092 ± 0.0005

Number of isotopes (including nuclear isomers): 6

Isotope mass range: $6 \rightarrow 11$

Nuclear properties

Key isotopes

Nuclide	Atomic mass	Natural abundance (%)	Half life $T_{1/2}$	Decay mode and energy (MeV)	Nuclear spin I	Nuclear magnetic moment μ	Uses
^{7}Be	7.0169	0	53.37d	EC(0.862); γ			tracer
^{9}Be	9.01218	100	stable		3/2	-1.17745	NMR
^{10}Be	10.0135	0	2.5×10^6y	β^-(0.555); no γ			

NMR

^{9}Be

Relative sensitivity (^{1}H = 1.00)	1.39×10^{-2}
Absolute sensitivity (^{1}H = 1.00)	1.39×10^{-2}
Receptivity (^{13}C = 1.00)	78.8
Magnetogyric ratio/rad T^{-1} s^{-1}	3.7589×10^7
Quadrupole moment/m^2	5.2×10^{-30}
Frequency (^{1}H = 100 MHz; 2.3488 T)/MHz	14.053

Reference: $Be(NO_3)_2$ (aq)

Ground state electron configuration: [He]2s^2

Term symbol: $^1S_0$

Electron affinity $(M \rightarrow M^-)$/kJ mol^{-1}: -18

Electron shell properties

Ionization energies/kJ mol^{-1}

1. $M \rightarrow M^+$ 899.4
2. $M^+ \rightarrow M^{2+}$ 1757.1
3. $M^{2+} \rightarrow M^{3+}$ 14848
4. $M^{3+} \rightarrow M^{4+}$ 21006

Principal lines in atomic spectrum

Wavelength/nm	Species	Sensitivity	Application
234.861	I	U1	AA AE
332.134	I	U2	AE
332.109	I	U3	AE
332.101	I	U4	AE
467.342	II		

Abundance: Earth's crust 2 p.p.m.; seawater 5×10^{-5} p.p.m.

Biological role: Toxic; carcinogenic

Bi

Atomic number: 83

Relative atomic mass ($^{12}C = 12.0000$): 208.9804

Chemical properties

Brittle metal, silvery lustre with pink tinge. Stable to oxygen and water. Dissolves in concentrated nitric acid. Basic oxide. Used in alloys and medicines.

Radii/pm: Bi^{5+} 74; Bi^{3+} 96; covalent 152; atomic 155; van der Waals 240

Electronegativity: 2.02 (Pauling); 1.67 (Allred–Rochow)

Effective nuclear charge: 6.30 (Slater); 13.34 (Clementi); 16.90 (Froese–Fischer)

Standard reduction potentials $E^{\ominus}/V$

V		III		0		−III
Bi^{5+}	$\xrightarrow{ca.\ 2}$	Bi^{3+}	$\xrightarrow{0.317}$	Bi	$\xrightarrow{-0.97}$	BiH_3

Covalent bonds

	r/pm	E/kJ mol^{-1}
Bi–H	n.a.	194
Bi–C	230	143
Bi–O	232	339
Bi–F	235	314
Bi–Cl	248	285
Bi–Bi	309	200

Oxidation states

Bi^{-III}	BiH_3
Bi^{I}	Bi^+, cation clusters Bi_3^+, Bi_5^{3+}, Bi_9^{5+} etc.
Bi^{III}	Bi_2O_3, $Bi(OH)_3$, Bi^{3+}(aq), BiOCl, BiF_3, $BiCl_3$ etc. $[BiBr_6]^{3-}$, salts
Bi^{V}	Bi_2O_5 unstable, $[Bi(OH)_6]^-$(aq), $NaBiO_3$, BiF_5, $KBiF_6$

Physical properties

Melting point/K: 544.5
Boiling point/K: 1833 ± 5
ΔH_{fusion}/kJ mol^{-1}: 10.48
ΔH_{vap}/kJ mol^{-1}: 179.1

Thermodynamic properties (298.15 K, 0.1 MPa)

State	$\Delta_f H^{\ominus}$/kJ mol^{-1}	$\Delta_f G^{\ominus}$/kJ mol^{-1}	$S^{\ominus}$/J K^{-1} mol^{-1}	C_p/J K^{-1} mol^{-1}
Solid	0	0	56.74	25.52
Gas	207.1	168.2	187.005	20.786

Density/kg m^{-3}: 9747 [293 K]; 10050 [liquid at m.p.]
Thermal conductivity/W m^{-1} K^{-1}: 7.87 [300 K]
Electrical resistivity/Ω m: 106.8×10^{-8} [273 K]
Mass magnetic susceptibility/kg^{-1} m^3: -1.684×10^{-8} (s)
Molar volume/cm^3: 21.44
Coefficient of linear thermal expansion/K^{-1}: 13.4×10^{-6}

Lattice structure (cell dimensions/pm), space group

Rhombohedral ($a = 454.950$, $c = 1186.225$), R$\bar{3}$m

X-ray diffraction: mass absorption coefficients (μ/ρ)/cm^2 g^{-1}:
CuK$_\alpha$ 240   MoK$_\alpha$ 120

Bismuth

Atomic number: 83
Thermal neutron capture cross-section/barns: 0.034
Number of isotopes (including nuclear isomers): 19
Isotope mass range: 199→215

Key isotopes

Nuclide	Atomic mass	Natural abundance (%)	Half life $T_{1/2}$	Decay mode and energy (MeV)	Nuclear spin I	Nuclear magnetic moment μ	Uses
^{206}Bi		0	6.3d	EC(0.98); γ	6	+4.56	tracer
^{207}Bi		0	30.2y	EC(2.40); γ			
^{209}Bi	208.9804	100	stable		9/2	+4.080	NMR
^{210m}Bi		0	3×10^6y	α(4.96); β^-; γ			
^{210}Bi		trace	5.01d	β^-(1.16); α; no γ	1	−0.0442	tracer
^{212}Bi		trace	60.6m	β^-(2.25); α(6.206); γ			
^{214}Bi		trace	19.7m	β^-(3.28); α(5.616); γ			

NMR

	^{209}Bi
Relative sensitivity (^{1}H = 1.00)	0.13
Absolute sensitivity (^{1}H = 1.00)	0.13
Receptivity (^{13}C = 1.00)	777
Magnetogyric ratio/rad T^{-1} s^{-1}	4.2986×10^7
Quadrupole moment/m^2	-0.4×10^{-28}

Frequency (^{1}H = 100 MHz; 2.3488 T)/MHz 16.069
Reference: $KBiF_6$

Ground state electron configuration: $[Xe]4f^{14}5d^{10}6s^26p^3$
Term symbol: $^4S_{3/2}$
Electron affinity (M→M$^-$)/kJ mol^{-1}: 101

Ionization energies/kJ mol^{-1}

1. M → M$^+$	703.2	6. M^{5+} → M^{6+}	8 520
2. M$^+$ → M^{2+}	1610	7. M^{6+} → M^{7+}	(10 300)
3. M^{2+} → M^{3+}	2466	8. M^{7+} → M^{8+}	(12 300)
4. M^{3+} → M^{4+}	4372	9. M^{8+} → M^{9+}	(14 300)
5. M^{4+} → M^{5+}	5400	10. M^{9+} → M^{10+}	(16 300)

Principal lines in atomic spectrum

Wavelength/nm	Species	Sensitivity	Application
202.121	I		AA
206.170	I		AA AE
222.825	I		AA
289.798	I	U2	AA AE
306.772	I	U1	AA AE

Abundance: Earth's crust 0.008 p.p.m.; seawater 0.0002 p.p.m.
Biological role: None; non-toxic

Chemical properties

Occurs as large deposits of borax in US and Turkey. Amorphous boron is dark powder unreactive to oxygen, water, acids, and alkalis. Forms metal borides with most metals. Used in glass, bleaches and fireproofing.

Radii/pm: B^{3+} 23; covalent 88; atomic 83;
van der Waals 208

Electronegativity: 2.04 (Pauling); 2.01 (Allred-Rochow)

Effective nuclear charge: 2.60 (Slater); 2.42 (Clementi);
2.27 (Froese–Fischer)

Standard reduction potentials $E^{\ominus}/V$

III		0
$B(OH)_3$	$\xrightarrow{-0.890}$	B
BF_4^-	$\xrightarrow{-1.284}$	B

Covalent bonds

Covalent bonds	r/pm	E/kJ mol^{-1}
B–H	119	381
B–H–B	132	439
B–C	156	372
B–O	136	523
B–F	129	644
B–Cl	174	444
B–B	175	335

Oxidation states

B^{III} B_2O_3, H_3BO_3
($=B(OH)_3$),
borates e.g. borax
$Na_2[B_4O_5(OH)_4] \cdot 8H_2O$,
B_2H_6, B_4H_{10} etc.,
$NaBH_4$, BF_3,
BCl_3 etc.

Physical properties

Melting point/K: 2573

Boiling point/K: 3931

ΔH_{fusion}/kJ mol^{-1}: 22.2

ΔH_{vap}/kJ mol^{-1}: 504.5

Thermodynamic properties (298.15 K, 0.1 MPa)

State	$\Delta_f H^{\ominus}$/kJ mol^{-1}	$\Delta_f G^{\ominus}$/kJ mol^{-1}	$S^{\ominus}$/J K^{-1} mol^{-1}	C_p/J K^{-1} mol^{-1}
Solid (α)	0	0	5.86	11.09
Gas	562.7	518.8	153.45	20.799

Density/kg m^{-3}: 2340 (β-rhomb.) [293 K]

Thermal conductivity/W m^{-1} K^{-1}: 27.0 [300 K]

Electrical resistivity/Ω m: 18 000 [273 K]

Mass magnetic susceptibility/kg^{-1} m^3: -7.8×10^{-9} (s)

Molar volume/cm^3: 4.62

Coefficient of linear thermal expansion/K^{-1}: 5×10^{-6}

Lattice structure (cell dimensions/pm), space group

Tetragonal ($a = 874.0$; $c = 506$), $P4_2/nnm$
α-B rhombohedral ($a = 506.7$, $\alpha = 58°4'$), R$\bar{3}$m
β-B rhombohedral ($a = 1014.5$, $\alpha = 65°12'$), R$\bar{3}$m, R32, R3m
Rhombic ($a = 1015$, $b = 895$, $c = 1790$)
Monoclinic ($a = 1013$, $b = 893$, $c = 1786$, $\alpha \simeq 90°$, $\beta \simeq 90°$, $\gamma \simeq 90°$) or triclinic
Hexagonal ($a = 1198$, $c = 954$)

X-ray diffraction: mass absorption coefficients (μ/ρ)/cm^2 g^{-1}:
CuK$_\alpha$ 2.39   MoK$_\alpha$ 0.392

Atomic number: 5

**Thermal neutron capture
cross-section**/barns: 3837 ^{10}B; 0.005 ^{11}B

Number of isotopes (including nuclear isomers): 6

Isotope mass range: $8 \rightarrow 13$

Key isotopes

Nuclide	Atomic mass	Natural abundance (%)	Half life $T_{1/2}$	Decay mode and energy (MeV)	Nuclear spin I	Nuclear magnetic moment μ	Uses
^{10}B	10.0129	20.0	stable		3	+1.8006	NMR
^{11}B	11.009 31	80.0	stable		3/2	+2.6885	NMR

NMR

	$[^{10}B]$	^{11}B
Relative sensitivity (^{1}H = 1.00)	1.99×10^{-2}	0.17
Absolute sensitivity (^{1}H = 1.00)	3.90×10^{-3}	0.13
Receptivity (^{13}C = 1.00)	22.1	754
Magnetogyric ratio/rad T^{-1} s^{-1}	2.8740×10^7	8.5794×10^7
Quadrupole moment/m^2	7.4×10^{-30}	3.55×10^{-30}
Frequency (^{1}H = 100 MHz; 2.3488 T)/MHz	10.746	32.084

Reference: Et_2O/BF_3

Ground state electron configuration: $[He]2s^2 2p^1$

Term symbol: $^2P_{1/2}$

Electron affinity ($M \rightarrow M^-$)/kJ mol^{-1}: 23

Ionization energies/kJ mol^{-1}

1. M	$\rightarrow M^+$	800.6
2. M$^+$	$\rightarrow M^{2+}$	2 427
3. M^{2+}	$\rightarrow M^{3+}$	3 660
4. M^{3+}	$\rightarrow M^{4+}$	25 025
5. M^{4+}	$\rightarrow M^{5+}$	32 822

Principal lines in atomic spectrum

Wavelength/nm	Species	Sensitivity	Application
249.678*	I	U2	AA AE
249.773	I	U1	AA AE
345.129	II	V2	AE
1116.004	I		

* Lowest energy transition from ground state to nearest empty orbital

Abundance: Earth's crust 9 p.p.m.; seawater 4.8 p.p.m.

Biological role: Essential for plants: toxic in excess

Br

Chemical properties

Deep red, dense, sharp smelling liquid, Br_2, obtained from natural brines or seawater by treatment with Cl_2. Compounds used in fuel additives, pesticides, flame-retardants, and photography.

Radii/pm: covalent 114.2; Br^- 196; van der Waals 195
Electronegativity: 2.96 (Pauling); 2.74 (Allred-Rochow)
Effective nuclear charge: 7.60 (Slater); 9.03 (Clementi); 10.89 (Froese–Fischer)

Standard reduction potentials $E^{\ominus}$/V

	VII	V	I	0	$-$I

Acid solution

$$BrO_4^- \xrightarrow{1.853} BrO_3^- \xrightarrow{1.447} HBrO \xrightarrow{1.604} Br_2 \xrightarrow{1.0652} Br^-$$

(1.478 over BrO_3^- to Br_2; Br_2 (aq) $\xrightarrow{1.0874}$; 1.341)

Alkaline solution

$$BrO_4^- \xrightarrow{1.025} BrO_3^- \xrightarrow{0.492} BrO^- \xrightarrow{0.455} Br_2 \xrightarrow{1.0652} Br^-$$

(0.766; 0.584)

Covalent bonds r/pm E/kJ mol^{-1}

	r/pm	E/kJ mol^{-1}
Br–H	140.8	366
Br–C	194	285
Br–O	160	234
Br–F	176	285
Br–Br	229	193
Br–B	187	410
Br–Si	215	310
Br–P	218	264

Oxidation states

Br^{-I} ([Kr])	Br^-(aq), HBr, KBr etc.
Br^I (s^2p^4)	Br_2O, $BrCl_2^-$
Br^{III} (s^2p^2)	BrF_3, BrF_4^-
Br^{IV} (s^2p^1)	BrO_2
Br^V (s^2)	BrO_3^-(aq), BrF_5, BrF_6^-
Br^{VII} (d^{10})	$KBrO_4$, BrF_6^+

Physical properties

Melting point/K: 265.9
Boiling point/K: 331.93
ΔH_{fusion}/kJ mol^{-1}: 10.8
ΔH_{vap}/kJ mol^{-1}: 30.5

Thermodynamic properties (298.15 K, 0.1 MPa)

State	$\Delta_f H^{\ominus}$/kJ mol^{-1}	$\Delta_f G^{\ominus}$/kJ mol^{-1}	$S^{\ominus}$/J K^{-1} mol^{-1}	C_p/J K^{-1} mol^{-1}
Liquid	0	0	152.231	75.689
Gas (atom)	111.884	82.396	175.022	20.786

Density/kg m^{-3}: 4050 [123 K]; 3122.6 [293 K]; 7.59 [gas]
Thermal conductivity/W m^{-1} K^{-1}: 0.122 [300 K]$_l$
Mass magnetic susceptibility/kg^{-1} m^3: -4.44×10^{-9} (l)
Molar volume/cm^3: 19.73 [123 K]

Lattice structure (cell dimensions/pm), space group

Orthorhombic (120 K) ($a = 673.7$; $b = 454.8$; $c = 876.1$), Cmca

X-ray diffraction: mass absorption coefficients (μ/ρ)/cm^2 g^{-1}:
CuK$_\alpha$ 99.6   MoK$_\alpha$ 79.8

Bromine

Atomic number: 35

Thermal neutron capture cross-section/barns: 6.8 ± 0.1

Number of isotopes (including nuclear isomers): 19

Isotope mass range: $74 \rightarrow 90$

**Nuclear
properties**

Key isotopes

Nuclide	Atomic mass	Natural abundance (%)	Half life $T_{1/2}$	Decay mode and energy (MeV)	Nuclear spin I	Nuclear magnetic moment μ	Uses
^{77}Br		0	57h	β^+, EC(1.365); γ	3/2		tracer
^{79}Br	79.9183	50.69	stable		3/2	$+2.1055$	NMR
^{81}Br	80.9163	49.31	stable		3/2	$+2.2696$	NMR
^{82}Br	81.917	0	35.5h	β^-(3.092); γ	5	± 1.626	tracer

NMR

	$[^{79}$Br$]$	^{81}Br
Relative sensitivity (^{1}H$=1.00$)	7.86×10^{-2}	9.85×10^{-2}
Absolute sensitivity (^{1}H$=1.00$)	3.97×10^{-2}	4.87×10^{-2}
Receptivity (^{13}C$=1.00$)	226	277
Magnetogyric ratio/rad T^{-1} s^{-1}	6.7023×10^7	7.2246×10^7
Quadrupole moment/m^2	0.33×10^{-28}	0.28×10^{-28}
Frequency (^{1}H$=100$ MHz; 2.3488 T)/MHz	25.053	27.006

Reference: NaBr(aq)

Ground state electron configuration: $[Ar]3d^{10}4s^24p^5$

Term symbol: $^2P_{3/2}$

Electron affinity (M$\rightarrow$M$^-$)/kJ mol^{-1}: 324.5

**Electron
shell
properties**

Ionization energies/kJ mol^{-1}

1.	M $\rightarrow$ M$^+$	1139.9	6. M$^{5+}\rightarrow$M^{6+}	8 550
2.	M$^+\rightarrow$M^{2+}	2104	7. M$^{6+}\rightarrow$M^{7+}	9 940
3.	M$^{2+}\rightarrow$M^{3+}	3500	8. M$^{7+}\rightarrow$M^{8+}	18 600
4.	M$^{3+}\rightarrow$M^{4+}	4560	9. M$^{8+}\rightarrow$M^{9+}	(23 900)
5.	M$^{4+}\rightarrow$M^{5+}	5760	10. M$^{9+}\rightarrow$M^{10+}	(28 100)

Principal lines in atomic spectrum

Wavelength/nm	Species	Sensitivity	Application
154.065* (vac)	I		AE
470.486	II	V1	AE
478.550	II	V2	AE
481.671	II	V3	AE
635.073	I		

* Lowest energy transition from ground state to nearest empty orbital

Abundance: Earth's crust 2.5 p.p.m.; seawater 65 p.p.m.

Biological role: None; Br$^-$ slightly toxic; Br$_2$ very toxic

Cd

Atomic number: 48
Relative atomic mass ($^{12}C = 12.0000$): 112.411

Chemical properties

Silvery metal obtained as by-product from zone refining. Tarnishes in air, soluble in acids but not alkalis. Some used in rechargeable batteries and alloys.

Radii/pm: Cd^{2+} 103; Cd^{+} 114; covalent 141; atomic 148.9
Electronegativity: 1.69 (Pauling); 1.46 (Allred-Rochow)
Effective nuclear charge: 4.35 (Slater); 8.19 (Clementi); 11.58 (Froese–Fischer)

Standard reduction potentials $E^{\ominus}$/V

	II		0
Acid solution	Cd^{2+}	-0.4025	Cd
Alkaline solution	$Cd(OH)_2$	-0.824	Cd
	$[Cd(NH_3)_4]^{2+}$	-0.622	Cd
	$[Cd(CN)_4]^{2-}$	-1.09	Cd

Oxidation states

Cd^I ($d^{10} s^1$)	rare $Cd_2[AlCl_4]_2$
Cd^{II} (d^{10})	CdO (basic), CdS, $Cd(OH)_2$, CdF_2, $CdCl_2$ etc., many salts, $[Cd(H_2O)_6]^{2+}$ (aq), many complexes, e.g. $[Cd(SCN)_4]^{2-}$

Physical properties

Melting point/K: 594.1
Boiling point/K: 1038
ΔH_{fusion}/kJ mol^{-1}: 6.11
ΔH_{vap}/kJ mol^{-1}: 100.0

Thermodynamic properties (298.15 K, 0.1 MPa)

State	$\Delta_f H^{\ominus}$/kJ mol^{-1}	$\Delta_f G^{\ominus}$/kJ mol^{-1}	$S^{\ominus}$/J K^{-1} mol^{-1}	C_p/J K^{-1} mol^{-1}
Solid	0	0	51.76	25.98
Gas	112.01	77.41	167.746	20.786

Density/kg m^{-3}: 8650 [293 K]; 7996 [liquid at m.p.]
Thermal conductivity/W m^{-1} K^{-1}: 96.8 [300 K]
Electrical resistivity/Ω m: 6.83×10^{-8} [273 K]
Mass magnetic susceptibility/kg^{-1} m^3: -2.21×10^{-9} (s)
Molar volume/cm^3: 13.00
Coefficient of linear thermal expansion/K^{-1}: 29.8×10^{-6}

Lattice structure (cell dimensions/pm), space group

h.c.p. ($a = 297.94$; $c = 561.86$), P6$_3$/mmc

X-ray diffraction: mass absorption coefficients (μ/ρ)/cm^2 g^{-1}:
CuK$_\alpha$ 231   MoK$_\alpha$ 27.5

Cadmium

Atomic number: 48
Thermal neutron capture cross-section/barns: 2450 ± 20
Number of isotopes (including nuclear isomers): 22
Isotope mass range: $103 \rightarrow 119$

Key isotopes

Nuclide	Atomic mass	Natural abundance (%)	Half life $T_{1/2}$	Decay mode and energy (MeV)	Nuclear spin I	Nuclear magnetic moment μ	Uses
^{106}Cd	105.907	1.25	stable				
^{108}Cd	107.9040	0.89	stable		0		
^{109}Cd		0	450d	EC(0.16); γ	5/2	-0.8270	tracer
^{110}Cd	109.9030	12.51	stable		0		
^{111}Cd	110.9042	12.81	stable		1/2	-0.5943	NMR
^{112}Cd	111.9028	24.13	stable		0		
^{113}Cd	112.9046	12.22	stable		1/2	-0.6217	NMR
^{114}Cd	113.9036	28.72	stable		0		
^{115m}Cd		0	43d	β^-(1.62); γ	11/2	-1.0400	tracer
^{115}Cd		0	53.5h	β^-(1.45); γ	1/2	-0.648	tracer
^{116}Cd	115.9050	7.47	stable		0		

NMR

	$[^{111}$Cd$]$	^{113}Cd
Relative sensitivity (^{1}H = 1.00)	9.54×10^{-3}	1.09×10^{-2}
Absolute sensitivity (^{1}H = 1.00)	1.21×10^{-3}	1.33×10^{-3}
Receptivity (^{13}C = 1.00)	6.93	7.6
Magnetogyric ratio/rad T^{-1} s^{-1}	-5.6714×10^7	-5.9328×10^7
Frequency (^{1}H = 100 MHz; 2.3488 T)/MHz	21.205	22.182

References: $Cd(ClO_4)_2$ (aq) and $Cd(CH_3)_2$

Ground state electron configuration: $[Kr]4d^{10}5s^2$
Term symbol: $^1S_0$
Electron affinity ($M \rightarrow M^-$)/kJ mol^{-1}: -26

Ionization energies/kJ mol^{-1}

1. $M \rightarrow M^+$	867.6	6. $M^{5+} \rightarrow M^{6+}$	(9 100)
2. $M^+ \rightarrow M^{2+}$	1631	7. $M^{6+} \rightarrow M^{7+}$	(11 100)
3. $M^{2+} \rightarrow M^{3+}$	3616	8. $M^{7+} \rightarrow M^{8+}$	(14 100)
4. $M^{3+} \rightarrow M^{4+}$	(5300)	9. $M^{8+} \rightarrow M^{9+}$	(16 400)
5. $M^{4+} \rightarrow M^{5+}$	(7000)	10. $M^{9+} \rightarrow M^{10+}$	(18 800)

Principal lines in atomic spectrum

Wavelength/nm	Species	Sensitivity	Application
214.441	II	V1	AE
226.502	II	V2	AE
228.802	I	U1	AA, AE
326.106*	I		AA, AE
643.847	I		AE

* Lowest energy transition from ground state to nearest empty orbital

Abundance: Earth's crust 0.16 p.p.m.; seawater 0.0001 p.p.m.
Biological role: Toxic; stimulatory; carcinogenic; teratogenic

Cs

Atomic number: 55
Relative atomic mass ($^{12}C=12.0000$): 132.9054

Chemical properties

Silvery-white, very soft metal; reacts rapidly with oxygen and explosively with water. Found as pollucite $Cs_4Al_4Si_9O_{26}.H_2O$ but most obtained as by-product of lithium production. Little used.

Radii/pm: Cs^+ 165; atomic 265.4; covalent 235; van der Waals 262
Electronegativity: 0.79 (Pauling); 0.86 (Allred-Rochow)
Effective nuclear charge: 2.20 (Slater); 6.36 (Clementi); 8.56 (Froese–Fischer)

Standard reduction potentials $E^\ominus$/V

I	0
Cs^+ $\xrightarrow{-2.923}$	Cs

Oxidation states

Cs^{-1} (s^2)	caesium metal in liquid ammonia
Cs^I ([Xe])	Cs_2O, Cs_2O_2, CsO_2, CsOH, CsH, CsF, CsCl etc., $[Cs(H_2O)_x]^+$(aq), Cs_2CO_3, many salts and salt hydrates, some complexes with crown ethers etc.

Physical properties

Melting point/K: 301.55
Boiling point/K: 951.6
ΔH_{fusion}/kJ mol^{-1}: 2.09
ΔH_{vap}/kJ mol^{-1}: 66.5

Thermodynamic properties (298.15 K, 0.1 MPa)

State	$\Delta_f H^\ominus$/kJ mol^{-1}	$\Delta_f G^\ominus$/kJ mol^{-1}	$S^\ominus$/J K^{-1} mol^{-1}	C_p/J K^{-1} mol^{-1}
Solid	0	0	85.23	32.17
Gas	76.065	49.121	175.595	20.786

Density/kg m^{-3}: 1873 [293 K]; 1843 [liquid at m.p.]
Thermal conductivity/W m^{-1} K^{-1}: 35.9 [300 K]
Electrical resistivity/Ω m: 20.0×10^{-8} [293 K]
Mass magnetic susceptibility/kg^{-1} m^3: $+2.8 \times 10^{-9}$ (s)
Molar volume/cm^3: 70.96
Coefficient of linear thermal expansion/K^{-1}: 97×10^{-6}

Lattice structure (cell dimensions/pm), space group

b.c.c. (78 K) ($a=614$), Im3m

High pressure forms: ($a=598.4$), Fm3m
($a=580.0$), Fm3m

X-ray diffraction: mass absorption coefficients (μ/ρ)/cm^2 g^{-1}: CuK$_\alpha$ 318 MoK$_\alpha$ 41.3

Caesium (Cesium)

Atomic number: 55
Thermal neutron capture cross-section/barns: 30.0 ± 1.5
Number of isotopes (including nuclear isomers): 22
Isotope mass range: $123 \rightarrow 144$

Key isotopes

Nuclide	Atomic mass	Natural abundance (%)	Half life $T_{1/2}$	Decay mode and energy (MeV)	Nuclear spin I	Nuclear magnetic moment μ	Uses
^{133}Cs	132.9051	100	stable		7/2	+2.5779	NMR
^{134}Cs		0	2.046y	β^- (2.062); γ	4	+2.989	tracer
^{135}Cs		0	3×10^6y	β^- (0.210); no γ	7/2	+2.7280	
^{137}Cs		0	30.23y	β^- (1.176); γ	7/2	+2.8372	tracer, medical

NMR

^{133}Cs

Relative sensitivity ($^1H = 1.00$) 4.74×10^{-2}
Absolute sensitivity ($^1H = 1.00$) 4.74×10^{-2}
Receptivity ($^{13}C = 1.00$)   269
Magnetogyric ratio/rad T^{-1} s^{-1} 3.5087×10^7
Quadrupole moment/m^2 -3×10^{-31}
Frequency ($^1H = 100$ MHz; 2.3488 T)/MHz   13.117
Reference: 0.5M CsBr(aq)

Ground state electron configuration: $[Xe]6s^1$
Term symbol: $^2S_{1/2}$
Electron affinity $(M \rightarrow M^-)$/kJ mol^{-1}:   45.5

Ionization energies/kJ mol^{-1}

1. $M \rightarrow M^+$	375.7	6. $M^{5+} \rightarrow M^{6+}$	(7 100)
2. $M^+ \rightarrow M^{2+}$	2420	7. $M^{6+} \rightarrow M^{7+}$	(8 300)
3. $M^{2+} \rightarrow M^{3+}$	(3400)	8. $M^{7+} \rightarrow M^{8+}$	(11 300)
4. $M^{3+} \rightarrow M^{4+}$	(4400)	9. $M^{8+} \rightarrow M^{9+}$	(12 700)
5. $M^{4+} \rightarrow M^{5+}$	(6000)	10. $M^{9+} \rightarrow M^{10+}$	(23 700)

Principal lines in atomic spectrum

Wavelength/nm	Species	Sensitivity	Application
455.528	I	U3	AA AE
459.317	I	U4	AA AE
460.376	II		
852.110	I	U1	AA AE
894.350*	I	U2	AA AE

* Lowest energy transition from ground state to nearest empty orbital

Abundance: Earth's crust 2.6 p.p.m.; seawater 0.0005 p.p.m.
Biological role: None; non-toxic

<table>
<tr><td>**Ca**</td><td>Atomic number: 20
Relative atomic mass ($^{12}C = 12.0000$): **40.078**</td></tr>
</table>

Chemical properties	Silvery white, relatively soft metal obtained from fused calcium chloride by electrolysis. Protected by oxide/nitride film and can be worked as a metal. Attacked by oxygen and water. Used in alloys.

Radii/pm: Ca^{2+} 106; covalent 174; atom (α form) 197.3

Electronegativity: 1.00 (Pauling); 1.04. (Allred-Rochow)

Effective nuclear charge: 2.85 (Slater); 4.40 (Clementi);
5.69 (Froese–Fischer)

Standard reduction potentials $E^{\ominus}/V$

Oxidation states

Ca^{II} ([Ar])	CaO, CaO_2 (peroxide), $Ca(OH)_2$, CaH_2, CaF_2, $CaCl_2$ etc., Ca^{2+}(aq), $CaCO_3$, $CaSO_4 . 2H_2O$ (gypsum) $CaSO_4 . \frac{1}{2}H_2O$ (plaster of Paris) CaC_2 (calcium carbide), many salts, few complexes

Physical properties	**Melting point**/K: 1112 **Boiling point**/K: 1757 ΔH_{fusion}/kJ mol^{-1}: 9.33 ΔH_{vap}/kJ mol^{-1}: 150.6

Thermodynamic properties (298.15 K, 0.1 MPa)

State	$\Delta_f H^{\ominus}$/kJ mol^{-1}	$\Delta_f G^{\ominus}$/kJ mol^{-1}	$S^{\ominus}$/J K^{-1} mol^{-1}	C_p/J K^{-1} mol^{-1}
Solid (α)	0	0	41.42	25.31
Gas	178.2	144.3	154.884	20.786

Density/kg m^{-3}: 1550 [293 K]; 1365 [liquid at m.p.]

Thermal conductivity/W m^{-1} K^{-1}: 200 [300 K]

Electrical resistivity/Ω m: 3.43×10^{-8} [293 K]

Mass magnetic susceptibility/kg^{-1} m^3: $+1.4 \times 10^{-8}$ (s)

Molar volume/cm^3: 25.86

Coefficient of linear thermal expansion/K^{-1}: 22×10^{-6}

Lattice structure (cell dimensions/pm), space group

α-Ca f.c.c. ($a = 558.84$), Fm3m
β-Ca f.c.c. ($a = 448.0$), Im3m
γ-Ca h.c.p. ($a = 397$, $c = 649$), P6$_3$/mmc, [may contain H]
$T(\alpha \rightarrow \beta) = 573$ K
$T(\beta \rightarrow \gamma) = 723$ K

X-ray diffraction: mass absorption coefficients (μ/ρ)/cm^2 g^{-1}:
CuK$_\alpha$ 162   MoK$_\alpha$ 18.3

Calcium

Atomic number: 20
Thermal neutron capture cross-section/barns: 0.44 ± 0.02
Number of isotopes (including nuclear isomers): 14
Isotope mass range: 37→50

Nuclear properties

Key isotopes

Nuclide	Atomic mass	Natural abundance (%)	Half life $T_{1/2}$	Decay mode and energy (MeV)	Nuclear spin I	Nuclear magnetic moment μ	Uses
^{40}Ca	39.962 59	96.941	stable		0		
^{42}Ca	41.958 63	0.647	stable				
^{43}Ca	42.958 78	0.135	stable		7/2	−1.3172	NMR
^{44}Ca	43.955 49	2.086	stable				
^{45}Ca	44.956	0	165d	β^- (0.252); no γ			tracer
^{46}Ca	45.9537	0.004	stable				
^{47}Ca	46.954	0	4.53d	β^- (1.979); γ			tracer
^{48}Ca	47.9524	0.187	stable				

NMR

^{43}Ca

Relative sensitivity (^{1}H = 1.00) 6.40×10^{-3}
Absolute sensitivity (^{1}H = 1.00) 9.28×10^{-6}
Receptivity (^{13}C = 1.00) 0.0527
Magnetogyric ratio/rad T^{-1} s^{-1} -1.8001×10^7
Quadrupole moment/m^2 -0.05×10^{-28}
Frequency (^{1}H = 100 MHz; 2.3488 T)/MHz 6.728
Reference: $CaCl_2$(aq)

Ground state electron configuration: [Ar]$4s^2$
Term symbol: $^1S_0$
Electron affinity (M→M$^-$)/kJ mol^{-1}: −186

Electron shell properties

Ionization energies/kJ mol^{-1}

1. M →M$^+$	589.7	6. M^{5+}→M^{6+}	10 496
2. M$^+$→M^{2+}	1145	7. M^{6+}→M^{7+}	12 320
3. M^{2+}→M^{3+}	4910	8. M^{7+}→M^{8+}	14 207
4. M^{3+}→M^{4+}	6474	9. M^{8+}→M^{9+}	18 191
5. M^{4+}→M^{5+}	8144	10. M^{9+}→M^{10+}	20 385

Principal lines in atomic spectrum

Wavelength/nm	Species	Sensitivity	Application
239.856	I		AA
393.366	II	V1	AE
422.673	I	U1	AA AE
443.496	I	U3	AE
445.478	I	U2	AE
657.278*	I		

* Lowest energy transition from ground state to nearest empty orbital

Abundance: Earth's crust 46 600 p.p.m. (limestone and dolomite deposits); seawater 400 p.p.m.
Biological role: Essential constituent of cells, fluids, and bones

Cf	Atomic number: 98
	Relative atomic mass ($^{12}C = 12.0000$): **(251)**

Chemical properties

Radioactive, silvery metal obtained in mg quantities as ^{249}Cf and ^{252}Cf from neutron bombardment of ^{239}Pu. Attacked by oxygen, steam, and acids, but not alkalis. ^{252}Cf used in cancer therapy.

Radii/pm: Cf^{4+} 86; Cf^{3+} 98; Cf^{2+} 117

Electronegativity: 1.3 (Pauling); n.a. (Allred-Rochow)

Effective nuclear charge: 1.65 (Slater)

Standard reduction potentials $E^\ominus$/V

Acid solution

[Cf^{IV} is reduced by water within minutes]

Oxidation states

Cf^{II} (f^{10})	CfO ?, $CfBr_2$, CfI_2
Cf^{III} (f^9)	Cf_2O_3, CfF_3, $CfCl_3$ etc.
	Cf^{3+}(aq), [$Cf(C_5H_5)_3$]
Cf^{IV} (f^8)	CfO_2, CfF_4

Physical properties

Melting point/K: n.a.

Boiling point/K: n.a.

ΔH_{fusion}/kJ mol^{-1}: n.a.

ΔH_{vap}/kJ mol^{-1}: n.a.

Thermodynamic properties (298.15 K, 0.1 MPa)

State	$\Delta_f H^\ominus$/kJ mol^{-1}	$\Delta_f G^\ominus$/kJ mol^{-1}	$S^\ominus$/J K^{-1} mol^{-1}	C_p/J K^{-1} mol^{-1}
Solid	0	0	n.a.	25.98
Gas	n.a.	n.a.	n.a.	n.a.

Density/kg m^{-3}: n.a.

Thermal conductivity/W m^{-1} K^{-1}: 10 (est.) [300 K]

Electrical resistivity/Ω m: n.a.

Mass magnetic susceptibility/kg^{-1} m^3: n.a.

Molar volume/cm^3: n.a.

Coefficient of linear thermal expansion/K^{-1}: n.a.

Lattice structure (cell dimensions/pm), space group

Cubic

X-ray diffraction: mass absorption coefficients (μ/ρ)/cm^2 g^{-1}: n.a.

Produced in 1950 by S. G. Thompson, K. Street Jr, A. Ghiorso, and G. T. Seaborg at Berkeley, California, USA
[English, *California*]

Californium

Atomic number: 98

Thermal neutron capture cross-section/barns: 2100 ± 1000 (^{251}Cf)

Number of isotopes (including nuclear isomers): 12

Isotope mass range: $242 \rightarrow 254$

Key isotopes

Nuclide	Atomic mass	Natural abundance (%)	Half life $T_{1/2}$	Decay mode and energy (MeV)	Nuclear spin I	Nuclear magnetic moment μ	Uses
^{249}Cf	249.0748	0	360y	α(6.295)			
^{251}Cf		0	900y	α(5.94)			
^{252}Cf		0	2.65y	α(6.217); SF			tracer, medical

Ground state electron configuration: [Rn]$5f^{10}7s^2$

Term symbol: $^5I_8$

Electron affinity $(M \rightarrow M^-)$/kJ mol^{-1}: n.a.

Ionization energies/kJ mol^{-1}

1. $M \rightarrow M^+$	608	6. $M^{5+} \rightarrow M^{6+}$	
2. $M^+ \rightarrow M^{2+}$		7. $M^{6+} \rightarrow M^{7+}$	
3. $M^{2+} \rightarrow M^{3+}$		8. $M^{7+} \rightarrow M^{8+}$	
4. $M^{3+} \rightarrow M^{4+}$		9. $M^{8+} \rightarrow M^{9+}$	
5. $M^{4+} \rightarrow M^{5+}$		10. $M^{9+} \rightarrow M^{10+}$	

Selected lines in atomic spectrum

Wavelength/nm	Species	Sensitivity	Application
339.222	I		
353.149	I		
354.098	I		
359.877	I		
360.532	I		
361.749	I		
366.270	I		

Abundance: Nil

Biological role: Toxic due to radioactivity

Atomic number: 6

Relative atomic mass ($^{12}C = 12.0000$): 12.011

Found naturally as element (graphite and diamond) but mainly as hydrocarbons (methane gas, oil, and coal) and carbonates (limestone and dolomite). Used as coke (steel making), carbon black (printing), and activated charcoal (sugar refining, etc.).

Radii/pm: covalent bond 77; double 67; triple 60; C^{4-} 260; van der Waals 185

Electronegativity: 2.55 (Pauling); 2.50 (Allred-Rochow)

Effective nuclear charge: 3.25 (Slater); 3.14 (Clementi); 2.87 (Froese–Fischer)

Standard reduction potentials $E^{\ominus}/V$

	IV		II		0		−IV		
Acid solution	CO_2	$\xrightarrow{-0.106}$	CO	$\xrightarrow{0.517}$	C	$\xrightarrow{0.132}$	CH_4		
	CO_2	$\xrightarrow{-0.20}$	HCO_2H	$\xrightarrow{0.034}$	HCHO	$\xrightarrow{0.232}$	CH_3OH	$\xrightarrow{0.59}$	CH_4
Alkaline solution	CO_2	$\xrightarrow{-1.01}$	HCO_2^-	$\xrightarrow{-1.07}$	HCHO	$\xrightarrow{0.59}$	CH_3OH	$\xrightarrow{-0.2}$	CH_4

Covalent bonds*	r/pm	E/kJ mol^{-1}
C–H	109.3	339
C–C	154	348
C=C	134	614
C≡C	120	839
C=N	130	615
C≡N	116	891
C=O	123	745
C≡O	112.8	1074

*See also other elements for bonds to carbon

Oxidation states

This concept is rarely used in discussing carbon and its compounds because of their subtleties of bonding. However, for simple compounds with a single carbon we can use it:

C^{-IV} CH_4
C^{II} CO
C^{IV} CO_2, CO_3^{2-}, CF_4, etc.

Melting point/K: c. 3820 (diamond)

Boiling point/K: 5100 (sublimes)

ΔH_{fusion}/kJ mol^{-1}: 105.0

ΔH_{vap}/kJ mol^{-1}: 710.9

Thermodynamic properties (298.15 K, 0.1 MPa)

State	$\Delta_f H^{\ominus}$/kJ mol^{-1}	$\Delta_f G^{\ominus}$/kJ mol^{-1}	$S^{\ominus}$/J K^{-1} mol^{-1}	C_p/J K^{-1} mol^{-1}
Solid (graphite)	0	0	5.740	8.527
Solid (diamond)	1.895	2.900	2.377	6.113
Gas	716.682	671.257	158.096	20.838

Density/kg m^{-3}: 3513 (diam.); 2260 (graph.) [293 K]

Thermal conductivity/W m^{-1} K^{-1}: 990–2320 (diam.); $5.7^{\perp}$, $1960^{\parallel}$ (graph.) [298 K]

Electrical resistivity/Ω m: 10^{11} (diam.); 1.375×10^{-5} (graph.) [293 K]

Mass magnetic susceptibility/kg^{-1} m^3: -6.3×10^{-9} (graph.); -6.2×10^{-9} (diam.)

Molar volume/cm^3: 3.42 (diam.)

Coefficient of linear thermal expansion/K^{-1}: 1.19×10^{-6} (diam.)

Lattice structure (cell dimensions/pm), space group

Cubic diamond ($a = 356.703$), Fd3m
Hexagonal graphite ($a = 246.12$, $c = 670.78$), P6$_3$mc
Rhombohedral graphite ($a = 364.2$, $\alpha = 39°30'$), R3m
Hexagonal diamond ($a = 252$, $c = 412$), P6$_3$/mmc
Hexagonal ($a = 894.8$, $c = 1408$)

X-ray diffraction: mass absorption coefficients (μ/ρ)/cm^2 g^{-1}:
CuK$_\alpha$ 4.60   MoK$_\alpha$ 0.625

Atomic number: 6
Thermal neutron capture cross-section/barns: 0.0034 ± 0.0002
Number of isotopes (including nuclear isomers): 7
Isotope mass range: $9 \rightarrow 16$

Nuclear properties

Key isotopes

Nuclide	Atomic mass	Natural abundance (%)	Half life $T_{1/2}$	Decay mode and energy (MeV)	Nuclear spin I	Nuclear magnetic moment μ	Uses
^{12}C	12.0000	98.90	stable		0		
^{13}C	13.003 35	1.10	stable		1/2	+0.7024	NMR
^{14}C	14.0032	trace	5730y	β^- (0.156); no γ	0		tracer

NMR

^{13}C

Relative sensitivity ($^1H = 1.00$) 1.59×10^{-2}
Absolute sensitivity ($^1H = 1.00$) 1.76×10^{-4}
Receptivity ($^{13}C = 1.00$)   1.00 (by definition)
Magnetogyric ratio/rad T^{-1} s^{-1} 6.7263×10^7
Frequency ($^1H = 100$ MHz; 2.3488 T)/MHz   25.144
Reference: $Si(CH_3)_4$

Ground state electron configuration: [He]$2s^2 2p^2$
Term symbol: $^3P_0$
Electron affinity ($M \rightarrow M^-$)/kJ mol^{-1}: 122.5

Electron shell properties

Ionization energies/kJ mol^{-1}

1. $M \rightarrow M^+$ 1 086.2	6. $M^{5+} \rightarrow M^{6+}$ 47 270
2. $M^+ \rightarrow M^{2+}$ 2 352	
3. $M^{2+} \rightarrow M^{3+}$ 4 620	
4. $M^{3+} \rightarrow M^{4+}$ 6 222	
5. $M^{4+} \rightarrow M^{5+}$ 37 827	

Principal lines in atomic spectrum

Wavelength/nm	Species	Sensitivity	Application
247.856	I	U2	AE
283.671	II	V4	AE
283.760	II	V5	AE
296.485*	I		
426.700	II	V3	AE
426.726	II	V2	AE

* Lowest energy transition from ground state to nearest empty orbital

Abundance: Atmosphere CO_2 335 p.p.m., CH_4 1.7 p.p.m.; CO up to 0.02 p.p.m. (all by volume); Earth's crust 180 p.p.m.; seawater 28 p.p.m.
Biological role: Basis of all life (part of DNA molecule)

Ce

Atomic number: 58

Relative atomic mass ($^{12}C = 12.0000$): 140.115

Chemical properties

Reactive, grey metal; most abundant of the lanthanide metals. Obtained from bastnaesite and monazite. Tarnishes in air, burns when heated, reacts rapidly with water, dissolves in acids. Used in glass, flints, ceramics.

Radii/pm: Ce^{3+} 107; Ce^{4+} 94; atomic 182.5; covalent 165

Electronegativity: 1.12 (Pauling); 1.06 (Allred-Rochow)

Effective nuclear charge: 2.85 (Slater); 10.80 (Clementi); 10.57 (Froese–Fischer)

Standard reduction potentials $E^{\ominus}/V$

	IV		III		0
Acid solution	Ce^{4+}	$\underline{1.72}$	Ce^{3+}	$\underline{-2.34}$	Ce
Alkaline solution	CeO_2	$\underline{-0.7}$	$Ce(OH)_3$	$\underline{-2.78}$	Ce

Oxidation states

Ce^{III} ($4f^1$)	Ce_2O_3, $Ce(OH)_3$, CeF_3, $CeCl_3$ etc., Ce^{3+} salts, $[Ce(H_2O)_6]^{3+}$ (aq), complexes
Ce^{IV} ([Xe])	CeO_2, CeF_4, $CaCl_6^{2-}$, $[Ce(NO_3)_6]^{2-}$ (aq)

Physical properties

Melting point/K: 1072

Boiling point/K: 3699

ΔH_{fusion}/kJ mol^{-1}: 8.87

ΔH_{vap}/kJ mol^{-1}: 398

Thermodynamic properties (298.15 K, 0.1 MPa)

State	$\Delta_f H^{\ominus}$/kJ mol^{-1}	$\Delta_f G^{\ominus}$/kJ mol^{-1}	$S^{\ominus}$/J K^{-1} mol^{-1}	C_p/J K^{-1} mol^{-1}
Solid	0	0	72.0	26.94
Gas	423	385	191.776	23.075

Density/kg m^{-3}: 8240 (α); 6749 (β); 6773 (γ); 6700 (δ) [298 K]

Thermal conductivity/W m^{-1} K^{-1}: 11.4 [300 K]

Electrical resistivity/Ω m: 73×10^{-8} [273 K]

Mass magnetic susceptibility/kg^{-1} m^3: $+2.17 \times 10^{-7}$ (s)

Molar volume/cm^3: 17.00

Coefficient of linear thermal expansion/K^{-1}: 8.5×10^{-6}

Lattice structure (cell dimensions/pm), space group

α-Ce f.c.c. ($a = 485$), Fm3m
β-Ce hexagonal ($a = 367.3$, $c = 1180.2$), P6$_3$/mmc
γ-Ce f.c.c. ($a = 516.01$), Fm3m
δ-Ce f.c.c. ($a = 412$), Im3m
$T(\beta \rightarrow \gamma) = 441$ K

X-ray diffraction: mass absorption coefficients (μ/ρ)/cm^2 g^{-1}:
CuK$_\alpha$ 352   MoK$_\alpha$ 48.2

Discovered in 1803 by J. J. Berzelius and W. Hisinger at Vestmanland, Sweden. First isolated by W. F. Hillebrand and T. H. Norton in 1875 at Washington, DC, USA
[Named after asteroid Ceres, discovered in 1801]

Cerium

Atomic number: 58
Thermal neutron capture cross-section/barns: 0.73 ± 0.08
Number of isotopes (including nuclear isomers): 19
Isotope mass range: $132 \rightarrow 148$

Nuclear properties

Key isotopes

Nuclide	Atomic mass	Natural abundance (%)	Half life $T_{1/2}$	Decay mode and energy (MeV)	Nuclear spin I	Nuclear magnetic moment μ	Uses
^{136}Ce		0.19	stable				
^{138}Ce	137.9057	0.25	stable				
^{139}Ce		0	140d	EC(0.27); γ	3/2	± 0.9	tracer
^{140}Ce	139.9053	88.48	stable				
^{141}Ce		0	32.5d	β^- (0.581); γ	7/2	± 0.97	tracer
^{142}Ce	141.9090	11.08	stable				
^{143}Ce		0	33h		3/2	$c. \pm 1$	tracer
^{144}Ce		0	284.9h	β^- (0.31); γ			

Ground state electron configuration: $[Xe]4f^2 6s^2$
Term symbol: $^3H_4$
Electron affinity $(M \rightarrow M^-)$/kJ mol^{-1}: ≤ 50

Electron shell properties

Ionization energies/kJ mol^{-1}

1. $M \rightarrow M^+$	527.4	6. $M^{5+} \rightarrow M^{6+}$ (8 200)
2. $M^+ \rightarrow M^{2+}$	1047	7. $M^{6+} \rightarrow M^{7+}$ (9 700)
3. $M^{2+} \rightarrow M^{3+}$	1949	8. $M^{7+} \rightarrow M^{8+}$ (11 800)
4. $M^{3+} \rightarrow M^{4+}$	3547	9. $M^{8+} \rightarrow M^{9+}$ (13 200)
5. $M^{4+} \rightarrow M^{5+}$ (6800)		10. $M^{9+} \rightarrow M^{10+}$ (14 700)

Principal lines in atomic spectrum

Wavelength/nm	Species	Sensitivity	Application
399.924	II		AE
418.660	II		AE
522.349	I		AA
569.699	I		AA
739.777*	I		

* Lowest energy transition from ground state to nearest empty orbital

Abundance: Earth's crust 66 p.p.m.; seawater 0.0004 p.p.m.
Biological role: None; stimulatory

<table>
<tr><td>**Cl**</td><td>**Atomic number: 17**
Relative atomic mass ($^{12}C = 12.0000$): 35.4527</td></tr>
</table>

Chemical properties

Yellow-green, dense, sharp smelling gas, Cl_2; produced on large scale by electrolysis of sodium chloride. Used as bleaching agent and to make organo-chlorine solvents and polymers (PVC)

Radii/pm: Cl^- 181; covalent 99; van der Waals 181

Electronegativity: 3.16 (Pauling); 2.83 (Allred-Rochow)

Effective nuclear charge: 6.10 (Slater); 6.12 (Clementi); 6.79 (Froese–Fischer)

Standard reduction potentials $E^\ominus$/V

	VII	V	III	I	0	−I

Acid solution

$$ClO_4^- \xrightarrow{1.201} ClO_3^- \xrightarrow{1.181} HClO_2 \xrightarrow{1.701} HClO \xrightarrow{1.630} Cl_2 \xrightarrow{1.35828} Cl^-$$

with ClO_2 branches 1.175, 1.188; 1.659; 1.584; 1.277; 1.468; 1.287

Alkaline solution

$$ClO_4^- \xrightarrow{0.374} ClO_3^- \xrightarrow{0.295} ClO_2^- \xrightarrow{0.681} ClO^- \xrightarrow{0.421} Cl_2 \xrightarrow{1.35828} Cl^-$$

with ClO_2 branches −0.481, 1.071; 0.488; 0.890; 0.622

Covalent bonds r/pm E/kJ mol^{-1}

	r/pm	E/kJ mol^{-1}
Cl–O	170	206
Cl–Cl	198.8	242
Cl–F	163	257

For other bonds see other elements

Oxidation states

Cl^{-I}	Cl^-(aq), HCl, NaCl etc
Cl^{I}	Cl_2O, HOCl, salts, ClO^-(aq), ClF
Cl^{III}	$NaClO_2$, ClF_3
Cl^{IV}	ClO_2
Cl^{V}	$HClO_3$, salts, ClO_3^- (aq), ClF_5, F_3ClO
Cl^{VI}	Cl_2O_6
Cl^{VII}	Cl_2O_7, $HClO_4$, salts, ClO_4^- (aq), $FClO_3$

Physical properties

Melting point/K: 172.17
Boiling point/K: 239.18   **Triple point/K:** 172.17
ΔH_{fusion}/kJ mol^{-1}: 6.41 ΔH_{vap}/kJ mol^{-1}: 20.4033

Thermodynamic properties (298.15 K, 0.1 MPa)

State	$\Delta_f H^\ominus$/kJ mol^{-1}	$\Delta_f G^\ominus$/kJ mol^{-1}	$S^\ominus$/J K^{-1} mol^{-1}	C_p/J K^{-1} mol^{-1}
Gas (Cl_2)	0	0	223.066	33.907
Gas (atoms)	121.679	105.680	165.198	21.840

Density/kg m^{-3}: 2030 [113 K]; 1507 [239 K]; 3.214 [273 K]
Thermal conductivity/W m^{-1} K^{-1}: 0.0089 [300 K]$_g$
Critical temperature/K: 416.9   **Critical pressure/kPa:** 7977
Mass magnetic susceptibility/kg^{-1} m^3: -7.2×10^{-9} (g)
Molar volume/cm^3: 17.46 [113 K]
Coefficient of linear thermal expansion/K^{-1}: n.a.

Lattice structure (cell dimensions/pm), space group

Tetragonal ($a = 856$; $c = 612$), P4/ncm
Orthorhombic ($a = 624$, $b = 448$, $c = 826$), Cmca
T (tetragonal →orthorhombic) = 100 K

X-ray diffraction: mass absorption coefficients (μ/ρ)/cm^2 g^{-1}:
CuK_α 106 MoK_α 11.4

Discovered in 1774 by C. W. Scheele at Uppsala, Sweden

[Greek, *chloros* = pale green]

Chlorine

Atomic number: 17
Thermal neutron capture cross-section/barns: 44 ± 2 (^{35}Cl); 0.403 ± 0.1 (^{37}Cl)
Number of isotopes (including nuclear isomers): 11
Isotope mass range: $32 \rightarrow 40$

Nuclear properties

Key isotopes

Nuclide	Atomic mass	Natural abundance (%)	Half life $T_{1/2}$	Decay mode and energy (MeV)	Nuclear spin I	Nuclear magnetic moment μ	Uses
^{35}Cl	34.968 85	75.77	stable		3/2	+0.821 81	NMR
^{36}Cl	35.9797	0	3.1×10^5y	β^-(0.712); β^+, EC; (2%); no γ	2	+1.2853	tracer
^{37}Cl	36.9658	24.23	stable		3/2	+0.684 07	NMR

NMR

	^{35}Cl	^{37}Cl
Relative sensitivity (^{1}H = 1.00)	4.70×10^{-3}	2.71×10^{-3}
Absolute sensitivity (^{1}H = 1.00)	3.55×10^{-3}	6.63×10^{-4}
Receptivity (^{13}C = 1.00)	20.2	3.8
Magnetogyric ratio/rad T^{-1} s^{-1}	2.6210×10^7	2.1718×10^7
Quadrupole moment/m^2	-8.0×10^{-30}	-6.32×10^{-30}
Frequency (^{1}H = 100 MHz; 2.3488 T)/MHz	9.798	8.156

Reference: NaCl (aq)

Ground state electron configuration: [Ne] $3s^2 3p^5$
Term symbol: $^2P_{3/2}$
Electron affinity (M$\rightarrow$M$^-$)/kJ mol^{-1}: 348.7

Electron shell properties

Ionization energies/kJ mol^{-1}

1.	M $\rightarrow$ M$^+$	1251.1	6.	M$^{5+} \rightarrow$ M^{6+}	9 362
2.	M$^+ \rightarrow$ M^{2+}	2297	7.	M$^{6+} \rightarrow$ M^{7+}	11 020
3.	M$^{2+} \rightarrow$ M^{3+}	3826	8.	M$^{7+} \rightarrow$ M^{8+}	33 610
4.	M$^{3+} \rightarrow$ M^{4+}	5158	9.	M$^{8+} \rightarrow$ M^{9+}	38 600
5.	M$^{4+} \rightarrow$ M^{5+}	6540	10.	M$^{9+} \rightarrow$ M^{10+}	43 960

Principal lines in atomic spectrum

Wavelength/nm	Species	Sensitivity	Application
138.99* (vac)	I		
479.455	II	V2	AE
481.006	II	V3	AE
481.947	II	V4	AE
837.594	I		

* Lowest energy transition from ground state to nearest empty orbital

Abundance: Earth's crust 126 p.p.m.; seawater 19 000 p.p.m.
Biological role: Essential as Cl$^-$; toxic as Cl$_2$ gas

Cr

Atomic number: 24

Relative atomic mass ($^{12}C = 12.0000$): **51.9961**

Chemical properties

Hard blue-white metal; main ore chromite $FeCr_2O_4$. Soluble in HCl and H_2SO_4 but not HNO_3, H_3PO_4 or $HClO_4$ due to formation of protective layer. Resists oxidation in air. Main use in alloys and chrome plating.

Radii/pm: Cr^{2+} 84; Cr^{3+} 64; Cr^{4+} 56; atomic 124.9

Electronegativity: 1.66 (Pauling); 1.56 (Allred-Rochow)

Effective nuclear charge: 3.45 (Slater); 5.13 (Clementi); 6.92 (Froese–Fischer)

Standard reduction potentials $E^\ominus$/V

Oxidation states

Cr^{-II} (d^8)	$Na_2[Cr(CO)_5]$
Cr^{-I} (d^7)	$Na_2[Cr_2(CO)_{10}]$
Cr^0 (d^6)	$Cr(CO)_6$
Cr^I (d^5)	$[Cr(bipyridyl)_3]$
Cr^{II} (d^4)	CrO, CrF_2, $CrCl_2$ etc., CrS
$\mathbf{Cr^{III}}$ (d^3)	Cr_2O_3, CrF_3, $CrCl_3$ etc. $[Cr(H_2O)_6]^{3+}$ (aq), $Cr(OH)_3$, salts, complexes
Cr^{IV} (d^2)	CrO_2, CrF_4
Cr^V (d^1)	CrF_5
Cr^{VI} (d^0)	CrO_3, $Na_2Cr_2O_7$, CrO_4^{2-}, $CrOF_4$

Physical properties

Melting point/K: 2130 ± 20

Boiling point/K: 2945

ΔH_{fusion}/kJ mol^{-1}: 15.3

ΔH_{vap}/kJ mol^{-1}: 341.8

Thermodynamic properties (298.15 K, 0.1 MPa)

State	$\Delta_f H^\ominus$/kJ mol^{-1}	$\Delta_f G^\ominus$/kJ mol^{-1}	$S^\ominus$/J K^{-1} mol^{-1}	C_p/J K^{-1} mol^{-1}
Solid	0	0	23.47	23.35
Gas	396.6	351.8	174.50	20.79

Density/kg m^{-3}: 7190 [293 K]; 6460 [liquid at m.p.]

Thermal conductivity/W m^{-1} K^{-1}: 93.7 [300 K]

Electrical resistivity/Ω m: 12.7×10^{-8} [273 K]

Mass magnetic susceptibility/kg^{-1} m^3: $+4.45 \times 10^{-8}$ (s)

Molar volume/cm^3: 7.23

Coefficient of linear thermal expansion/K^{-1}: 6.2×10^{-6}

Lattice structure (cell dimensions/pm), space group

b.c.c. ($a = 288.46$), Im3m

X-ray diffraction: mass absorption coefficients (μ/ρ)/cm^2 g^{-1}: CuK_α 260 MoK_α 31.1

Atomic number: 24
Thermal neutron capture cross-section/barns: 3.1 ± 0.2
Number of isotopes (including nuclear isomers): 9
Isotope mass range: $48 \rightarrow 56$

Key isotopes

Nuclide	Atomic mass	Natural abundance (%)	Half life $T_{1/2}$	Decay mode and energy (MeV)	Nuclear spin I	Nuclear magnetic moment μ	Uses
^{50}Cr	49.9461	4.35	stable				
^{51}Cr	50.945	0	27.8d	EC (0.752); γ	7/2	± 0.934	tracer, medical
^{52}Cr	51.9405	83.79	stable				
^{53}Cr	52.9407	9.50	stable		3/2	-0.4735	NMR
^{54}Cr	53.9389	2.36	stable				

NMR

^{53}Cr

Relative sensitivity (^{1}H = 1.00) 9.03×10^{-4}
Absolute sensitivity (^{1}H = 1.00) 8.62×10^{-3}
Receptivity (^{13}C = 1.00)   0.49
Magnetogyric ratio/rad T^{-1} s^{-1} -1.5120×10^7
Quadrupole moment/m^2 $\pm 0.3 \times 10^{-28}$
Frequency (^{1}H = 100 MHz; 2.3488 T)/MHz   5.652
Reference: CrO_4^{2-} (aq)

Ground state electron configuration: [Ar]$3d^5 4s^1$
Term symbol: $^7S_3$
Electron affinity ($M \rightarrow M^-$)/kJ mol^{-1}:   94

Ionization energies/kJ mol^{-1}

1. $M \rightarrow M^+$	652.7	6. $M^{5+} \rightarrow M^{6+}$	8738
2. $M^+ \rightarrow M^{2+}$	1592	7. $M^{6+} \rightarrow M^{7+}$	15550
3. $M^{2+} \rightarrow M^{3+}$	2987	8. $M^{7+} \rightarrow M^{8+}$	17830
4. $M^{3+} \rightarrow M^{4+}$	4740	9. $M^{8+} \rightarrow M^{9+}$	20220
5. $M^{4+} \rightarrow M^{5+}$	6690	10. $M^{9+} \rightarrow M^{10+}$	23580

Principal lines in atomic spectrum

Wavelength/nm	Species	Sensitivity	Application
357.869	I		AA
359.349	I		AA
425.435	I	U1	AA, AE
427.480	I	U2	AA, AE
428.972*	I	U3	AA, AE
520.844	I	U4	AA, AE

* Lowest energy transition from ground state to nearest empty orbital

Abundance: Earth's crust 122 p.p.m.; seawater 5×10^{-5} p.p.m.
Biological role: Essential trace element; chromates toxic; stimulatory; carcinogenic

Atomic number: 27

Relative atomic mass ($^{12}C=12.0000$): 58.93320

Chemical properties

Lustrous, silvery-blue, hard metal; ferromagnetic and used to make magnetic alloys. Also used in ceramics, paints and as catalysts. Stable in air, slowly attacked by dilute acids. ^{60}Co useful radio-isotope.

Radii/pm: Co^{2+} 82; Co^{3+} 64; atomic 125.3; covalent 116

Electronegativity: 1.88 (Pauling); 1.70 (Allred-Rochow)

Effective nuclear charge: 3.90 (Slater); 5.58 (Clementi); 7.63 (Froese–Fischer)

Standard reduction potentials $E^{\ominus}/V$

	IV		III		II		0
Acid solution	CoO_2	$\xrightarrow{1.416}$	Co^{3+}	$\xrightarrow{1.92}$	Co^{2+}	$\xrightarrow{-0.277}$	Co
Alkaline solution	CoO_2	$\xrightarrow{0.7}$	$Co(OH)_3$	$\xrightarrow{0.17}$	$Co(OH)_2$	$\xrightarrow{-0.733}$	Co

Oxidation states

Co^{-I} (d^{10})	rare $[Co(CO)_4]^-$
Co^0 (d^9)	rare $[Co_2(CO)_8]$
Co^I (d^8)	rare $[Co(NCCH_2)_5]^+$
Co^{II} (d^7)	CoO, Co_3O_4 ($=Co^{II}Co^{III}_2O_4$), $Co(OH)_2$, CoF_2, $CoCl_2$ etc., $[Co(H_2O)_6]^{2+}$ (aq), many salts and complexes
Co^{III} (d^6)	$Co(OH)_3$, CoF_3, $[Co(H_2O)_6]^{3+}$ (aq), $[Co(NH_3)_6]^{3+}$, many complexes
Co^{IV} (d^5)	CoO_2?, CoS_2, $[CoF_6]^{2-}$
Co^V (d^4)	K_3CoO_4

Physical properties

Melting point/K: 1768

Boiling point/K: 3143

$\Delta H_{fusion}/kJ\ mol^{-1}$: 15.2

$\Delta H_{vap}/kJ\ mol^{-1}$: 382.4

Thermodynamic properties (298.15 K, 0.1 MPa)

State	$\Delta_f H^{\ominus}/kJ\ mol^{-1}$	$\Delta_f G^{\ominus}/kJ\ mol^{-1}$	$S^{\ominus}/J\ K^{-1}\ mol^{-1}$	$C_p/J\ K^{-1}\ mol^{-1}$
Solid	0	0	30.04	24.81
Gas	424.7	380.3	179.515	23.020

Density/kg m^{-3}: 8900 [293 K]; 7670 [liquid at m.p.]

Thermal conductivity/W m^{-1} K^{-1}: 100 [300 K]

Electrical resistivity/Ω m: 6.24×10^{-8} [293 K]

Mass magnetic susceptibility/kg^{-1} m^3: Ferromagnetic

Molar volume/cm^3: 6.62

Coefficient of linear thermal expansion/K^{-1}: 13.36×10^{-6}

Lattice structure (cell dimensions/pm), space group

α-Co f.c.c. ($a=354.41$), Fm3m
ε-Co h.c.p. ($a=250.7$; $c=406.9$), P6$_3$/mmc
T ($\alpha\rightarrow\varepsilon$)$=690$ K

X-ray diffraction: mass absorption coefficients $(\mu/\rho)/cm^2\ g^{-1}$:
CuK_{α} 313 MoK_{α} 42.5

Cobalt

Atomic number: 27
Thermal neutron capture cross-section/barns: 37.5 ± 0.2
Number of isotopes (including nuclear isomers): 14
Isotope mass range: $54 \rightarrow 63$

Nuclear properties

Key isotopes

Nuclide	Atomic mass	Natural abundance (%)	Half life $T_{1/2}$	Decay mode and energy (MeV)	Nuclear spin I	Nuclear magnetic moment μ	Uses
^{56}Co	55.940	0	77d	β^+ (4.57); EC; γ 4		± 3.83	tracer
^{57}Co	56.936	0	270d	EC (0.837); γ	7/2	$+4.72$	tracer, medical
^{58}Co	57.936	0	71.3d	β^+ (2.309); EC; γ	2	$+4.04$	tracer, medical
^{59}Co	58.9332	100	stable		7/2	$+4.616$	NMR
^{60}Co	59.934	0	5.26y	β^- (2.819); γ	5	$+3.79$	tracer, medical

NMR

^{59}Co

Relative sensitivity (^{1}H = 1.00) 0.28
Absolute sensitivity (^{1}H = 1.00) 0.28
Receptivity (^{13}C = 1.00) 1570
Magnetogyric ratio/rad T^{-1} s^{-1} 6.3472×10^7
Quadrupole moment/m^2 0.40×10^{-28}
Frequency (^{1}H = 100 MHz; 2.3488 T)/MHz 23.614
Reference: $K_3Co(CN)_6$

Ground state electron configuration: [Ar] $3d^74s^2$
Term symbol: $^4F_{9/2}$
Electron affinity (M→M$^-$)/kJ mol^{-1}: 102

Electron shell properties

Ionization energies/kJ mol^{-1}

1. M →M$^+$ 760.0	6. M^{5+}→M^{6+} 9 840
2. M$^+$→M^{2+} 1646	7. M^{6+}→M^{7+} 12 400
3. M^{2+}→M^{3+} 3232	8. M^{7+}→M^{8+} 15 100
4. M^{3+}→M^{4+} 4950	9. M^{8+}→M^{9+} 17 900
5. M^{4+}→M^{5+} 7670	10. M^{9+}→M^{10+} 26 600

Principal lines in atomic spectrum

Wavelength/nm	Species	Sensitivity	Application
240.725	I		AA
242.493	I		AA
345.350	I	U1	AE
346.580	I	U2	AA, AE
352.981	I	U3	AE
423.400*	I		

* Lowest energy transition from ground state to nearest empty orbital

Abundance: Earth's crust 29 p.p.m.; seawater 0.0005 p.p.m.
Biological role: Essential trace element (vitamin B$_{12}$); carcinogenic

Cu

Atomic number: 29

Relative atomic mass $(^{12}C = 12.0000)$: **63.546**

Chemical properties

Reddish metal, malleable and ductile, with high electrical and thermal conductivities. Main ore is chalcopyrite, $CuFeS_2$. Resistant to air and water but slowly weathers to green patina of carbonate. Historically important alloys.

Radii/pm: Cu^+ 96; Cu^{2+} 72; atomic 127.8; covalent 117

Electronegativity: 1.90 (Pauling); 1.75 (Allred–Rochow)

Effective nuclear charge: 4.20 (Slater); 5.84 (Clementi); 8.07 (Froese–Fischer)

Standard reduction potentials $E^{\ominus}/V$

II		I		0
		0.340		
Cu^{2+}	0.159	Cu^+	0.520	Cu
$Cu(NH_3)_4^{2+}$	0.10	$Cu(NH_3)_4^+$	−0.100	Cu
$Cu(CN)_2$	1.12	$Cu(CN)_2^-$	−0.44	Cu

Oxidation states

Cu^{-1} $(d^{10} s^2)$	—
Cu^0 $(d^{10} s^1)$	rare, $[Cu(CO)_3]$ at 10 K
Cu^I (d^{10})	Cu_2O, CuCl, $K[Cu(CN)_2]$
Cu^{II} (d^9)	CuO, $CuCl_2$, Cu^{2+} (aq), Cu^{2+} salts
Cu^{III} (d^8)	rare $K_3[CuF_6]$
Cu^{IV} (d^7)	rare $Cs_2[CuF_6]$

Physical properties

Melting point/K: 1356.6

Boiling point/K: 2840

ΔH_{fusion}/kJ mol^{-1}: 13.0

ΔH_{vap}/kJ mol^{-1}: 306.7

Thermodynamic properties (298.15 K, 0.1 MPa)

State	$\Delta_f H^{\ominus}$/kJ mol^{-1}	$\Delta_f G^{\ominus}$/kJ mol^{-1}	$S^{\ominus}$/J K^{-1} mol^{-1}	C_p/J K^{-1} mol^{-1}
Solid	0	0	33.150	24.435
Gas	338.32	298.58	166.38	20.786

Density/kg m^{-3}: 8960 [293 K]; 7940 [liquid at m.p.]

Thermal conductivity/W m^{-1} K^{-1}: 401 [300 K]

Electrical resistivity/Ω m: 1.6730×10^{-8} [293 K]

Mass magnetic susceptibility/kg^{-1} m^3: -1.081×10^{-9} (s)

Molar volume/cm^3: 7.09

Coefficient of linear thermal expansion/K^{-1}: 16.5×10^{-6}

Lattice structure (cell dimensions/pm), space group

f.c.c. $(a = 361.47)$, Fm3m

X-ray diffraction: mass absorption coefficients (μ/ρ)/cm^2 g^{-1}:

CuK_α 52.9 MoK_α 50.9

Copper

Atomic number: 29
Thermal neutron capture cross-section/barns: 3.8 ± 0.1
Number of isotopes (including nuclear isomers): 11
Isotope mass range: $58 \rightarrow 68$

Nuclear properties

Key isotopes

Nuclide	Atomic mass	Natural abundance (%)	Half life $T_{1/2}$	Decay mode and energy (MeV)	Nuclear spin I	Nuclear magnetic moment μ	Uses
^{63}Cu	62.9298	69.17	stable		3/2	+2.2228	NMR
^{64}Cu	63.930	0	12.9h	β^- (0.573) 40%;1 β^+, EC; γ		−0.216	tracer, medical
^{65}Cu	64.9278	30.83	stable		3/2	+2.3812	NMR
^{67}Cu			61.88h	β^- (0.576); γ			tracer

NMR

	^{63}Cu	^{65}Cu
Relative sensitivity (^{1}H = 1.00)	9.31×10^{-2}	0.11
Absolute sensitivity (^{1}H = 1.00)	6.43×10^{-2}	3.52×10^{-2}
Receptivity (^{13}C = 1.00)	365	201
Magnetogyric ratio/rad T^{-1} s^{-1}	7.0965×10^7	7.6018×10^7
Quadrupole moment/m^2	-0.211×10^{-28}	-0.195×10^{-28}
Frequency (^{1}H = 100 MHz; 2.3488 T)/MHz	26.505	28.394

Reference: $Cu(MeCN)_4^+ BF_4^-$ in MeCN

Ground state electron configuration: $[Ar]3d^{10}4s^1$
Term symbol: $^2S_{1/2}$
Electron affinity $(M \rightarrow M^-)$/kJ mol^{-1}: 118.3

Electron shell properties

Ionization energies/kJ mol^{-1}

1. $M \rightarrow M^+$	745.4	6. $M^{5+} \rightarrow M^{6+}$ (9 940)
2. $M^+ \rightarrow M^{2+}$	1958	7. $M^{6+} \rightarrow M^{7+}$ (13 400)
3. $M^{2+} \rightarrow M^{3+}$	3554	8. $M^{7+} \rightarrow M^{8+}$ (16 000)
4. $M^{3+} \rightarrow M^{4+}$	5326	9. $M^{8+} \rightarrow M^{9+}$ (19 200)
5. $M^{4+} \rightarrow M^{5+}$	7709	10. $M^{9+} \rightarrow M^{10+}$ (22 400)

Principal lines in atomic spectrum

Wavelength/nm	Species	Sensitivity	Application
216.509	I		AA
217.894	I		AA
324.754	I	U1	AA, AE
327.396*	I	U2	AA, AE
515.324	I	U4	AE
521.820	I	U3	AE

* Lowest energy transition from ground state to nearest empty orbital

Abundance: Earth's crust 68 p.p.m.; seawater 0.003 p.p.m.
Biological role: Essential; excess copper is toxic

Cm

Atomic number: 96

Relative atomic mass ($^{12}C = 12.0000$): (247)

Radioactive, silvery metal obtained in gramme quantities as ^{242}Cm and ^{244}Cm from neutron bombardment of ^{239}Pu. Attacked by oxygen, steam, and acids but not alkalis.

Radii/pm: Cm^{2+} 119; Cm^{3+} 99; Cm^{4+} 88

Electronegativity: 1.3 (Pauling); n.a. (Allred-Rochow)

Effective nuclear charge: 1.80 (Slater)

Standard reduction potentials $E^{\ominus}$/V

	IV		III		0
Acid solution	Cm^{4+}	—3.2—	Cm^{3+}	—2.06—	Cm
Alkaline solution	CmO_2	—0.7—	$Cm(OH)_3$	—2.5—	Cm

Oxidation states

Cm^{II} ($f^7 d^1$)	CmO
Cm^{III} (f^7)	Cm_2O_3, $Cm(OH)_3$, CmF_3, $CmCl_3$ etc., Cm^{3+} (aq)
Cm^{IV} (f^6)	CmO_2, CmF_4, Cm^{4+} (aq) very unstable

Melting point/K: 1610 ± 40
Boiling point/K: n.a.
ΔH_{fusion}/kJ mol^{-1}: n.a.
ΔH_{vap}/kJ mol^{-1}: n.a.

Thermodynamic properties (298.15 K, 0.1 MPa)

State	$\Delta_f H^{\ominus}$/kJ mol^{-1}	$\Delta_f G^{\ominus}$/kJ mol^{-1}	$S^{\ominus}$/J K^{-1} mol^{-1}	C_p/J K^{-1} mol^{-1}
Solid	0	0	n.a.	n.a.
Gas	n.a.	n.a.	n.a.	n.a.

Density/kg m^{-3}: 13 300 [293 K]
Thermal conductivity/W m^{-1} K^{-1}: 10 (est.) [300 K]
Electrical resistivity/Ω m: n.a.
Mass magnetic susceptibility/kg^{-1} m^3: n.a.
Molar volume/cm^3: 18.6
Coefficient of linear thermal expansion/K^{-1}: n.a.

Lattice structure (cell dimensions/pm), space group

n.a.

X-ray diffraction: mass absorption coefficients (μ/ρ)/cm^2 g^{-1}:
n.a.

Prepared by G. T. Seaborg, R. A. James, and A. Ghiorso in 1944
at Berkeley, California, USA
[Named after Pierre and Marie Curie]

Curium

Atomic number: 96
Thermal neutron capture cross-section/barns: 180 (^{247}Cm)
Number of isotopes (including nuclear isomers): 13
Isotope mass range: 238→250

Key isotopes

Nuclide	Atomic mass	Natural abundance (%)	Half life $T_{1/2}$	Decay mode and energy (MeV)	Nuclear spin I	Nuclear magnetic moment μ	Uses
^{242}Cm	242.0588	0	163d	α (6.12); γ	0		available in mg quantities
^{244}Cm	244.0629	0	17.6y	α (5.902); γ			
^{247}Cm	247.0704	0	1.6×10^7y	α (c. 5.3); γ	9/2	± 0.4	
^{248}Cm		0	4.7×10^5y	α (5.161); SF			

Ground state electron configuration: [Rn]$5f^76d^17s^2$
Term symbol: $^9D_2$
Electron affinity $(M \rightarrow M^-)$/kJ mol^{-1}: n.a.

Ionization energies/kJ mol^{-1}

1. $M \rightarrow M^+$ 581
2. $M^+ \rightarrow M^{2+}$
3. $M^{2+} \rightarrow M^{3+}$
4. $M^{3+} \rightarrow M^{4+}$
5. $M^{4+} \rightarrow M^{5+}$
6. $M^{5+} \rightarrow M^{6+}$
7. $M^{6+} \rightarrow M^{7+}$
8. $M^{7+} \rightarrow M^{8+}$
9. $M^{8+} \rightarrow M^{9+}$
10. $M^{9+} \rightarrow M^{10+}$

Selected lines in atomic spectrum

Wavelength/nm	Species	Sensitivity	Application
299.939	I		
310.969	I		
311.641	I		
313.716	I		
314.733	I		
315.510	I		
315.860	I		

Abundance: Nil
Biological role: None; toxic due to radioactivity

Dy

Atomic number: 66

Relative atomic mass ($^{12}C=12.0000$): 162.50

Chemical properties

Reactive, hard, silvery metal of the rare earth group. Obtained from monazite and bastnaesite ores. Oxidized by oxygen, reacts rapidly with water, dissolves in acids.

Radii/pm: Dy^{3+} 91; atomic 177.3; covalent 159

Electronegativity: 1.22 (Pauling); 1.10 (Allred-Rochow)

Effective nuclear charge: 2.85 (Slater); 8.34 (Clementi); 11.49 (Froese–Fischer)

Standard reduction potentials $E^{\ominus}/V$

	IV		III		II		0
Acid solution	Dy^{4+}	— 5.7 —	Dy^{3+}	— -2.5 —	Dy^{2+}	— -2.2 —	Dy
Alkaline solution	DyO_2	— 3.5 —	$Dy(OH)_3$		— -2.80 —		Dy

(Acid solution III→0: -2.29)

Oxidation states

Dy^{II} (f^{10})	$DyCl_2$, DyI_2
Dy^{III} (f^9)	Dy_2O_3, $Dy(OH)_3$, DyF_3, $DyCl_3$ etc., $[Dy(H_2O)_x]^{3+}$ (aq), Dy^{3+} salts, $DyCl_6^{3-}$
Dy^{IV} (f^8)	Cs_3DyF_7

Physical properties

Melting point/K: 1685
Boiling point/K: 2835
ΔH_{fusion}/kJ mol^{-1}: 17.2
ΔH_{vap}/kJ mol^{-1}: 293

Thermodynamic properties (298.15 K, 0.1 MPa)

State	$\Delta_f H^{\ominus}$/kJ mol^{-1}	$\Delta_f G^{\ominus}$/kJ mol^{-1}	$S^{\ominus}$/J K^{-1} mol^{-1}	C_p/J K^{-1} mol^{-1}
Solid	0	0	74.77	28.16
Gas	290.4	254.4	196.63	20.79

Density/kg m^{-3}: 8550 [293 K]
Thermal conductivity/W m^{-1} K^{-1}: 10.7 [300 K]
Electrical resistivity/Ω m: 57.0×10^{-8} [273 K]
Mass magnetic susceptibility/kg^{-1} m^3: $+8.00 \times 10^{-6}$ (s)
Molar volume/cm^3: 19.00
Coefficient of linear thermal expansion/K^{-1}: 10.0×10^{-6}

Lattice structure (cell dimensions/pm), space group

Rhombic ($a=359.5$; $b=618.3$; $c=567.7$), Cmcm
h.c.p. ($a=359.03$; $c=564.75$), P6$_3$/mmc
b.c.c. ($a=398$), Im3m

T (rhombic→h.c.p.) = 86 K
High pressure form: ($a=334$, $c=245$), R$\bar{3}$m

X-ray diffraction: mass absorption coefficients (μ/ρ)/cm^2 g^{-1}:
CuK$_\alpha$ 286   MoK$_\alpha$ 70.6

Dysprosium

Atomic number: 66

Thermal neutron capture cross-section/barns: 90 ± 20

Number of isotopes (including nuclear isomers): 21

Isotope mass range: $149 \rightarrow 167$

Nuclear properties

Key isotopes

Nuclide	Atomic mass	Natural abundance (%)	Half life $T_{1/2}$	Decay mode and energy (MeV)	Nuclear spin I	Nuclear magnetic moment μ	Uses
^{156}Dy	155.9238	0.06	stable				
^{158}Dy	157.9240	0.10	stable				
^{160}Dy	159.9248	2.34	stable				
^{161}Dy	160.9266	19.0	stable		5/2	−0.482	NMR
^{162}Dy	161.9265	25.5	stable				
^{163}Dy	162.9284	24.9	stable		5/2	+0.676	NMR
^{164}Dy	163.9288	28.1	stable				

NMR

	^{161}Dy	^{163}Dy
Relative sensitivity (^{1}H = 1.00)	4.17×10^{-4}	1.12×10^{-3}
Absolute sensitivity (^{1}H = 1.00)	7.87×10^{-5}	2.79×10^{-4}
Receptivity (^{13}C = 1.00)	0.509	1.79
Magnetogyric ratio/rad T^{-1} s^{-1}	-0.9206×10^{7}	1.2750×10^{7}
Quadrupole moment/m^{2}	1.4×10^{-28}	1.6×10^{-28}
Frequency (^{1}H = 100 MHz; 2.3488 T)/MHz	3.294	4.583

Ground state electron configuration: $[Xe]4f^{10}6s^{2}$

Term symbol: $^{5}I_{8}$

Electron affinity $(M \rightarrow M^{-})$/kJ mol^{-1}: n.a.

Electron shell properties

Ionization energies/kJ mol^{-1}

1. $M \rightarrow M^{+}$ 571.9	6. $M^{5+} \rightarrow M^{6+}$
2. $M^{+} \rightarrow M^{2+}$ 1126	7. $M^{6+} \rightarrow M^{7+}$
3. $M^{2+} \rightarrow M^{3+}$ 2200	8. $M^{7+} \rightarrow M^{8+}$
4. $M^{3+} \rightarrow M^{4+}$ 4001	9. $M^{8+} \rightarrow M^{9+}$
5. $M^{4+} \rightarrow M^{5+}$	10. $M^{9+} \rightarrow M^{10+}$

Principal lines in atomic spectrum

Wavelength/nm	Species	Sensitivity	Application
353.170	II		
400.045	II		AE
404.597	I		AA, AE
416.797	I		AA, AE
418.682	I		AA
421.172	I		AA, AE

Abundance: Earth's crust 4.5 p.p.m.; seawater 2×10^{-7} p.p.m.

Biological role: None; low toxicity; stimulatory

Atomic number: 99

Relative atomic mass ($^{12}C = 12.0000$): **(254)**

Chemical properties

Radioactive, silvery metal obtained in mg quantities as ^{254}Es from neutron bombardment of ^{239}Pu. Attacked by oxygen, steam, and acids, but not alkalis.

Radii/pm: Es^{2+} 116; Es^{3+} 98; Es^{4+} 85

Electronegativity: 1.3 (Pauling); n.a. (Allred-Rochow)

Effective nuclear charge: 1.65 (Slater)

Standard reduction potentials $E^{\ominus}$/V

	III		II		0
Acid solution	Es^{3+}	$\xrightarrow{-1.5}$	Es^{2+}	$\xrightarrow{-2.2}$	Es
			-2.0		

Oxidation states

Es^{II} (f^{11})	transient state
Es^{III} (f^{10})	Es_2O_3, $EsCl_3$, $EsBr_3$, Es^{3+} (aq), EsOCl

Physical properties

Melting point/K: n.a.

Boiling point/K: n.a.

ΔH_{fusion}/kJ mol^{-1}: n.a.

ΔH_{vap}/kJ mol^{-1}: n.a.

Thermodynamic properties (298.15 K, 0.1 MPa)

State	$\Delta_f H^{\ominus}$/kJ mol^{-1}	$\Delta_f G^{\ominus}$/kJ mol^{-1}	$S^{\ominus}$/J K^{-1} mol^{-1}	C_p/J K^{-1} mol^{-1}
Solid	0	0	n.a.	n.a.
Gas	n.a.	n.a.	n.a.	n.a.

Density/kg m^{-3}: n.a.

Thermal conductivity/W m^{-1} K^{-1}: 10 (est.) [300 K]

Electrical resistivity/Ω m: n.a.

Mass magnetic susceptibility/kg^{-1} m^3: n.a.

Molar volume/cm^3: n.a.

Coefficient of linear thermal expansion/K^{-1}: n.a.

Lattice structure (cell dimensions/pm), space group

n.a.

X-ray diffraction: mass absorption coefficients (μ/ρ)/cm^2 g^{-1}:
n.a.

Einsteinium

Atomic number: 99

Thermal neutron capture cross-section/barns: < 40 (^{254}Es)

Number of isotopes (including nuclear isomers): 12

Isotope mass range: $246 \rightarrow 256$

Nuclear properties

Key isotopes

Nuclide	Atomic mass	Natural abundance (%)	Half life $T_{1/2}$	Decay mode and energy (MeV)	Nuclear spin I	Nuclear magnetic moment μ	Uses
^{253}Es		0	20.7d	α			Available
^{254}Es	254.0881	0	207d	α (6.623); γ			

Ground state electron configuration: [Rn]$4f^{11}7s^2$

Term symbol: $^5I_{15/2}$

Electron affinity $(M \rightarrow M^-)$/kJ mol^{-1}: ≤ 50

Electron shell properties

Ionization energies/kJ mol^{-1}

1. M $\rightarrow$ M$^+$ 619	6. M$^{5+} \rightarrow$ M^{6+}
2. M$^+ \rightarrow$ M^{2+}	7. M$^{6+} \rightarrow$ M^{7+}
3. M$^{2+} \rightarrow$ M^{3+}	8. M$^{7+} \rightarrow$ M^{8+}
4. M$^{3+} \rightarrow$ M^{4+}	9. M$^{8+} \rightarrow$ M^{9+}
5. M$^{4+} \rightarrow$ M^{5+}	10. M$^{9+} \rightarrow$ M^{10+}

Selected lines in atomic spectrum

Wavelength/nm	Species	Sensitivity	Application
342.848	I		
349.811	I		
351.433	I		
352.349	I		
355.534	I		
516.174	I		
520.440	I		

Abundance: Nil

Biological role: None; toxic due to radioactivity

Er	Atomic number: **68**
	Relative atomic mass (^{12}C = 12.0000): **167.26**

Chemical properties

Silver-grey metal of the rare earth group. Obtained from bastnaesite, monazite, and euxenite. Slowly tarnishes in air, slowly reacts with water, dissolves in acids. Some used in infrared absorbing glass.

Radii/pm: Er^{3+} 89; atomic 175.7; covalent 157

Electronegativity: 1.24 (Pauling); 1.11 (Allred-Rochow)

Effective nuclear charge: 2.85 (Slater); 8.48 (Clementi); 11.70 (Froese–Fischer)

Standard reduction potentials $E^{\ominus}$/V

	III		0
Acid solution	Er^{3+}	$\underline{-2.32}$	Er
Alkaline solution	$Er(OH)_3$	$\underline{-2.84}$	Er

Oxidation states

Er^{III} (f^{11})	Er_2O_3, $Er(OH)_3$, ErF_3, $ErCl_3$ etc., $[Er(H_2O)_x]^{3+}$ (aq), Er^{3+} salts, $ErCl_6^{3-}$, complexes etc.

Physical properties

Melting point/K: 1802
Boiling point/K: 3136
ΔH_{fusion}/kJ mol^{-1}: 17.2
ΔH_{vap}/kJ mol^{-1}: 280

Thermodynamic properties (298.15 K, 0.1 MPa)

State	$\Delta_f H^{\ominus}$/kJ mol^{-1}	$\Delta_f G^{\ominus}$/kJ mol^{-1}	$S^{\ominus}$/J K^{-1} mol^{-1}	C_p/J K^{-1} mol^{-1}
Solid	0	0	73.18	28.12
Gas	317.1	280.7	195.59	20.79

Density/kg m^{-3}: 9066 [298 K]
Thermal conductivity/W m^{-1} K^{-1}: 14.3 [300 K]
Electrical resistivity/Ω m: 87×10^{-8} [298 K]
Mass magnetic susceptibility/kg^{-1} m^3: $+3.33 \times 10^{-6}$ (s)
Molar volume/cm^3: 18.44
Coefficient of linear thermal expansion/K^{-1}: 9.2×10^{-6}

Lattice structure (cell dimensions/pm), space group

α-Er h.c.p. ($a = 355.88$; $c = 558.74$), P6$_3$/mmc
β-Er b.c.c. ($a = 394$), Im3m

$T(\alpha \rightarrow \beta) = 1640$ K

X-ray diffraction: mass absorption coefficients (μ/ρ)/cm^2 g^{-1}:
CuK$_\alpha$ 134   MoK$_\alpha$ 77.3

Discovered in 1842 by C. G. Mosander at Stockholm, Sweden

[Named after Ytterby, Sweden]

Erbium

Atomic number: 68
Thermal neutron capture cross-section/barns: 0.16 ± 0.03
Number of isotopes (including nuclear isomers): 16
Isotope mass range: $158 \rightarrow 172$

Nuclear properties

Key isotopes

Nuclide	Atomic mass	Natural abundance (%)	Half life $T_{1/2}$	Decay mode and energy (MeV)	Nuclear spin I	Nuclear magnetic moment μ	Uses
^{162}Er	161.9288	0.14	stable				
^{164}Er	163.9293	1.56	stable				
^{166}Er	165.9304	33.4	stable				
^{167}Er	166.9320	22.9	stable		7/2	-0.564	NMR
^{168}Er	167.9324	27.1	stable				
^{169}Er		0	9.4d	β^- (0.34); γ	1/2	$+0.513$	tracer
^{170}Er	169.9355	14.9	stable				
^{171}Er		0	7.52h	β^- (1.490); γ	5/2	± 0.70	tracer

NMR

^{167}Er

Relative sensitivity (^{1}H = 1.00)	5.07×10^{-4}
Absolute sensitivity (^{1}H = 1.00)	1.16×10^{-4}
Receptivity (^{13}C = 1.00)	0.665
Magnetogyric ratio/rad T^{-1} s^{-1}	-0.7752×10^7
Quadrupole moment/m^2	2.83×10^{-28}
Frequency (^{1}H = 100 MHz; 2.3488 T)/MHz	2.890

Ground state electron configuration: $[Xe]4f^{12}6s^2$
Term symbol: $^3H_6$
Electron affinity $(M \rightarrow M^-)$/kJ mol^{-1}: ≤ 50

Electron shell properties

Ionization energies/kJ mol^{-1}

1. $M \rightarrow M^+$	588.7	6. $M^{5+} \rightarrow M^{6+}$	
2. $M^+ \rightarrow M^{2+}$	1151	7. $M^{6+} \rightarrow M^{7+}$	
3. $M^{2+} \rightarrow M^{3+}$	2194	8. $M^{7+} \rightarrow M^{8+}$	
4. $M^{3+} \rightarrow M^{4+}$	4115	9. $M^{8+} \rightarrow M^{9+}$	
5. $M^{4+} \rightarrow M^{5+}$		10. $M^{9+} \rightarrow M^{10+}$	

Principal lines in atomic spectrum

Wavelength/nm	Species	Sensitivity	Application
349.910	II		AE
369.265	II		AE
386.285	I		AA
390.632	II		AE
400.796	I		AA

Abundance: Earth's crust 3.5 p.p.m.; seawater 2×10^{-7} p.p.m.
Biological role: None; low toxicity; stimulatory

Eu

Atomic number: 63
Relative atomic mass ($^{12}C = 12.0000$): **151.965**

Chemical properties

Rare, and most reactive of the rare earth metals. Obtained from monazite and bastnaesite ores. Soft, silvery metal which reacts quickly with oxygen and water. Little used.

Radii/pm: Eu^{2+} 112; Eu^{3+} 98; atomic 204.2; covalent 185
Electronegativity: n.a. (Pauling); 1.01 (Allred-Rochow)
Effective nuclear charge: 2.85 (Slater); 8.11 (Clementi); 11.17 (Froese–Fischer)

Standard reduction potentials $E^\ominus$/V

	III		II		0
Acid solution	Eu^{3+}	$\xrightarrow{-0.35}$	Eu^{2+}	$\xrightarrow{-2.80}$	Eu
Alkaline solution	$Eu(OH)_3$		$\xrightarrow{-2.51}$		Eu

(Acid solution III to II: -1.99)

Oxidation states

Eu^{II} (f^7)	EuO, EuS, EuF_2, $EuCl_2$ etc.
Eu^{III} (f^6)	Eu_2O_3, $Eu(OH)_3$, EuF_3, $EuCl_3$ etc., $[Eu(H_2O)_x]^{3+}$ (aq), Eu^{3+} salts, complexes

Physical properties

Melting point/K: 1095
Boiling point/K: 1870
ΔH_{fusion}/kJ mol^{-1}: 10.5
ΔH_{vap}/kJ mol^{-1}: 176

Thermodynamic properties (298.15 K, 0.1 MPa)

State	$\Delta_f H^\ominus$/kJ mol^{-1}	$\Delta_f G^\ominus$/kJ mol^{-1}	$S^\ominus$/J K^{-1} mol^{-1}	C_p/J K^{-1} mol^{-1}
Solid	0	0	77.78	27.66
Gas	175.3	142.2	188.795	20.786

Density/kg m^{-3}: 5243 [293 K]
Thermal conductivity/W m^{-1} K^{-1}: 13.9 [300 K]
Electrical resistivity/Ω m: 90.0×10^{-8} [298 K]
Mass magnetic susceptibility/kg^{-1} m^3: $+2.81 \times 10^{-6}$ (s)
Molar volume/cm^3: 28.98
Coefficient of linear thermal expansion/K^{-1}: 32×10^{-6}

Lattice structure (cell dimensions/pm), space group

b.c.c. ($a = 458.20$), Im3m

X-ray diffraction: mass absorption coefficients (μ/ρ)/cm^2 g^{-1}:
CuK$_\alpha$ 425   MoK$_\alpha$ 61.5

Discovered in 1901 by E.-A. Demarçay at Paris,
France
[Named after Europe]

Europium

Atomic number: 63

Thermal neutron capture cross-section/barns: 4100 ± 100

Number of isotopes (including nuclear isomers): 21

Isotope mass range: $144 \rightarrow 160$

Nuclear properties

Key isotopes

Nuclide	Atomic mass	Natural abundance (%)	Half life $T_{1/2}$	Decay mode and energy (MeV)	Nuclear spin I	Nuclear magnetic moment μ	Uses
^{151}Eu	150.9196	47.8	stable		5/2	$+3.463$	NMR
^{151}Eu		0	12.7y	EC, β^- (1.82), β^+; γ	3	-1.937	tracer
^{152m_1}Eu		0	9.2h	β^-; γ	0		tracer
^{153}Eu	152.9209	52.2	stable		5/2		NMR

NMR

	^{151}Eu	^{153}Eu
Relative sensitivity (^{1}H = 1.00)	0.18	1.52×10^{-2}
Absolute sensitivity (^{1}H = 1.00)	8.51×10^{-2}	7.98×10^{-3}
Receptivity (^{13}C = 1.00)	4.64×10^2	45.7
Magnetogyric ratio/rad T^{-1} s^{-1}	6.5477×10^7	2.9371×10^7
Quadrupole moment/m^2	1.16×10^{-28}	2.9×10^{-28}
Frequency (^{1}H = 100 MHz; 2.3488 T)/MHz	24.801	10.951

Ground state electron configuration: $[Xe]4f^7 6s^2$

Term symbol: $^8S_{7/2}$

Electron affinity ($M \rightarrow M^-$)/kJ mol^{-1}: ≤ 50

Electron shell properties

Ionization energies/kJ mol^{-1}

1. $M \rightarrow M^+$ 546.7	6. $M^{5+} \rightarrow M^{6+}$
2. $M^+ \rightarrow M^{2+}$ 1085	7. $M^{6+} \rightarrow M^{7+}$
3. $M^{2+} \rightarrow M^{3+}$ 2404	8. $M^{7+} \rightarrow M^{8+}$
4. $M^{3+} \rightarrow M^{4+}$ 4110	9. $M^{8+} \rightarrow M^{9+}$
5. $M^{4+} \rightarrow M^{5+}$	10. $M^{9+} \rightarrow M^{10+}$

Principal lines in atomic spectrum

Wavelength/nm	Species	Sensitivity	Application
412.974	II		AE
420.505	II		AE
459.402	I		AA, AE
462.722	I		AA, AE
466.187	I		AA, AE

Abundance: Earth's crust 2.1 p.p.m.; seawater 4×10^{-8} p.p.m.

Biological role: None; low toxicity

Fm

Chemical properties

Radioactive metal obtained in microgramme quantities as ^{253}Fm from neutron bombardment of ^{239}Pu.

Radii/pm: Fm^{2+} 115; Fm^{3+} 97; Fm^{4+} 84

Electronegativity: 1.3 (Pauling); n.a. (Allred-Rochow)

Effective nuclear charge: 1.65 (Slater)

Standard reduction potentials $E^{\ominus}/V$

	III	II	0
Acid solution	Fm^{3+} $\xrightarrow{-1.15}$	Fm^{2+} $\xrightarrow{-2.37}$	Fm

-1.96

Oxidation states

Fm^{II} (f^{12}) ?

Fm^{III} (f^{11}) $[Fm(H_2O)_x]^{3+}$ (aq)

Physical properties

Melting point/K: n.a.

Boiling point/K: n.a.

ΔH_{fusion}/kJ mol^{-1}: n.a.

ΔH_{vap}/kJ mol^{-1}: n.a.

Thermodynamic properties (298.15 K, 0.1 MPa)

State	$\Delta_f H^{\ominus}$/kJ mol^{-1}	$\Delta_f G^{\ominus}$/kJ mol^{-1}	$S^{\ominus}$/J K^{-1} mol^{-1}	C_p/J K^{-1} mol^{-1}
Solid	0	0	n.a.	n.a.
Gas	n.a.	n.a.	n.a.	n.a.

Density/kg m^{-3}: n.a.

Thermal conductivity/W m^{-1} K^{-1}: 10 (est.) [300 K]

Electrical resistivity/Ω m: n.a.

Mass magnetic susceptibility/kg^{-1} m^3: n.a.

Molar volume/cm^3: n.a.

Coefficient of linear thermal expansion/K^{-1}: n.a.

Lattice structure (cell dimensions/pm), space group

n.a.

X-ray diffraction: mass absorption coefficients (μ/ρ)/cm^2 g^{-1}:

n.a.

Fermium

Atomic number: 100

Thermal neutron capture cross-section/barns: 26 ± 3 (^{255}Fm)

Number of isotopes (including nuclear isomers): 10

Isotope mass range: $248 \rightarrow 257$

Nuclear properties

Key isotopes

Nuclide	Atomic mass	Natural abundance (%)	Half life $T_{1/2}$	Decay mode and energy (MeV)	Nuclear spin I	Nuclear magnetic moment μ	Uses
^{254}Fm		0	3.24h	α			
^{255}Fm		0	20.0h	α			
^{257}Fm		0	80d	α (6.871); γ			

Ground state electron configuration: $[\text{Rn}]5f^{12}7s^2$

Term symbol: $^3H_6$

Electron affinity $(M \rightarrow M^-)$/kJ mol^{-1}: n.a.

Electron shell properties

Ionization energies/kJ mol^{-1}

1. $M \rightarrow M^+$ 627	6. $M^{5+} \rightarrow M^{6+}$
2. $M^+ \rightarrow M^{2+}$	7. $M^{6+} \rightarrow M^{7+}$
3. $M^{2+} \rightarrow M^{3+}$	8. $M^{7+} \rightarrow M^{8+}$
4. $M^{3+} \rightarrow M^{4+}$	9. $M^{8+} \rightarrow M^{9+}$
5. $M^{4+} \rightarrow M^{5+}$	10. $M^{9+} \rightarrow M^{10+}$

Principal lines in atomic spectrum

Wavelength/nm	Species	Sensitivity	Application
n.a.			

Abundance: Nil

Biological role: None; toxic due to radioactivity

F

Atomic number: 9
Relative atomic mass ($^{12}C = 12.0000$): **19.998 4032**

Chemical properties

Pale yellow gas, F_2, most reactive of all elements. Produced by electrolysis of molten KF.2HF. Used to make UF_6, SF_6, and fluorinating agents such as ClF_3. Organic compounds, polymers, and salts all used, notably CaF_2 as flux in metallurgy, and AlF_3 in aluminium production.

Radii/pm: F^- 133; covalent 64; atomic 71.7; van der Waals 135
Electronegativity: 3.98 (Pauling); 4.10 (Allred-Rochow)
Effective nuclear charge: 5.20 (Slater); 5.10 (Clementi); 4.61 (Froese–Fischer)

Standard reduction potentials $E^{\ominus}/V$

```
    0                   −I
F₂   ──2.866──   F⁻ (aq)
F₂   ──2.979──   HF₂⁻
F₂   ──3.053──   HF (aq)
```

Oxidation states

F^{-I} ([Ne]) F^- (aq), HF, KHF_2, CaF_2
many salts and derivatives
of other elements

Covalent bonds r/pm E/kJ mol^{-1}

	r/pm	E/kJ mol^{-1}
F–F	141.7	159
F–O	147	190
F–N	137	272

For other bonds see other elements

Physical properties

Melting point/K: 53.53
Boiling point/K: 85.01
ΔH_{fusion}/kJ mol^{-1}: 1.02
ΔH_{vap}/kJ mol^{-1}: 3.26

Thermodynamic properties (298.15 K, 0.1 MPa)

State	$\Delta_f H^{\ominus}$/kJ mol^{-1}	$\Delta_f G^{\ominus}$/kJ mol^{-1}	$S^{\ominus}$/J K^{-1} mol^{-1}	C_p/J K^{-1} mol^{-1}
Gas (F_2)	0	0	202.78	31.30
Gas atoms	78.99	61.91	158.754	22.744

Density/kg m^{-3}: n.a. [solid]; 1516 [liquid, 85 K]; 1.696 [gas, 273 K]
Thermal conductivity/W m^{-1} K^{-1}: 0.0279 [300 K]
Mass magnetic susceptibility/kg^{-1} m^3: n.a.
Molar volume/cm^3: 18.05 [85 K]

Lattice structure (cell dimensions/pm), space group

α-F_2 rhombic (cell dimensions n.a.), C2/m
β-F_2 cubic ($a = 667$), Pm3n

$T(\alpha \rightarrow \beta) = 45.6$ K

X-ray diffraction: mass absorption coefficients (μ/ρ)/cm^2 g^{-1}:
CuK$_\alpha$ 16.4   MoK$_\alpha$ 1.80

Fluorine

Atomic number: 9

Thermal neutron capture cross-section/barns: 0.098 ± 0.007

Number of isotopes (including nuclear isomers): 6

Isotope mass range: $17 \rightarrow 22$

Nuclear properties

Key isotopes

Nuclide	Atomic mass	Natural abundance (%)	Half life $T_{1/2}$	Decay mode and energy (MeV)	Nuclear spin I	Nuclear magnetic moment μ	Uses
^{18}F		0	109.7m	β^+ (1.65), EC (3%), no γ			tracer, medical
^{19}F	18.998 40	100	stable		1/2	+2.6283	NMR

NMR

^{19}F

Relative sensitivity (^{1}H = 1.00)   0.83

Absolute sensitivity (^{1}H = 1.00)   0.83

Receptivity (^{13}C = 1.00)   4730

Magnetogyric ratio/rad T^{-1} s^{-1} 25.1665×10^7

Frequency (^{1}H = 100 MHz; 2.3488 T)/MHz   94.077

Reference: CFCl$_3$

Ground state electron configuration: [He]$2s^2 2p^5$

Term symbol: $^2P_{3/2}$

Electron affinity (M$\rightarrow$M$^-$)/kJ mol^{-1}: 322

Electron shell properties

Ionization energies/kJ mol^{-1}

1. M $\rightarrow$M$^+$	1 681	6. M$^{5+}\rightarrow$M^{6+}	15 164
2. M$^+\rightarrow$M^{2+}	3 374	7. M$^{6+}\rightarrow$M^{7+}	17 867
3. M$^{2+}\rightarrow$M^{3+}	6 050	8. M$^{7+}\rightarrow$M^{8+}	92 036
4. M$^{3+}\rightarrow$M^{4+}	8 408	9. M$^{8+}\rightarrow$M^{9+}	106 432
5. M$^{4+}\rightarrow$M^{5+}	11 023		

Principal lines in atomic spectrum

Wavelength/nm	Species	Sensitivity	Application
95.482* (vac)	I		
685.603	I	U3	AE
690.248	I	U2	AE
703.747	I		
712.787			

* Lowest energy transition from ground state to nearest empty orbital

Abundance: Earth's crust 544 p.p.m.; seawater 1.3 p.p.m.

Biological role: Essential; excess fluoride toxic; HF and F$_2$ very toxic

Fr	**Atomic number:** 87
	Relative atomic mass ($^{12}C = 12.0000$): **(223)**

Chemical properties

Intensely radioactive, short-lived metal element obtained in minute quantities from actinium obtained from the neutron bombardment of radium.

Radii/pm: Fr^+ 180; atomic *c.* 270
Electronegativity: 0.7 (Pauling); 0.86 (Allred-Rochow)
Effective nuclear charge: 2.20 (Slater)

Standard reduction potentials $E^{\ominus}/V$

I		0
Fr^+	$\xrightarrow{c. -2.9}$	Fr

Oxidation states

Fr^I ([Rn])	little studied, Fr^+ (aq), $FrClO_4$ insoluble

Physical properties

Melting point/K: 300
Boiling point/K: 950
ΔH_{fusion}/kJ mol^{-1}: n.a.
ΔH_{vap}/kJ mol^{-1}: n.a.

Thermodynamic properties (298.15 K, 0.1 MPa)

State	$\Delta_f H^{\ominus}$/kJ mol^{-1}	$\Delta_f G^{\ominus}$/kJ mol^{-1}	$S^{\ominus}$/J K^{-1} mol^{-1}	C_p/J K^{-1} mol^{-1}
Solid	0	0	95.4	n.a.
Gas	72.8	n.a.	n.a.	n.a.

Density/kg m^{-3}: n.a.
Thermal conductivity/W m^{-1} K^{-1}: 15 (est.) [300 K]
Electrical resistivity/Ω m: n.a.
Mass magnetic susceptibility/kg^{-1} m^3: n.a.
Molar volume/cm^3: n.a.
Coefficient of linear thermal expansion/K^{-1}: n.a.

Lattice structure (cell dimensions/pm), space group

n.a.

X-ray diffraction: mass absorption coefficients (μ/ρ)/cm^2 g^{-1}:
n.a.

Francium

Atomic number: 87
Thermal neutron capture cross-section/barns: n.a.
Number of isotopes (including nuclear isomers): 21
Isotope mass range: $204 \rightarrow 224$

Key isotopes

Nuclide	Atomic mass	Natural abundance (%)	Half life $T_{1/2}$	Decay mode and energy (MeV)	Nuclear spin I	Nuclear magnetic moment μ	Uses
^{212}Fr		0	19m				
^{223}Fr	223.0198	some	22m*	β^- (1.15); γ			

* longest lived isotope

Ground state electron configuration: $[Rn]7s^1$
Term symbol: $^2S_{1/2}$
Electron affinity $(M \rightarrow M^-)$/kJ mol^{-1}: 44 (calc.)

Ionization energies/kJ mol^{-1}

1. $M \rightarrow M^+$	400	6. $M^{5+} \rightarrow M^{6+}$	(6 900)	
2. $M^+ \rightarrow M^{2+}$	(2100)	7. $M^{6+} \rightarrow M^{7+}$	(8 100)	
3. $M^{2+} \rightarrow M^{3+}$	(3100)	8. $M^{7+} \rightarrow M^{8+}$	(12 300)	
4. $M^{3+} \rightarrow M^{4+}$	(4100)	9. $M^{8+} \rightarrow M^{9+}$	(12 800)	
5. $M^{4+} \rightarrow M^{5+}$	(5700)	10. $M^{9+} \rightarrow M^{10+}$	(29 300)	

Principal lines in atomic spectrum

Wavelength/nm	Species	Sensitivity	Application
n.a.			

Abundance: Minute traces in uranium ores
Biological role: None; toxic due to radioactivity

Gd

Atomic number: 64
Relative atomic mass ($^{12}C = 12.0000$): 157.25

Chemical properties

Silvery-white metal of rare earth group. Obtained from monazite and bastnae-site. Reacts slowly with oxygen and water, dissolves in acids. Limited use in electronics, refractories, and alloys.

Radii/pm: Gd^{3+} 97; atomic 180.2; covalent 161
Electronegativity: 1.20 (Pauling); 1.11 (Allred-Rochow)
Effective nuclear charge: 2.85 (Slater); 8.22 (Clementi);
11.28 (Froese–Fischer)

Standard reduction potentials $E^{\ominus}$/V

	III		0
Acid solution	Gd^{3+}	$\xrightarrow{-2.28}$	Gd
Alkaline solution	$Gd(OH)_3$	$\xrightarrow{-2.82}$	Gd

Oxidation states

Gd^{II} (f^8)	GdI_2
Gd^{III} (f^7)	Gd_2O_3, $Gd(OH)_3$, GdF_3,
	$GdCl_3$ etc., $[Gd(H_2O)_x]^{3+}$ (aq),
	Gd^{3+} salts and complexes

Physical properties

Melting point/K: 1586
Boiling point/K: 3539
ΔH_{fusion}/kJ mol^{-1}: 15.5
ΔH_{vap}/kJ mol^{-1}: 301

Thermodynamic properties (298.15 K, 0.1 MPa)

State	$\Delta_f H^{\ominus}$/kJ mol^{-1}	$\Delta_f G^{\ominus}$/kJ mol^{-1}	$S^{\ominus}$/J K^{-1} mol^{-1}	C_p/J K^{-1} mol^{-1}
Solid	0	0	68.07	37.03
Gas	397.5	359.8	193.314	27.547

Density/kg m^{-3}: 7900.4 [298 K]
Thermal conductivity/W m^{-1} K^{-1}: 10.6 [300 K]
Electrical resistivity/Ω m: 134.0×10^{-8} [298 K]
Mass magnetic susceptibility/kg^{-1} m^3: $+6.030 \times 10^{-5}$ (s)
Molar volume/cm^3: 19.90
Coefficient of linear thermal expansion/K^{-1}: 8.6×10^{-6}

Lattice structure (cell dimensions/pm), space group

α-Gd h.c.p. ($a = 363.60$, $c = 578.26$), P6$_3$/mmc
β-Gd b.c.c. ($a = 405$), Im3m

$T(\alpha \rightarrow \beta) = 1535$ K

high pressure form: ($a = 361$, $c = 2603$), R$\bar{3}$m

X-ray diffraction: mass absorption coefficients (μ/ρ)/cm^2 g^{-1}:
CuK$_\alpha$ 439   MoK$_\alpha$ 64.4

Discovered in 1880 by J.-C. Galissard de Marignac at Geneva,
Switzerland. Isolated in 1886 by P.-E. Lecoq de Boisbaudran at
Paris, France
[Named after J. Gadolin, a Finnish chemist]

Gadolinium

Atomic number: 64

Thermal neutron capture cross-section/barns: $49\,000 \pm 2000$

Number of isotopes (including nuclear isomers): 17

Isotope mass range: $145 \rightarrow 161$

Nuclear properties

Key isotopes

Nuclide	Atomic mass	Natural abundance (%)	Half life $T_{1/2}$	Decay mode and energy (MeV)	Nuclear spin I	Nuclear magnetic moment μ	Uses
^{152}Gd	151.9195	0.20	1.1×10^{14}y	α (2.24)			
^{153}Gd		0	242d	EC (0.243); γ	3/2		tracer
^{154}Gd	153.9207	2.1	stable				
^{155}Gd	154.9226	14.8	stable		3/2	-0.2584	NMR
^{156}Gd	155.9221	20.6	stable				
^{157}Gd	156.9339	15.7	stable		3/2	-0.3388	NMR
^{158}Gd	157.9241	24.8	stable				
^{160}Gd	159.9071	21.8	stable				

NMR

	^{155}Gd	^{157}Gd
Relative sensitivity (^{1}H $= 1.00$)	2.79×10^{-4}	5.44×10^{-4}
Absolute sensitivity (^{1}H $= 1.00$)	4.11×10^{-5}	8.53×10^{-5}
Receptivity (^{13}C $= 1.00$)	0.124	0.292
Magnetogyric ratio/rad T^{-1} s^{-1}	-0.8273×10^7	-1.0792×10^7
Quadrupole moment/m^2	1.6×10^{-28}	2×10^{-28}
Frequency (^{1}H $= 100$ MHz; 2.3488 T)/MHz	3.819	4.774

Ground state electron configuration: $[Xe]4f^75d^16s^2$

Term symbol: $^9D_2$

Electron affinity $(M \rightarrow M^-)$/kJ mol^{-1}: ≤ 50

Electron shell properties

Ionization energies/kJ mol^{-1}

1. $M \rightarrow M^+$	592.5	6. $M^{5+} \rightarrow M^{6+}$
2. $M^+ \rightarrow M^{2+}$	1167	7. $M^{6+} \rightarrow M^{7+}$
3. $M^{2+} \rightarrow M^{3+}$	1990	8. $M^{7+} \rightarrow M^{8+}$
4. $M^{3+} \rightarrow M^{4+}$	4250	9. $M^{8+} \rightarrow M^{9+}$
5. $M^{4+} \rightarrow M^{5+}$		10. $M^{9+} \rightarrow M^{10+}$

Principal lines in atomic spectrum

Wavelength/nm	Species	Sensitivity	Application
364.620	II		AE
368.413	I		AA
376.841	II		AE
378.305	I		AA
404.870	I		AA

Abundance: Earth's crust 6.1 p.p.m.; seawater 2×10^{-7} p.p.m.

Biological role: None; toxicity unknown, probably low

<table>
<tr><td>**Ga**</td><td>**Atomic number: 31**
Relative atomic mass ($^{12}C = 12.0000$): **69.723**</td></tr>
</table>

Chemical properties

Soft, silvery-white metal, stable in air and with water. Soluble in acids and alkalis. Longest liquid range of all elements. Semiconductor properties with phosphorus, arsenic, and antimony.

Radii/pm: Ga^{3+} 62; Ga^+ 113; atomic 122.1; covalent 125

Electronegativity: 1.81 (Pauling); 1.82 (Allred-Rochow)

Effective nuclear charge: 5.00 (Slater); 6.22 (Clementi); 6.72 (Froese–Fischer)

Standard reduction potentials $E^{\ominus}$/V

	III	II	0
Acid solution	Ga^{3+}	Ga^{2+}	Ga

-0.53 (III to 0); $c.\,-0.65$ (III to II); $c.\,-0.45$ (II to 0)

Oxidation states

Ga^I (s^2)	Ga_2O, $GaCl_2$ etc., $Ga_2Cl_4 = Ga^I[Ga^{III}Cl_4]$
Ga^{II} (s^1)	$[Ga_2Cl_6]^{2-}$
Ga^{III} ($[d^{10}]$)	Ga_2O_3, $Ga(OH)_3$, $[Ga(H_2O)_6]^{3+}$ (aq) GaF_3, Ga_2Cl_6, $GaCl_6^{3-}$

Physical properties

Melting point/K: 302.93

Boiling point/K: 2676

ΔH_{fusion}/kJ mol^{-1}: 5.59

ΔH_{vap}/kJ mol^{-1}: 270.3

Thermodynamic properties (298.15 K, 0.1 MPa)

State	$\Delta_f H^{\ominus}$/kJ mol^{-1}	$\Delta_f G^{\ominus}$/kJ mol^{-1}	$S^{\ominus}$/J K^{-1} mol^{-1}	C_p/J K^{-1} mol^{-1}
Solid	0	0	40.88	25.86
Gas	277.0	238.9	169.06	25.36

Density/kg m^{-3}: 5907 [293 K]; 6113.6 [liquid at m.p.]

Thermal conductivity/W m^{-1} K^{-1}: 40.6 [300 K]

Electrical resistivity/Ω m: 27×10^{-8} [273 K] varies with axis

Mass magnetic susceptibility/kg^{-1} m^3: -3.9×10^{-9} (s)

Molar volume/cm^3: 11.81

Coefficient of linear thermal expansion/K^{-1}:
11.5×10^{-6} (*a* axis); 31.5×10^{-6} (*b* axis); 16.5×10^{-6} (*c* axis)

Lattice structure (cell dimensions/pm), space group

α-Ga orthorhombic ($a = 451.86$, $b = 765.70$, $c = 452.58$), Cmca
β-Ga rhombic ($a = 290$, $b = 813$, $c = 317$), Cmcm (metastable form)
γ-Ga rhombic ($a = 1060$, $b = 1356$, $c = 519$), Cmc2$_1$

$T(\gamma \rightarrow \alpha) = 238$ K
High pressure form: ($a = 279$, $c = 438$), I4/mmm

X-ray diffraction: mass absorption coefficients (μ/ρ)/cm^2 g^{-1}:
CuK$_\alpha$ 67.9   MoK$_\alpha$ 60.1

Discovered in 1875 by Paul-Émile Lecoq de Boisbaudran at Paris, France

[Latin, *Gallia* = France]

Gallium

Atomic number: 31
Thermal neutron capture cross-section/barns: 3.1 ± 0.3
Number of isotopes (including nuclear isomers): 14
Isotope mass range: $63 \rightarrow 76$

Key isotopes

Nuclide	Atomic mass	Natural abundance (%)	Half life $T_{1/2}$	Decay mode and energy (MeV)	Nuclear spin I	Nuclear magnetic moment μ	Uses
^{67}Ga		0	78.1h	EC(1.0003); γ	3/2	+1.849	medical tracer
^{69}Ga	68.9257	60.1	stable		3/2	+2.0145	NMR
^{71}Ga	70.9249	39.9	stable		3/2	+2.5597	NMR
^{72}Ga		0	14.1h	β^-(4.00); γ	3	-0.1321	tracer

NMR

	$[^{69}$Ga$]$	^{71}Ga
Relative sensitivity (^{1}H = 1.00)	6.91×10^{-2}	0.14
Absolute sensitivity (^{1}H = 1.00)	4.17×10^{-2}	5.62×10^{-2}
Receptivity (^{13}C = 1.00)	237	319
Magnetogyric ratio/rad T^{-1} s^{-1}	6.420×10^7	8.158×10^7
Quadrupole moment/m^2	0.178×10^{-28}	0.112×10^{-28}
Frequency (^{1}H = 100 MHz; 2.3488 T)/MHz	24.003	30.495

Reference: $Ga(H_2O)_6^{3+}$

Ground state electron configuration: $[Ar]3d^{10}4s^24p^1$
Term symbol: $^2P_{1/2}$
Electron affinity $(M \rightarrow M^-)$/kJ mol^{-1}: 36 (calc.)

Ionization energies/kJ mol^{-1}

1.	$M \rightarrow M^+$	578.8	6.	$M^{5+} \rightarrow M^{6+}$	(11 400)
2.	$M^+ \rightarrow M^{2+}$	1979	7.	$M^{6+} \rightarrow M^{7+}$	(14 400)
3.	$M^{2+} \rightarrow M^{3+}$	2963	8.	$M^{7+} \rightarrow M^{8+}$	(17 700)
4.	$M^{3+} \rightarrow M^{4+}$	6200	9.	$M^{8+} \rightarrow M^{9+}$	(22 300)
5.	$M^{4+} \rightarrow M^{5+}$	(8700)	10.	$M^{9+} \rightarrow M^{10+}$	(26 100)

Principal lines in atomic spectrum

Wavelength/nm	Species	Sensitivity	Application
287.424	I	U4	AA, AE
294.364	I	U3	AA, AE
403.298*	I	U2	AA, AE
417.206	I	U1	AA, AE
639.656	I		

* Lowest energy transition from ground state to nearest empty orbital

Abundance: Earth's crust 19 p.p.m.; seawater 0.0005 p.p.m.
Biological role: None; non-toxic; stimulatory

Ge	Atomic number: 32
	Relative atomic mass ($^{12}C = 12.0000$): **72.61**

Chemical properties

Ultrapure it is a silvery-white brittle metalloid element. Minerals are rare; obtained from zinc smelter flue dust. Stable in air and water, unaffected by acids, except nitric, and alkalis. Used in semiconductors, alloys, and special glasses.

Radii/pm: Ge^{2+} 90; atomic 122.5; covalent 122; Ge^{4-} 272

Electronegativity: 2.01 (Pauling); 2.02 (Allred-Rochow)

Effective nuclear charge: 5.65 (Slater); 6.78 (Clementi); 7.92 (Froese–Fischer)

Standard reduction potentials $E^{\ominus}$/V

	IV		II		0		−IV
Acid solution	GeO_2	$\xrightarrow{-0.370}$	GeO	$\xrightarrow{0.255}$	Ge	$\xrightarrow{-0.29}$	GeH_4
	Ge^{4+}	$\xrightarrow{0.00}$	Ge^{2+}	$\xrightarrow{-0.247}$			

[Alkaline solutions contain many different forms]

Covalent bonds r/pm E/kJ mol^{-1}

	r/pm	E/kJ mol^{-1}
Ge–H	152.9	288
Ge–C	194	237
Ge–O	165	363
Ge–F	170	464
Ge–Cl	210	340
Ge–Ge	241	163

Oxidation states

Ge^{II} (s^2) GeO, GeS, GeF_2, $GeCl_2$ etc.

Ge^{IV} (d^{10}) GeO_2, GeH_4 etc., GeF_4, $GeCl_4$ etc., GeF_6^{2-}, $GeCl_6^{2-}$, GeS_2, $[GeO(OH)_3]^-$ (aq)

Physical properties

Melting point/K: 1210.6
Boiling point/K: 3103
ΔH_{fusion}/kJ mol^{-1}: 34.7
ΔH_{vap}/kJ mol^{-1}: 327.6

Thermodynamic properties (298.15 K, 0.1 MPa)

State	$\Delta_f H^{\ominus}$/kJ mol^{-1}	$\Delta_f G^{\ominus}$/kJ mol^{-1}	$S^{\ominus}$/J K^{-1} mol^{-1}	C_p/J K^{-1} mol^{-1}
Solid	0	0	31.09	23.347
Gas	376.6	335.9	167.900	30.731

Density/kg m^{-3}: 5323 [293 K]; 5490 [liquid at m.p.]
Thermal conductivity/W m^{-1} K^{-1}: 59.9 [300 K]
Electrical resistivity/Ω m: 0.46 [295 K]
Mass magnetic susceptibility/kg^{-1} m^3: -1.328×10^{-9} (s)
Molar volume/cm^3: 13.64
Coefficient of linear thermal expansion/K^{-1}: 5.57×10^{-6}

Lattice structure (cell dimensions/pm), space group

Cubic ($a = 565.754$), Fd3m, diamond structure
High pressure forms: ($a = 488.4$; $c = 269.2$), I4$_1$/amd
($a = 593$; $c = 698$), P4$_3$2$_1$2
($a = 692$), b.c.c.

X-ray diffraction: mass absorption coefficients (μ/ρ)/cm^2 g^{-1}:
CuK$_\alpha$ 75.6   MoK$_\alpha$ 64.8

Germanium

Atomic number: 32
Thermal neutron capture cross-section/barns: 2.3 ± 0.3
Number of isotopes (including nuclear isomers): 17
Isotope mass range: $65 \rightarrow 78$

Key isotopes

Nuclide	Atomic mass	Natural abundance (%)	Half life $T_{1/2}$	Decay mode and energy (MeV)	Nuclear spin I	Nuclear magnetic moment μ	Uses
^{68}Ge		0	287d	EC($c.$ 0.7); γ			tracer
^{70}Ge	69.9243	20.5	stable		0		
^{71}Ge	70.925	0	11.4d	EC(0.235); γ	1/2	+0.546	tracer
^{72}Ge	71.9217	27.4	stable		0		
^{73}Ge	72.9234	7.8	stable		9/2	−0.8792	NMR
^{74}Ge	73.9219	36.5	stable		0		
^{76}Ge	75.9214	7.8	stable		0		
^{77}Ge		0	11.3h	β^- (2.75); γ			tracer

NMR

^{73}Ge

Relative sensitivity (^{1}H = 1.00)	1.4×10^{-3}
Absolute sensitivity (^{1}H = 1.00)	1.08×10^{-4}
Receptivity (^{13}C = 1.00)	0.617
Magnetogyric ratio/rad T^{-1} s^{-1}	-0.9331×10^7
Quadrupole moment/m^2	-0.2×10^{-28}
Frequency (^{1}H = 100 MHz; 2.3488 T)/MHz	3.488

Reference: $Ge(CH_3)_4$

Ground state electron configuration: [Ar] $3d^{10}4s^24p^2$
Term symbol: $^3P_0$
Electron affinity (M→M$^-$)/kJ mol^{-1}: 116

Ionization energies/kJ mol^{-1}

1. M $\rightarrow$M$^+$	762.1	6. M^{5+}$\rightarrow$M^{6+}	(11 900)
2. M$^+$$\rightarrow$M^{2+}	1537	7. M^{6+}$\rightarrow$M^{7+}	(15 000)
3. M^{2+}$\rightarrow$M^{3+}	3302	8. M^{7+}$\rightarrow$M^{8+}	(18 200)
4. M^{3+}$\rightarrow$M^{4+}	4410	9. M^{8+}$\rightarrow$M^{9+}	(21 800)
5. M^{4+}$\rightarrow$M^{5+}	9020	10. M^{9+}$\rightarrow$M^{10+}	(27 000)

Principal lines in atomic spectrum

Wavelength/nm	Species	Sensitivity	Application
206.866	I		
259.253	I		AA
265.158*	I		AA, AE
303.906	I	U2	AA, AE
326.949	I	U3	AE

* Lowest energy transition from ground state to nearest empty orbital

Abundance: Earth's crust 1.5 p.p.m.; seawater 1×10^{-4} p.p.m.
Biological role: None; non-toxic; stimulatory

<table>
<tr><td>

Au

</td><td>

Atomic number: 79

Relative atomic mass ($^{12}C = 12.0000$): **196.96654**

</td></tr>
</table>

<table>
<tr><td>

Chemical properties

</td><td>

Soft metal with characteristic yellow colour. Highest malleability and ductility of any element. Found in free state. Unaffected by air, water, acids (except HNO_3–HCl), and alkalis. Used as bullion, in jewellery, electronics, and glass (colouring and heat reflecting).

</td></tr>
</table>

Radii/pm: Au^+ 137; Au^{3+} 91; atomic 144.2; covalent 134

Electronegativity: 2.54 (Pauling); 1.42 (Allred-Rochow)

Effective nuclear charge: 4.20 (Slater); 10.94 (Clementi); 15.94 (Froese–Fischer)

Standard reduction potentials $E^\ominus$/V

	III		I		0
Acid solution	Au^{3+}	$\frac{1.36}{1.52}$	Au^+	1.83	Au
	$AuCl_4^-$	$\frac{0.926}{1.002}$	$AuCl_2^-$	1.154	Au
	$Au(SCN)_4^-$	$\frac{0.623}{0.636}$	$Au(SCN)_2^-$	0.662	Au

Oxidation states

Au^{-1} ($d^{10} s^2$)	$[Au(NH_3)_n]^-$ in liquid NH_3
Au^0 ($d^{10} s^1$)	gold clusters eg. $[Au_8(PPh_3)_8]^{2+}$
Au^I (d^{10})	Au_2S, $[Au(CN)_2]^-$ and other complexes
Au^{II} (d^9)	rare but some complexes known
Au^{III} (d^8)	Au_2O_3, $Au(OH)_4^-$ (aq), $AuCl_4^-$ (aq), $AuCl_3(OH)^-$ (aq), Au_2S_3, AuF_3, Au_2Cl_6, $AuBr_3$, complexes
Au^V (d^6)	AuF_5
Au^{VII} (d^4)	AuF_7

<table>
<tr><td>

Physical properties

</td><td>

Melting point/K: 1337.58

Boiling point/K: 3080

ΔH_{fusion}/kJ mol^{-1}: 12.7

ΔH_{vap}/kJ mol^{-1}: 343.1

</td></tr>
</table>

Thermodynamic properties (298.15 K, 0.1 MPa)

State	$\Delta_f H^\ominus$/kJ mol^{-1}	$\Delta_f G^\ominus$/kJ mol^{-1}	$S^\ominus$/J K^{-1} mol^{-1}	C_p/J K^{-1} mol^{-1}
Solid	0	0	47.40	25.418
Gas	336.1	326.3	180.503	20.786

Density/kg m^{-3}: 19 320 [293 K]; 17 280 [liquid at m.p.]

Thermal conductivity/W m^{-1} K^{-1}: 317 [300 K]

Electrical resistivity/Ω m: 2.35×10^{-8} [293 K]

Mass magnetic susceptibility/kg^{-1} m^3: -1.78×10^{-9} (s)

Molar volume/cm^3: 10.19

Coefficient of linear thermal expansion/K^{-1}: 14.16×10^{-6}

Lattice structure (cell dimensions/pm), space group

f.c.c. ($a = 407.833$), Fm3m

X-ray diffraction: mass absorption coefficients (μ/ρ)/cm^2 g^{-1}:
CuK$_\alpha$ 208 MoK$_\alpha$ 115

Gold

Atomic number: 79
Thermal neutron capture cross-section/barns: 98.8 ± 0.3
Number of isotopes (including nuclear isomers): 21
Isotope mass range: $185 \rightarrow 203$

Key isotopes

Nuclide	Atomic mass	Natural abundance (%)	Half life $T_{1/2}$	Decay mode and energy (MeV)	Nuclear spin I	Nuclear magnetic moment μ	Uses
^{195}Au		0	183d	EC(0.227); γ	3/2	± 0.147	tracer
^{197}Au	196.9666	100	stable		3/2	$+0.144\,86$	NMR
^{198}Au		0	2.693d	β^-(1.374); γ	2	$+0.590$	tracer, medical
^{199}Au		0	3.15d	β^-(0.46); γ	3/2	$+0.270$	tracer

NMR (difficult to detect) ^{197}Au

Relative sensitivity (^{1}H $= 1.00$)	2.51×10^{-5}
Absolute sensitivity (^{1}H $= 1.00$)	2.51×10^{-5}
Receptivity (^{13}C $= 1.00$)	0.06
Magnetogyric ratio/rad T^{-1} s^{-1}	0.357×10^7
Quadrupole moment/m^2	0.58×10^{-28}
Frequency (^{1}H $= 100$ MHz; 2.3488 T)/MHz	1.712

Ground state electron configuration: [Xe]4f^{14}5d^{10}6s^1
Term symbol: $^2S_{1/2}$
Electron affinity (M$\rightarrow$M$^-$)/kJ mol^{-1}: 223

Ionization energies/kJ mol^{-1}

1. M $\rightarrow$ M$^+$ 890.1	6. M$^{5+}\rightarrow$M^{6+} (7 000)
2. M$^+\rightarrow$M^{2+} 1980	7. M$^{6+}\rightarrow$M^{7+} (9 300)
3. M$^{2+}\rightarrow$M^{3+} (2900)	8. M$^{7+}\rightarrow$M^{8+} (11 000)
4. M$^{3+}\rightarrow$M^{4+} (4200)	9. M$^{8+}\rightarrow$M^{9+} (12 800)
5. M$^{4+}\rightarrow$M^{5+} (5600)	10. M$^{9+}\rightarrow$M^{10+} (14 800)

Principal lines in atomic spectrum

Wavelength/nm	Species	Sensitivity	Application
201.200	I		
242.795	I	U1	AA, AE
267.595*	I	U2	AA, AE
274.826	I		AA
312.278	I		AA

* Lowest energy transition from ground state to nearest empty orbital

Abundance: Earth's crust 0.004 p.p.m.; seawater 4×10^{-6} p.p.m.
Biological role: None; non-toxic; stimulatory

Hf

Atomic number: 72

Relative atomic mass ($^{12}C = 12.0000$): **178.49**

Chemical properties

Lustrous, silvery, ductile metal. Obtained from zirconium ores. Resists corrosion due to oxide film, but powdered Hf will burn in air. Unaffected by acids (except HF) and alkalis. Used in control rods for nuclear reactors.

Radii/pm: Hf^{4+} 84; atomic 156.4; covalent 144

Electronegativity: 1.3 (Pauling); 1.23 (Allred–Rochow)

Effective nuclear charge: 3.15 (Slater); 9.16 (Clementi); 13.27 (Froese–Fischer)

Standard reduction potentials $E^{\ominus}/V$

IV		0
Hf^{4+}	$\xrightarrow{-1.70}$	Hf
HfO_2	$\xrightarrow{-1.57}$	Hf

Oxidation states

Hf^I (d^3)	HfCl?
Hf^{II} (d^2)	$HfCl_2$?
Hf^{III} (d^1)	$HfCl_3$, $HfBr_3$, HfI_3, Hf^{3+} reduces water
Hf^{IV} (d^0, f^{14})	HfO_2, $Hf(OH)^{3+}$ (aq), HfF_4, $HfCl_4$ etc., HfF_6^{2-}, HfF_7^{3-}, HfF_8^{4-}

Physical properties

Melting point/K: 2503

Boiling point/K: 5470

ΔH_{fusion}/kJ mol^{-1}: 25.5

ΔH_{vap}/kJ mol^{-1}: 570.7

Thermodynamic properties (298.15 K, 0.1 MPa)

State	$\Delta_f H^{\ominus}$/kJ mol^{-1}	$\Delta_f G^{\ominus}$/kJ mol^{-1}	$S^{\ominus}$/J K^{-1} mol^{-1}	C_p/J K^{-1} mol^{-1}
Solid	0	0	43.56	25.73
Gas	619.2	576.5	186.892	20.803

Density/kg m^{-3}: 13 310 [293 K]; 12 000 [liquid at m.p.]

Thermal conductivity/W m^{-1} K^{-1}: 23.0 [300 K]

Electrical resistivity/Ω m: 35.1×10^{-8} [293 K]

Mass magnetic susceptibility/kg^{-1} m^3: $+5.3 \times 10^{-9}$ (s)

Molar volume/cm^3: 13.41

Coefficient of linear thermal expansion/K^{-1}: 5.9×10^{-6}

Lattice structure (cell dimensions/pm), space group

α-Hf h.c.p. ($a = 319.46$; $c = 505.10$), P6$_3$/mmc

β-Hf cubic ($a = 362$)

$T(\alpha \rightarrow \beta) = 2033$ K

X-ray diffraction: mass absorption coefficients (μ/ρ)/cm^2 g^{-1}:
CuK$_\alpha$ 159   MoK$_\alpha$ 91.7

Hafnium

Atomic number: 72

Thermal neutron capture cross-section/barns: 103 ± 3

Number of isotopes (including nuclear isomers): 19

Isotope mass range: $169 \rightarrow 183$

Nuclear properties

Key isotopes

Nuclide	Atomic mass	Natural abundance (%)	Half life $T_{1/2}$	Decay mode and energy (MeV)	Nuclear spin I	Nuclear magnetic moment μ	Uses
^{172}Hf		0	$c.$ 5y	EC($c.$ 1); γ			tracer
^{174}Hf	173.9403	0.2	2×10^{15}y	α(2.55)			
^{175}Hf		0	70d	EC(0.59); γ			tracer
^{176}Hf	175.941 65	5.2	stable				
^{177}Hf	176.9435	18.6	stable		7/2	$+0.7902$	NMR
^{178}Hf	177.9439	27.1	stable		0		
^{179}Hf	178.9460	13.7	stable		9/2	-0.638	NMR
^{180}Hf	179.9468	35.2	stable		0		
^{181}Hf		0	42.5d	β^-(1.023); γ			tracer
^{182}Hf		0	9×10^6y	β^-($c.$ 0.5)			tracer

NMR (difficult to observe)	^{177}Hf	^{179}Hf
Relative sensitivity (^{1}H = 1.00)	6.38×10^{-4}	2.16×10^{-4}
Absolute sensitivity (^{1}H = 1.00)	1.18×10^{-4}	2.97×10^{-5}
Receptivity (^{13}C = 1.00)	0.88	0.27
Magnetogyric ratio/rad T^{-1} s^{-1}	$+0.945 \times 10^7$	-0.609×10^7
Quadrupole moment/m^2	4.5×10^{-28}	5.1×10^{-28}
Frequency (^{1}H = 100 MHz; 2.3488 T)/MHz	3.120	1.869

Ground state electron configuration: $[Xe]4f^{14}5d^26s^2$

Term symbol: $^3F_2$

Electron affinity $(M \rightarrow M^-)$/kJ mol^{-1}: -61

Electron shell properties

Ionization energies/kJ mol^{-1}

1. $M \rightarrow M^+$ 642	6. $M^{5+} \rightarrow M^{6+}$
2. $M^+ \rightarrow M^{2+}$ 1440	7. $M^{6+} \rightarrow M^{7+}$
3. $M^{2+} \rightarrow M^{3+}$ 2250	8. $M^{7+} \rightarrow M^{8+}$
4. $M^{3+} \rightarrow M^{4+}$ 3216	9. $M^{8+} \rightarrow M^{9+}$
5. $M^{4+} \rightarrow M^{5+}$	10. $M^{9+} \rightarrow M^{10+}$

Principal lines in atomic spectrum

Wavelength/nm	Species	Sensitivity	Application
202.818	II		
289.826	I		AA, AE
307.288	I		AA, AE
368.224	I		AA, AE
713.182*	I		AE

* Lowest energy transition from ground state to nearest empty orbital

Abundance: Earth's crust 2.8 p.p.m.; seawater <0.008 p.p.m.

Biological role: None; non-toxic

Atomic number: 2

Relative atomic mass ($^{12}C = 12.0000$): **4.002 602**

Colourless, odourless gas obtained from natural gas wells. Unreactive chemically. Used in deep-sea diving, weather balloons, and low temperature research instruments.

Radii/pm: atomic 128; van der Waals 122

Electronegativity: n.a. (Pauling); n.a. (Allred-Rochow)

Effective nuclear charge: 1.70 (Slater); 1.69 (Clementi); 1.62 (Froese–Fischer)

Standard reduction potentials $E^{\ominus}/V$

n.a.

Covalent bonds r/pm E/kJ mol^{-1}	**Oxidation states**
n.a.	n.a.

Melting point/K: 0.95 (under pressure)

Boiling point/K: 4.216

ΔH_{fusion}/kJ mol^{-1}: 0.021

ΔH_{vap}/kJ mol^{-1}: 0.082

Thermodynamic properties (298.15 K, 0.1 MPa)

State	$\Delta_f H^{\ominus}$/kJ mol^{-1}	$\Delta_f G^{\ominus}$/kJ mol^{-1}	$S^{\ominus}$/J K^{-1} mol^{-1}	C_p/J K^{-1} mol^{-1}
Gas	0	0	126.150	20.786

Density/kg m^{-3}: n.a. [solid]; 124.8 [liquid at b.p.]; 0.1785 [gas, 273 K]

Thermal conductivity/W m^{-1} K^{-1}: 0.152 [300 K]$_g$

Mass magnetic susceptibility/kg^{-1} m^3: -5.9×10^{-9} (g)

Molar volume/cm^3: 32.07 [4 K]

Lattice structure (cell dimensions/pm), space group [T, p]

α-He h.c.p. ($a = 353.1$; $c = 569.3$), P6$_3$/mmc. [$T = 1.15$ K, $p = 6.69$ MPa]

β-He f.c.c. ($a = 424.0$), Fm3m. [$T = 16$ K, $p = 127$ MPa]

γ-He b.c.c. ($a = 411$), Im3m. [$T = 1.73$ K, $p = 2.94$ MPa]

X-ray diffraction: mass absorption coefficients (μ/ρ)/cm^2 g^{-1}:
CuK$_\alpha$ 0.383 MoK$_\alpha$ 0.207

Helium

Atomic number: 2
Thermal neutron capture cross-section/barns: *c.* 0.007
Number of isotopes (including nuclear isomers): 5
Isotope mass range: $3 \rightarrow 8$ except 7

Nuclear properties

Key isotopes

Nuclide	Atomic mass	Natural abundance (%)	Half life $T_{1/2}$	Decay mode and energy (MeV)	Nuclear spin I	Nuclear magnetic moment μ	Uses
^{3}He	3.016 03	1.38×10^{-4}	stable		1/2	−2.1276	NMR, tracer
^{4}He	4.002 60	99.999 862	stable	0			

NMR (no compounds known)

	^{3}He
Relative sensitivity (^{1}H = 1.00)	0.44
Absolute sensitivity (^{1}H = 1.00)	5.75×10^{-7}
Receptivity (^{13}C = 1.00)	0.003 26
Magnetogyric ratio/rad T^{-1} s^{-1}	-20.378×10^7
Frequency (^{1}H = 100 MHz; 2.3488 T)/MHz	76.178

Ground state electron configuration: $1s^2$
Term symbol: $^1S_0$
Electron affinity ($M \rightarrow M^-$)/kJ mol^{-1}: -21 (calc.)

Electron shell properties

Ionization energies/kJ mol^{-1}

1. $M \rightarrow M^+$ 2372.3
2. $M^+ \rightarrow M^{2+}$ 5250.4

Principal lines in atomic spectrum

Wavelength/nm	Species	Sensitivity	Application
59.1412*(vac)	I		
388.864	I	U2	AE
468.575	II		AE
587.562	I	U3	AE
1083.034	I		

* Lowest energy transition from ground state to nearest empty orbital

Abundance: Atmosphere 5.2 p.p.m. (volume); Earth's crust 0.003 p.p.m.; seawater 7.2×10^{-6} p.p.m.
Biological role: None; non-toxic

Ho

Atomic number: 67
Relative atomic mass ($^{12}C = 12.0000$): **164.93032**

Chemical properties

Silvery metal of the rare earth group, obtained from monazite and bastnaesite ores. Slowly attacked by oxygen and water, dissolves in acid. Little used.

Radii/pm: Ho^{3+} 89; atomic 176.6; covalent 158
Electronegativity: 1.23 (Pauling); 1.10 (Allred-Rochow)
Effective nuclear charge: 2.85 (Slater); 8.44 (Clementi); 11.60 (Froese–Fischer)

Standard reduction potentials $E^{\ominus}$/V

	III		0
Acid solution	Ho^{3+}	$\xrightarrow{-2.33}$	Ho
Alkaline solution	$Ho(OH)_3$	$\xrightarrow{-2.85}$	Ho

Oxidation states

Ho^{III} (f^{10})	Ho_2O_3, $Ho(OH)_3$, $[Ho(H_2O)_x]^{3+}$ (aq), Ho^{3+} salts, HoF_3, $HoCl_3$ etc., $HoCl_6^{3-}$, complexes

Physical properties

Melting point/K: 1747
Boiling point/K: 2968
ΔH_{fusion}/kJ mol^{-1}: 17.2
ΔH_{vap}/kJ mol^{-1}: 303

Thermodynamic properties (298.15 K, 0.1 MPa)

State	$\Delta_f H^{\ominus}$/kJ mol^{-1}	$\Delta_f G^{\ominus}$/kJ mol^{-1}	$S^{\ominus}$/J K^{-1} mol^{-1}	C_p/J K^{-1} mol^{-1}
Solid	0	0	75.3	27.15
Gas	300.8	264.8	195.59	20.79

Density/kg m^{-3}: 8795 [298 K]
Thermal conductivity/W m^{-1} K^{-1}: 16.2 [300 K]
Electrical resistivity/Ω m: 87.0×10^{-8} [298 K]
Mass magnetic susceptibility/kg^{-1} m^3: $+5.49 \times 10^{-6}$ (s)
Molar volume/cm^3: 18.75
Coefficient of linear thermal expansion/K^{-1}: 9.5×10^{-6}

Lattice structure (cell dimensions/pm), space group

α-Ho h.c.p. ($a = 357.73$; $c = 561.58$), P6$_3$/mmc
β-Ho b.c.c. ($a = 396$), Im3m
$T(\alpha \rightarrow \beta)$ just below melting point
High pressure form: ($a = 334$; $c = 2410$), R$\bar{3}$m

X-ray diffraction: mass absorption coefficients (μ/ρ)/cm^2 g^{-1}:
CuK$_{\alpha}$ 128   MoK$_{\alpha}$ 73.9

Discovered in 1878 by P. T. Cleve at Uppsala, Sweden and independently by M. Delafontaine and J.-L. Soret at Geneva, Switzerland
[Latin, *Holmia* = Stockholm]

Holmium

Atomic number: 67

Thermal neutron capture cross-section/barns: 65 ± 2

Number of isotopes (including nuclear isomers): 29

Isotope mass range: $151 \rightarrow 170$

Nuclear properties

Key isotopes

Nuclide	Atomic mass	Natural abundance (%)	Half life $T_{1/2}$	Decay mode and energy (MeV)	Nuclear spin I	Nuclear magnetic moment μ	Uses
^{165}Ho	164.9303	100	stable		7/2	+4.12	NMR
^{166}Ho		0	26.9h	β^- (1.847)	0		tracer

NMR

	^{165}Ho
Relative sensitivity (^{1}H = 1.00)	0.18
Absolute sensitivity (^{1}H = 1.00)	0.18
Receptivity (^{13}C = 1.00)	1.16×10^3
Magnetogyric ratio/rad T^{-1} s^{-1}	5.710×10^7
Quadrupole moment/m^2	2.82×10^{-28}
Frequency (^{1}H = 100 MHz; 2.3488 T)/MHz	20.513

Ground state electron configuration: $[Xe]4f^{11}6s^2$

Term symbol: $^4I_{15/2}$

Electron affinity $(M \rightarrow M^-)$/kJ mol^{-1}: ≤ 50

Electron shell properties

Ionization energies/kJ mol^{-1}

1. $M \rightarrow M^+$	580.7	6. $M^{5+} \rightarrow M^{6+}$	
2. $M^+ \rightarrow M^{2+}$	1139	7. $M^{6+} \rightarrow M^{7+}$	
3. $M^{2+} \rightarrow M^{3+}$	2204	8. $M^{7+} \rightarrow M^{8+}$	
4. $M^{3+} \rightarrow M^{4+}$	4100	9. $M^{8+} \rightarrow M^{9+}$	
5. $M^{4+} \rightarrow M^{5+}$		10. $M^{9+} \rightarrow M^{10+}$	

Principal lines in atomic spectrum

Wavelength/nm	Species	Sensitivity	Application
345.600	II		
374.817	II		AE
405.393	I		AA
410.384	I		AA, AE
416.303	I		

Abundance: Earth's crust 1.3 p.p.m.; seawater 8×10^{-8} p.p.m.

Biological role: None; non-toxic; stimulatory

Chemical properties

Colourless, odourless gas, insoluble in water. Burns in air, forms explosive mixtures with air. Used in industry for making ammonia, cyclohexane, methanol etc. Obtained from methane CH_4.

Radii/pm: H^- 154; atomic 78; covalent 30;
van der Waals 120; H^+ 10^{-5}

Electronegativity: 2.20 (Pauling); n.a. (Allred–Rochow)

Effective nuclear charge: 1.00 (Slater); 1.00 (Clementi);
1.00 (Froese–Fischer)

Standard reduction potentials $E^\ominus$/V

	I		0		$-$I
Acid solution	H_3O^+	_0.00_	H_2	_-2.25_	H^-
Alkaline solution	H_2O	_0.828_	H_2	_-2.25_	H^-

Covalent bonds

	r/pm	E/kJ mol^{-1}
H—H	74.14	453.6
H—F	91.7	566
H—Cl	127.4	431
H—Br	140.8	366
H—I	160.9	299

For covalent bonds to hydrogen
see other elements

D—D	74.14	447.3

Oxidation states

H^{-1}	NaH, CaH_2, etc.
H^0	H_2
H^I	H_2O, H_3O^+, etc..
	OH^-, HF, HCl, etc.,
	other acids, NH_3 etc.,
	CH_4 etc.

Hydrogen is the most versatile of elements in its range of chemical bonds

Physical properties

Melting point/K: 14.01
Boiling point/K: 20.28
ΔH_{fusion}/kJ mol^{-1}: 0.12
ΔH_{vap}/kJ mol^{-1}: 0.46

Thermodynamic properties (298.15 K, 0.1 MPa)

State	$\Delta_f H^\ominus$/kJ mol^{-1}	$\Delta_f G^\ominus$/kJ mol^{-1}	$S^\ominus$/J K^{-1} mol^{-1}	C_p/J K^{-1} mol^{-1}
Gas (H_2)	0	0	130.684	28.824
Gas (atoms)	217.965	203.247	114.713	20.784

Density/kg m^{-3}: 76.0 [solid, 11 K]; 70.8 [liquid, b.p.];
0.089 88 [gas, 273 K]

Thermal conductivity/W m^{-1} K^{-1}: 0.1815 [300 K] g

Mass magnetic susceptibility/kg^{-1} m^3: -2.50×10^{-8} (g)

Molar volume/cm^3: 13.26 [11 K]

Lattice structure (cell dimensions/pm), space group

H_2 h.c.p. ($a = 377.6$, $c = 616.2$), $P6_3$/mmc
D_2 h.c.p. ($a = 360.0$, $c = 585.8$), $P6_3$/mmc
H_2 cubic ($a = 533.8$), Fm3m
D_2 cubic ($a = 509.2$), Fm3m
H_2 tetragonal ($a = 450$, $c = 368$), I4
D_2 tetragonal ($a = 338$, $c = 560$)
T (h.c.p.→cubic) = 4.5 K

X-ray diffraction: mass absorption coefficients (μ/ρ)/cm^2 g^{-1}:
CuK_α 0.435 MoK_α 0.380

Atomic number: 1
Thermal neutron capture cross-section/barns: 0.332 ± 0.002
Number of isotopes (including nuclear isomers): 3
Isotope mass range: $1 \rightarrow 3$

Nuclear properties

Key isotopes

Nuclide	Atomic mass	Natural abundance (%)	Half life $T_{1/2}$	Decay mode and energy (MeV)	Nuclear spin I	Nuclear magnetic moment μ	Uses
^{1}H	1.007825	99.985	stable		1/2	+2.79278	NMR
^{2}H	2.01400	0.015	stable		1	+0.85742	NMR
^{3}H	3.01605	0	12.262y	β^-(0.01861); no γ	1/2	+2.9789	tracer, NMR

NMR

	^{1}H	^{2}H	^{3}H
Relative sensitivity (^{1}H $= 1.00$)	1.00 (by defn)	9.65×10^{-3}	1.21
Absolute sensitivity (^{1}H $= 1.00$)	1.00 (by defn)	1.45×10^{-6}	0
Receptivity (^{13}C $= 1.00$)	5680	8.2×10^{-6}	—
Magnetogyric ratio/rad T^{-1} s^{-1}	26.7510×10^7	4.1064×10^7	28.5335×10^7
Quadrupole moment/m^2	—	2.73×10^{-31}	—
Frequency (^{1}H $= 100$ MHz; 2.3488 T)/MHz	100.000	15.351	106.663

Reference: $Si(CH_3)_4$

Ground state electron configuration: $1s^1$
Term symbol: $^2S_{1/2}$
Electron affinity $(M \rightarrow M^-)$/kJ mol^{-1}: 72.8

Electron shell properties

Ionization energies/kJ mol^{-1}

1. $M \rightarrow M^+$ 1312.0

Principal lines in atomic spectrum

Wavelength/nm	Species	Sensitivity	Application
121.568* (vac)	I		
486.133	I	U3	AE
656.272	I	U2	AE
656.285	I		
1875.10	I		

* Lowest energy transition from ground state to nearest empty orbital

Abundance: Atmosphere 0.53 p.p.m. (volume); Earth's crust 1520 p.p.m.; seawater, constituent
Biological role: Basis of all life (part of DNA molecule)

<table>
<tr><td>**In**</td><td>**Atomic number: 49**
Relative atomic mass (^{12}C $= 12.0000$): **114.82**</td></tr>
</table>

<table>
<tr><td>**Chemical properties**</td><td>Soft, silvery-white metal. Stable in air and with water, dissolves in acids. Used in low melting alloys in safety devices. Semiconductor uses as InAs and InSb in transistors, thermistors etc.</td></tr>
</table>

Radii/pm: In^{3+} 92; In^+ 132; atomic 162.6; covalent 150

Electronegativity: 1.78 (Pauling); 1.49 (Allred-Rochow)

Effective nuclear charge: 5.00 (Slater); 8.47 (Clementi); 9.66 (Froese–Fischer)

Standard reduction potentials $E^{\ominus}$/V

	III		I		0
Acid solution	In^{3+}	$\underline{-0.444}$	In^+	$\underline{-0.126}$	In
			-0.3382		

Covalent bonds r/pm E/kJ mol^{-1}

	r/pm	E/kJ mol^{-1}
In–H	185	243
In–C	216	165
In–O	213	109
IN–F	199	c.525
In–Cl	240	439
In–In	325.1	c. 85

Oxidation states

In^I (s^2)	InCl, InBr, InI
In^{II} (s^1)	$[In_2Cl_6]^{2-}$
	$[In_2Br_2]^{2-}$, $[In_2I_6]^{2-}$
In^{III} (d^{10})	In_2O_3, $In(OH)_3$,
	$[In(H_2O)_6]^{3+}$ (aq)
	InF_3, $InCl_3$ etc.,
	$InCl_5^{2-}$, $InCl_6^{3-}$,
	complexes

<table>
<tr><td>**Physical properties**</td><td>**Melting point**/K: 429.32
Boiling point/K: 2353
ΔH_{fusion}/kJ mol^{-1}: 3.27
ΔH_{vap}/kJ mol^{-1}: 231.8</td></tr>
</table>

Thermodynamic properties (298.15 K, 0.1 MPa)

State	$\Delta_f H^{\ominus}$/kJ mol^{-1}	$\Delta_f G^{\ominus}$/kJ mol^{-1}	$S^{\ominus}$/J K^{-1} mol^{-1}	C_p/J K^{-1} mol^{-1}
Solid	0	0	57.82	26.74
Gas	243.30	208.71	173.79	20.84

Density/kg m^{-3}: 7310 [298 K]; 7032 [liquid at m.p.]

Thermal conductivity/W m^{-1} K^{-1}: 81.6 [300 K]

Electrical resistivity/Ω m: 8.37×10^{-8} [293 K]

Mass magnetic susceptibility/kg^{-1} m^3: -7.0×10^{-9} (s)

Molar volume/cm^3: 15.71

Coefficient of linear thermal expansion/K^{-1}: 33×10^{-6}

Lattice structure (cell dimensions/pm), space group

Face centred tetragonal ($a = 325.30$, $c = 494.55$), I4/mmm

X-ray diffraction: mass absorption coefficients (μ/ρ)/cm^2 g^{-1}:
CuK$_\alpha$ 243 MoK$_\alpha$ 29.3

Discovered in 1863 by Ferdinand Reich and Hieronymous Richter
at Freiberg, Germany
[Named after the indigo line in its spectrum]

Indium

Atomic number: 49

Thermal neutron capture cross-section/barns: 194 ± 2

Number of isotopes (including nuclear isomers): 34

Isotope mass range: $106 \rightarrow 124$

Nuclear properties

Key isotopes

Nuclide	Atomic mass	Natural abundance (%)	Half life $T_{1/2}$	Decay mode and energy (MeV)	Nuclear spin I	Nuclear magnetic moment μ	Uses
^{111}In			2.81d	EC; γ	9/2	+5.53	medical tracer
^{113}In	112.9043	4.3	stable		9/2	+5.5229	NMR
^{113m}In			100m	IT(0.3916)	1/2	−0.210	medical tracer
^{114m_1}In			50d	IT, EC(0.191); γ 5		+4.7	tracer
^{115}In	114.9041	95.7	6×10^{14}y	β^-; no γ	9/2	+5.5348	NMR

NMR

	$[^{113}$In$]$	^{115}In
Relative sensitivity (^{1}H = 1.00)	0.34	0.34
Absolute sensitivity (^{1}H = 1.00)	1.47×10^{-2}	0.33
Receptivity (^{13}C = 1.00)	83.8	1890
Magnetogyric ratio/rad T^{-1} s^{-1}	5.8493×10^7	5.8618×10^7
Quadrupole moment/m^2	1.14×10^{-28}	1.16×10^{-28}
Frequency (^{1}H = 100 MHz; 2.3488 T)/MHz	21.866	21.914

Reference: $In(H_2O)_6^{3+}$

Ground state electron configuration: $[Kr]4d^{10}5s^25p^1$

Term symbol: $^2P_{1/2}$

Electron affinity (M→M$^-$)/kJ mol^{-1}: 34

Electron shell properties

Ionization energies/kJ mol^{-1}

1. M $\rightarrow$ M$^+$	558.3	6. M$^{5+} \rightarrow$ M^{6+} (9 500)
2. M$^+ \rightarrow$ M^{2+}	1820.6	7. M$^{6+} \rightarrow$ M^{7+} (11 700)
3. M$^{2+} \rightarrow$ M^{3+}	2704	8. M$^{7+} \rightarrow$ M^{8+} (13 900)
4. M$^{3+} \rightarrow$ M^{4+}	5200	9. M$^{8+} \rightarrow$ M^{9+} (17 200)
5. M$^{4+} \rightarrow$ M^{5+}	(7400)	10. M$^{9+} \rightarrow$ M^{10+} (19 700)

Principal lines in atomic spectrum

Wavelength/nm	Species	Sensitivity	Application
303.936	I	U4	AA, AE
325.609	I	U3	AA, AE
325.856	I	U5	AE
410.177*	I	U2	AA, AE
451.132	I	U1	AA, AE

* Lowest energy transition from ground state to nearest empty orbital

Abundance: Earth's crust 0.24 p.p.m.; seawater 0.02 p.p.m.

Biological role: None; teratogenic; stimulatory

I

Atomic number: 53

Relative atomic mass ($^{12}C = 12.0000$): 126.90447

Chemical properties

Black, shiny solid, I_2, obtained from certain brines rich in iodide or from iodates. Sublimes easily. Compounds used in food supplements, dyes, catalysts and photography.

Radii/pm: I^- 220; Covalent 133.3; van der Waals 215

Electronegativity: 2.66 (Pauling); 2.21 (Allred-Rochow)

Effective nuclear charge: 7.60 (Slater); 11.61 (Clementi); 14.59 (Froese–Fischer)

Standard reduction potentials $E^{\ominus}$/V

Covalent bonds r/pm E/kJ mol^{-1}

	r/pm	E/kJ mol^{-1}
I–H	160.9	299
I–C	213	218
I–O	195	234
I–F	191	280
I–Cl	232	208
I–I	266.6	151
I–Si	243	234
I–P	252	184

Oxidation states

I^{-I}	I^-(aq), HI, KI, etc
I^0	I_2, I_3^-, I_5^-, etc.
I^{I}	I_n^+, ICl_2^-, etc.
I^{III}	I_4O_9 ($= I^{3+}(IO_3^-)_3$))
	ICl_3
I^{V}	I_2O_5, HIO_3, IO_3^-(aq)
	IF_5, IF_6^-
I^{VII}	H_5IO_6, $H_4IO_6^-$(aq) etc.,
	HIO_4, IO_4^-(aq), IF_7

Physical properties

Melting point/K: 386.7

Boiling point/K: 457.50

ΔH_{fusion}/kJ mol^{-1}: 15.27   **ΔH_{vap}/kJ mol^{-1}:** 41.67

Thermodynamic properties (298.15 K, 0.1 MPa)

State	$\Delta_f H^{\ominus}$/kJ mol^{-1}	$\Delta_f G^{\ominus}$/kJ mol^{-1}	$S^{\ominus}$/J K^{-1} mol^{-1}	C_p/J K^{-1} mol^{-1}
Solid	0	0	116.135	54.438
Gas (I_2)	62.438	19.327	260.69	36.90
Gas (atoms)	106.838	70.250	180.791	20.786

Density/kg m^{-3}: 4930 [293 K]

Thermal conductivity/W m^{-1} K^{-1}: 0.449 [300 K]

Electrical resistivity/Ω m: 1.37×10^7 [293 K]

Mass magnetic susceptibility/kg^{-1} m^3: -4.40×10^{-9} (s)

Molar volume/cm^3: 25.74

Coefficient of linear thermal expansion/K^{-1}: n.a.

Lattice structure (cell dimensions/pm), space group

Orthorhombic ($a = 7.26.47$, $b = 478.57$, $c = 979.08$), Cmca

X-ray diffraction: mass absorption coefficients (μ/ρ)/cm^2 g^{-1}:
CuK$_{\alpha}$ 294   MoK$_{\alpha}$ 37.1

Iodine

Atomic number: 53
Thermal neutron capture cross-section/barns: 6.2 ± 0.2
Number of isotopes (including nuclear isomers): 24
Isotope mass range: $117 \rightarrow 139$

Nuclear properties

Key isotopes

Nuclide	Atomic mass	Natural abundance (%)	Half life $T_{1/2}$	Decay mode and energy (MeV)	Nuclear spin I	Nuclear magnetic moment μ	Uses
^{123}I		0	13.3h	EC(*c*. 1.4); γ	5/2		tracer
^{125}I		0	60.2d	EC(0.149); γ	5/2	$+3.0$	tracer medical
^{127}I	126.9004	100	stable		5/2	$+2.8091$	NMR
^{129}I		0	1.7×10^7y	β^-(0.189); γ	7/2	$+2.6174$	
^{131}I		0	8.070d	β^-(0.970); γ	7/2	$+2.738$	tracer, medical

NMR

^{127}I

Relative sensitivity ($^1H = 1.00$) 9.34×10^{-2}
Absolute sensitivity ($^1H = 1.00$) 9.34×10^{-2}
Receptivity ($^{13}C = 1.00$)   530
Magnetogyric ratio/rad T^{-1} s^{-1} 5.3525×10^7
Quadrupole moment/m^2 -0.79×10^{-28}
Frequency ($^1H = 100$ MHz; 2.3488 T)/MHz   20.007
Reference: NaI (aq)

Ground state electron configuration: $[Kr]4d^{10}5s^25p^5$
Term symbol: $^2P_{3/2}$
Electron affinity ($M \rightarrow M^-$)/kJ mol^{-1}: 295.3

Electron shell properties

Ionization energies/kJ mol^{-1}

1. $M \rightarrow M^+$ 1008.4	6. $M^{5+} \rightarrow M^{6+}$ (7 400)
2. $M^+ \rightarrow M^{2+}$ 1845.9	7. $M^{6+} \rightarrow M^{7+}$ (8 700)
3. $M^{2+} \rightarrow M^{3+}$ 3200	8. $M^{7+} \rightarrow M^{8+}$ (16 400)
4. $M^{3+} \rightarrow M^{4+}$ (4100)	9. $M^{8+} \rightarrow M^{9+}$ (19 300)
5. $M^{4+} \rightarrow M^{5+}$ (5000)	10. $M^{9+} \rightarrow M^{10+}$ (22 100)

Principal lines in atomic spectrum

Wavelength/nm	Species	Sensitivity	Application
183.036* (vac)	I		
516.119	II		AE
546.461	II		AE
804.374	I		

* Lowest energy transition from ground state to nearest empty orbital

Abundance: Earth's crust 0.46 p.p.m.; seawater 0.05 p.p.m.
Biological role: Essential; I_2 vapour harmful

Atomic number: 77
Relative atomic mass ($^{12}C = 12.0000$): **192.22**

Chemical properties

Hard, lustrous, silvery metal of the platinum group. Stable to air and water, inert to all acids but fused NaOH will attack it. Used in special alloys and spark plugs.

Radii/pm: Ir^{2+} 89; Ir^{3+} 75; Ir^{4+} 66; atomic 135.7; covalent 126
Electronegativity: 2.20 (Pauling); 1.55 (Allred–Rochow)
Effective nuclear charge: 3.90 (Slater); 10.57 (Clementi); 15.33 (Froese–Fischer)

Standard reduction potentials $E^{\ominus}$/V

IV		III		0
IrO_2	$\xrightarrow{0.223}$	Ir^{3+}	$\xrightarrow{1.156}$	Ir
		$\xrightarrow{0.926}$		

$IrCl_6^{2-}$	$\xrightarrow{0.867}$	$IrCl_6^{3-}$	$\xrightarrow{0.86}$	Ir

Oxidation states

Ir^{-I}	(d^{10})	rare $[Ir(CO)_3(PPh_3)]^-$
Ir^0	(d^9)	rare $[Ir_4(CO)_{12}]$
Ir^I	(d^8)	rare $[Ir(CO)Cl(PPh_3)_2]$
Ir^{II}	(d^7)	$IrCl_2$
Ir^{III}	(d^6)	IrF_3, $IrCl_3$ etc., $[IrCl_6]^{3-}$(aq)
Ir^{IV}	(d^5)	IrO_2, IrF_4, IrS_2, $[IrCl_6]^{2}$(aq)
Ir^V	(d^4)	IrF_5, $[IrF_6]^-$
Ir^{VI}	(d^3)	IrF_6

Physical properties

Melting point/K: 2683
Boiling point/K: 4403
ΔH_{fusion}/kJ mol^{-1}: 26.4
ΔH_{vap}/kJ mol^{-1}: 612.1

Thermodynamic properties (298.15 K, 0.1 MPa)

State	$\Delta_f H^{\ominus}$/kJ mol^{-1}	$\Delta_f G^{\ominus}$/kJ mol^{-1}	$S^{\ominus}$/J K^{-1} mol^{-1}	C_p/J K^{-1} mol^{-1}
Solid	0	0	35.48	25.10
Gas	665.3	617.9	193.578	20.786

Density/kg m^{-3}: 22 420 [290 K]; 20 000 [liquid at m.p.]
Thermal conductivity/W m^{-1} K^{-1}: 147 [300 K]
Electrical resistivity/Ω m: 5.3×10^{-8} [293 K]
Mass magnetic susceptibility/kg^{-1} m^3: $+1.67 \times 10^{-9}$ (s)
Molar volume/cm^3: 8.57
Coefficient of linear thermal expansion/K^{-1}: 6.8×10^{-6}

Lattice structure (cell dimensions/pm), space group

f.c.c. ($a = 383.89$), Fm3m

X-ray diffraction: mass absorption coefficients (μ/ρ)/cm^2 g^{-1}:
CuK$_\alpha$ 193   MoK$_\alpha$ 110

Discovered in 1803 by S. Tennant at London, UK

[Latin, *iris* = rainbow]

Iridium

Atomic number: 77
Thermal neutron capture cross-section/barns: 425 ± 15
Number of isotopes (including nuclear isomers): 25
Isotope mass range: $182 \rightarrow 198$

Key isotopes

Nuclide	Atomic mass	Natural abundance (%)	Half life $T_{1/2}$	Decay mode and energy (MeV)	Nuclear spin I	Nuclear magnetic moment μ	Uses
^{191}Ir	190.9609	37.3	stable		3/2	+0.1454	NMR
^{192}Ir		0	74.2d	β^- (1.453); EC; β^+; γ	4	+1.90	tracer medical
^{193}Ir	192.9633	62.7	stable		3/2	+0.1583	NMR

NMR (never used)	^{191}Ir	^{193}Ir
Relative sensitivity (^{1}H = 1.00)	2.53×10^{-5}	3.27×10^{-5}
Absolute sensitivity (^{1}H = 1.00)	9.43×10^{-6}	2.05×10^{-5}
Receptivity (^{13}C = 1.00)	0.023	0.050
Magnetogyric ratio/rad T^{-1} s^{-1}	0.539×10^7	0.391×10^7
Quadrupole moment/m^2	1.5×10^{-28}	1.4×10^{-28}
Frequency (^{1}H = 100 MHz; 2.3488 T)/MHz	1.718	1.871

Ground state electron configuration: [Xe]$4f^{14}5d^76s^2$
Term symbol: $^4F_{9/2}$
Electron affinity (M$\rightarrow$M$^-$)/kJ mol^{-1}: 190

Ionization energies/kJ mol^{-1}

1. M $\rightarrow$M$^+$	880	6. M$^{5+}\rightarrow$M^{6+}	(6 900)
2. M$^+\rightarrow$M^{2+}	(1680)	7. M$^{6+}\rightarrow$M^{7+}	(8 500)
3. M$^{2+}\rightarrow$M^{3+}	(2600)	8. M$^{7+}\rightarrow$M^{8+}	(10 000)
4. M$^{3+}\rightarrow$M^{4+}	(3800)	9. M$^{8+}\rightarrow$M^{9+}	(11 700)
5. M$^{4+}\rightarrow$M^{5+}	(5500)	10. M$^{9+}\rightarrow$M^{10+}	

Principal lines in atomic spectrum

Wavelength/nm	Species	Sensitivity	Application
208.882	I		AA
263.971	I		AA
322.078	I	U1	AE
351.365	I	U2	AE
380).012*			

* Lowest energy transition from ground state to nearest empty orbital

Abundance: Earth's crust 0.001 p.p.m.; seawater n.a. (minute)
Biological role: None; low toxicity

<table>
<tr><td>**Fe**</td><td>**Atomic number:** 26
Relative atomic mass ($^{12}C = 12.0000$): **55.847**</td></tr>
</table>

Chemical properties	When pure, iron is lustrous, silvery, and soft (workable). Chief ores are haematite (Fe_2O_3), magnesite (Fe_3O_4), and siderite ($FeCO_3$). Most important of all metals, used principally as steels. Rusts in damp air, dissolves in dilute acids.

Radii/pm: Fe^{2+} 82; Fe^{3+} 67: atomic (α form) 124.1; covalent 116.5

Electronegativity: 1.83 (Pauling); 1.64 (Allred-Rochow)

Effective nuclear charge: 3.75 (Slater); 5.43 (Clementi); 7.40 (Froese–Fischer)

Standard reduction potentials $E^{\ominus}/V$

	VI	III	II	0
Acid solution (pH 0)		Fe^{3+} —0.771→ Fe^{2+} —-0.44→ Fe $[Fe(CN)_6]^{3-}$ —0.361→ $[Fe(CN)_6]^{2-}$ —-1.16→ Fe		
Alkaline solution (pH14)	FeO_4^{2-} —c.0.55→	FeO_2^- —c.-0.69→	$HFeO_2^-$ —c.-0.8→ Fe	

Oxidation states

Fe^{-II}	(d^{10})	rare $Fe(CO)_4^{2-}$
Fe^{-I}	(d^9)	rare $Fe_2(CO)_8^{2-}$
Fe^0	(d^8)	rare $Fe(CO)_5$
Fe^I	(d^7)	rare $[Fe(NO)(H_2O)_5]^{2+}$
Fe^{II}	(d^6)	FeO, FeS_2 ($=Fe^{II}S_2^{2-}$) $Fe(OH)_2$, $[Fe(H_2O)_6]^{2+}$(aq), FeF_2, $Fe(C_5H_5)_2$, etc.
Fe^{III}	(d^5)	Fe_2O_3, Fe_3O_4 ($=Fe^{II}O.Fe_2^{III}O_3$), FeF_3, $FeCl_3$, FeO(OH), $[Fe(H_2O)_6]^{3+}$(aq), etc.
Fe^{IV}	(d^4)	rare, some complexes
Fe^V	(d^3)	FeO_4^{3-} ?
Fe^{IV}	(d^2)	FeO_4^{2-}

Physical properties*	**Melting point**/K: 1808 **Boiling point**/K: 3023 ΔH_{fusion}/kJ mol^{-1}: 14.9 $\qquad$ ΔH_{vap}/kJ mol^{-1}: 340.2

*Very much affected by impurities such as carbon

Thermodynamic properties (298.15 K, 0.1 MPa)

State	$\Delta_f H^{\ominus}$/kJ mol^{-1}	$\Delta_f G^{\ominus}$/kJ mol^{-1}	$S^{\ominus}$/J K^{-1} mol^{-1}	C_p/J K^{-1} mol^{-1}
Solid	0	0	27.28	25.10
Gas	416.3	370.7	180.490	25.677

Density/kg m^{-3}: 7874 [293 K]; 7035 [liquid at m.p.]

Thermal conductivity/W m^{-1} K^{-1}: 80.2 [300 K]

Electrical resistivity/Ω m: 9.71×10^{-8} [293 K]

Mass magnetic susceptibility/kg^{-1} m^3: ferromagnetic

Molar volume/cm^3: 7.09

Coefficient of linear thermal expansion/K^{-1}: 12.3×10^{-6}

Lattice structure (cell dimensions/pm), space group

α-Fe b.c.c. ($a = 286.645$), Im3m
β-Fe not true allotrope
γ-Fe c.c.p. ($a = 364.68$), Fm3m
δ-Fe b.c.c. ($a = 293.22$), Im3m

$T(\alpha \rightarrow \gamma) = 1183$ K
$T(\gamma \rightarrow \delta) = 1663$ K

X-ray diffraction: mass absorption coefficients (μ/ρ)/cm^2 g^{-1}:
CuK_α 308 $\quad$ MoK_α 38.5

Iron

Atomic number: 26
Thermal neutron capture cross-section/barns: 2.56 ± 0.05
Number of isotopes (including nuclear isomers): 10
Isotope mass range: $52 \rightarrow 61$

Key isotopes

Nuclide	Atomic mass	Natural abundance (%)	Half life $T_{1/2}$	Decay mode and energy (MeV)	Nuclear spin I	Nuclear magnetic moment μ	Uses
^{52}Fe		0	8.2h	β^+, EC(2.37); γ			tracer
^{54}Fe	53.9396	5.8	stable				
^{55}Fe	54.938	0	2.6y	EC(0.232); no γ			tracer medical
^{56}Fe	55.9349	91.7	stable				
^{57}Fe	56.9354	2.2	stable		1/2	+0.090 42	NMR
^{58}Fe	57.9333	0.3	stable				
^{59}Fe	58.935	0	45.1d	β^-(1.57 max); γ 3/2		± 1.1	tracer, medical
^{60}Fe		0	3×10^5y	β^-(0.14)			

NMR

^{57}Fe

Relative sensitivity (^{1}H $= 1.00$) 3.37×10^{-5}
Absolute sensitivity (^{1}H $= 1.00$) 7.38×10^{-7}
Receptivity (^{13}C $= 1.00$) 4.2×10^{-3}
Magnetogyric ratio/rad T^{-1} s^{-1} 0.8661×10^7
Frequency (^{1}H $= 100$ MHz; 2.3488 T)/MHz   3.231
Reference: Fe(CO)$_5$

Ground state electron configuration: [Ar]3d^{6}4s^2
Term symbol: ^{5}D$_4$
Electron affinity (M$\rightarrow$M$^-$)/kJ mol^{-1}:   44

Ionization energies/kJ mol^{-1}

1. M $\rightarrow$ M$^+$	759.3	6. M$^{5+} \rightarrow$ M^{6+}	9 600
2. M$^+ \rightarrow$ M^{2+}	1561	7. M$^{6+} \rightarrow$ M^{7+}	12 100
3. M$^{2+} \rightarrow$ M^{3+}	2957	8. M$^{7+} \rightarrow$ M^{8+}	14 575
4. M$^{3+} \rightarrow$ M^{4+}	5290	9. M$^{8+} \rightarrow$ M^{9+}	22 678
5. M$^{4+} \rightarrow$ M^{5+}	7240	10. M$^{9+} \rightarrow$ M^{10+}	25 290

Principal lines in atomic spectrum

Wavelength/nm	Species	Sensitivity	Application
248.327	I		AA
248.814	I		AA
371.994	I	U1	AE
373.713	I	U2	AAAE
374.556	I	U3	AE
835.991	I		AA

* Lowest energy transition from ground state to nearest empty orbital

Abundance: Earth's crust 62 000 p.p.m.; seawater 0.01 p.p.m.
Biological role: Essential; non-toxic

Kr

Atomic number: 36

Relative atomic mass ($^{12}C = 12.0000$): 83.80

<table>
<tr><td>**Chemical
properties**</td><td>Colourless, odourless gas obtained from liquid air. Chemically inert to everything but fluorine. ^{86}Kr has orange-red line in atomic spectrum which is fundamental standard of length: 1 metre = 1 650 763.73 wavelengths.</td></tr>
</table>

Radii/pm: covalent 189; van der Waals 198; Kr^+ 169

Electronegativity: n.a. (Pauling); n.a. (Allred-Rochow)

Effective nuclear charge: 8.25 (Slater); 9.77 (Clementi);
11.79 (Froese–Fischer)

Standard reduction potentials $E^{\ominus}$/V

n.a. [KrF_2 decomposes in water]

Covalent bonds r/pm E/kJ mol^{-1}

Kr–F	188.9	50

Oxidation states

Kr^0	clathrates $Kr_8(H_2O)_{46}$, Kr (quinol)$_3$
Kr^{II}	KrF_2, $[KrF]^+[AsF_6]^-$

<table>
<tr><td>**Physical
properties**</td><td>**Melting point**/K: 116.6
Boiling point/K: 120.85
ΔH_{fusion}/kJ mol^{-1}: 1.64
ΔH_{vap}/kJ mol^{-1}: 9.05</td></tr>
</table>

Thermodynamic properties (298.15 K, 0.1 MPa)

State	$\Delta_f H^{\ominus}$/kJ mol^{-1}	$\Delta_f G^{\ominus}$/kJ mol^{-1}	$S^{\ominus}$/J K^{-1} mol^{-1}	C_p/J K^{-1} mol^{-1}
Gas	0	0	164.082	20.786

Density/kg m^{-3}: 2823 [solid, m.p.]; 2413 [liquid b.p.];
3.7493 [gas, 273 K]

Thermal conductivity/W m^{-1} K^{-1}: 0.009 49 [300 K]$_g$

Mass magnetic susceptibility/kg^{-1} m^3: -4.32×10^{-9} (g)

Molar volume/cm^3: 29.68 [116 K]

Lattice structure (cell dimensions/pm), space group

f.c.c. (80 K) ($a = 572.1$), Fm3m

X-ray diffraction: mass absorption coefficients (μ/ρ)/cm^2 g^{-1}:
CuK$_\alpha$ 108   MoK$_\alpha$ 84.9

Krypton

Atomic number: 36
Thermal neutron capture cross-section/barns: 24.1 ± 1.0
Number of isotopes (including nuclear isomers): 23
Isotope mass range: $74 \rightarrow 94$

**Nuclear
properties**

Key isotopes

Nuclide	Atomic mass	Natural abundance (%)	Half life $T_{1/2}$	Decay mode and energy (MeV)	Nuclear spin I	Nuclear magnetic moment μ	Uses
^{78}Kr	77.9204	0.35	stable				
^{80}Kr	79.9164	2.25	stable				
^{82}Kr	81.9135	11.6	stable		0		
^{83}Kr	82.914	11.5	stable		9/2	-0.970	NMR
^{84}Kr	83.912	57.0	stable		0		
^{85}Kr	84.913	0	10.76y	β^-(0.67); γ	9/2	± 1.005	tracer
^{86}Kr	85.911	17.3	stable		0		

NMR (few compounds) ^{83}Kr
Relative sensitivity (^{1}H = 1.00) 1.88×10^{-3}
Absolute sensitivity (^{1}H = 1.00) 2.17×10^{-4}
Receptivity (^{13}C = 1.00)   1.23
Magnetogyric ratio/rad T^{-1} s^{-1} -1.029×10^7
Quadrupole moment/m^2 0.15×10^{-28}
Frequency (^{1}H = 100 MHz; 2.3488 T)/MHz   3.847

Ground state electron configuration: [Ar]$3d^{10}4s^24p^6$
Term symbol: $^1S_0$
Electron affinity (M$\rightarrow$M$^-$)/kJ mol^{-1}: -39 (calc.)

**Electron
shell
properties**

Ionization energies/kJ mol^{-1}

1. M $\rightarrow$M$^+$ 1350.7	6. M$^{5+}\rightarrow$M^{6+} 7 570
2. M $^+\rightarrow$M^{2+} 2350	7. M$^{6+}\rightarrow$M^{7+} 10 710
3. M$^{2+}\rightarrow$M^{3+} 3565	8. M$^{7+}\rightarrow$M^{8+} 12 200
4. M$^{3+}\rightarrow$M^{4+} 5070	9. M$^{8+}\rightarrow$M^{9+} 22 229
5. M$^{4+}\rightarrow$M^{5+} 6240	10. M$^{9+}\rightarrow$M^{10+} (28 900)

Principal lines in atomic spectrum

Wavelength/nm	Species	Sensitivity	Application
123.584* (vac)	I		
557.029	I	U3	AE
587.092	I	U2	AE
877.675	I		

* Lowest energy transition from ground state to nearest empty orbital

Abundance: Atmosphere 1.14 p.p.m. (volume); Earth's crust c. 0;
seawater 0.0003 p.p.m.
Biological role: None; non-toxic

La

Chemical properties

Soft, silvery-white metal; rapidly tarnishes in air and burns easily. Reacts with water to give hydrogen gas. Used in optical glass and for flints. La^{3+} is used as a biological tracer for calcium Ca^{2+}.

Radii/pm: La^{3+} 122; atomic 187.7; covalent 169

Electronegativity: 1.10 (Pauling); 1.08 (Allred-Rochow)

Effective nuclear charge: 2.85 (Slater); 9.31 (Clementi); 10.43 (Froese–Fischer)

Standard reduction potentials $E^{\ominus}$/V

	III		0
Acid solution	La^{3+}	$\xrightarrow{-2.38}$	La
Alkaline solution	$La(OH)_3$	$\xrightarrow{-2.80}$	La

Oxidation states

La^{III} ([Xe])	La_2O_3, $La(OH)_3$, $[La(H_2O)_x]^{3+}$(aq)
	LaF_3, $LaCl_3$ etc., La^{3+} salts,
	$LaOCl$, $[La(NCS)_6]^{3-}$, complexes
	LaH_2–LaH_3 is probably $La^{3+}H^-$

Physical properties

Melting point/K: 1194
Boiling point/K: 3730
ΔH_{fusion}/kJ mol^{-1}: 10.04
ΔH_{vap}/kJ mol^{-1}: 402.1

Thermodynamic properties (298.15 K, 0.1 MPa)

State	$\Delta_f H^{\ominus}$/kJ mol^{-1}	$\Delta_f G^{\ominus}$/kJ mol^{-1}	$S^{\ominus}$/J K^{-1} mol^{-1}	C_p/J K^{-1} mol^{-1}
Solid	0	0	56.9	27.11
Gas	431.0	393.56	182.377	22.753

Density/kg m^{-3}: 6145 [298 K]
Thermal conductivity/W m^{-1} K^{-1}: 13.5 [300 K]
Electrical resistivity/Ω m: 57×10^{-8} [298 K]
Mass magnetic susceptibility/kg^{-1} m^3: $+1.1 \times 10^{-8}$ (s)
Molar volume/cm^3: 22.60
Coefficient of linear thermal expansion/K^{-1}: 4.9×10^{-6}

Lattice structure (cell dimensions/pm), space group

α-La hexagonal ($a = 377.0$, $c = 121.59$), P6$_3$/mmc
β-La f.c.c. ($a = 529.6$), Fm3m
γ-La b.c.c. ($a = 426$), Im3m

$T(\alpha \rightarrow \beta) = 583$ K
$T(b \rightarrow \gamma) = 1137$ K

X-ray diffraction: mass absorption coefficients (μ/ρ)/cm^2 g^{-1}:
CuK$_\alpha$ 341   MoK$_\alpha$ 45.8

Lanthanum

Atomic number: 57
Thermal neutron capture cross-section/barns: 8.9 ± 0.2
Number of isotopes (including nuclear isomers): 19
Isotope mass range: $126 \rightarrow 144$

Nuclear properties

Key isotopes

Nuclide	Atomic mass	Natural abundance (%)	Half life $T_{1/2}$	Decay mode and energy (MeV)	Nuclear spin I	Nuclear magnetic moment μ	Uses
^{138}La	137.9068	0.09	stable		5	$+3.707$	NMR
^{139}La	138.9061	99.91			7/2	$+2.778$	NMR
^{140}La		0	40.22h	β^- (3.769); γ	3	$+0.73$	tracer

NMR

	$[^{138}$La$]$	^{139}La
Relative sensitivity (^{1}H = 1.00)	9.19×10^{-2}	5.92×10^{-2}
Absolute sensitivity (^{1}H = 1.00)	8.18×10^{-5}	5.91×10^{-2}
Receptivity (^{13}C = 1.00)	0.43	336
Magnetogyric ratio/rad T^{-1} s^{-1}	3.5295×10^7	3.7787×10^7
Quadrupole moment/m^2	-0.47×10^{-28}	0.21×10^{-28}
Frequency (^{1}H = 100 MHz; 2.3488 T)/MHz	13.193	14.126

References: 0.01M LaCl$_3$

Ground state electron configuration: [Xe]$5d^1 6s^2$
Term symbol: $^2D_{3/2}$
Electron affinity (M$\rightarrow$M$^-$)/kJ mol^{-1}: 53

Electron shell properties

Ionization energies/kJ mol^{-1}

1.	M $\rightarrow$M$^+$	538.1	6.	M$^{5+}\rightarrow$M^{6+}	(7 600)
2.	M$^+\rightarrow$M^{2+}	1067	7.	M$^{6+}\rightarrow$M^{7+}	(9 600)
3.	M$^{2+}\rightarrow$M^{3+}	1850	8.	M$^{7+}\rightarrow$M^{8+}	(11 000)
4.	M$^{3+}\rightarrow$M^{4+}	4819	9.	M$^{8+}\rightarrow$M^{9+}	(12 400)
5.	M$^{4+}\rightarrow$M^{5+}	(6400)	10.	M$^{9+}\rightarrow$M^{10+}	(15 900)

Principal lines in atomic spectrum

Wavelength/nm	Species	Sensitivity	Application
394.910	II	V2	
418.732	I		AA
550.134	I		AA
593.065	I	U2	AE
624.993	I	U1	AE
645.599*	I		

* Lowest energy transition from ground state to nearest empty orbital

Abundance: Earth's crust 35 p.p.m.; seawater 0.0003 p.p.m.
Biological role: None; moderately toxic

Lr

Atomic number: **103**

Relative atomic mass ($^{12}C = 12.0000$): **(260)**

Radioactive metal, of which only a few atoms have ever been produced by bombarding ^{252}Cf with boron nuclei.

Radii/pm: Lr^{2+} 112; Lr^{3+} 94; Lr^{4+} 83

Electronegativity: 1.3 (Pauling)

Effective nuclear charge: 1.80 (Slater)

Standard reduction potentials $E^{\ominus}/V$

III		0
Lr^{3+}	$\underline{\quad -2.06 \quad}$	Lr

Oxidation states

Lr^{III} (f^{14})	Lr^{3+} (aq)

Physical properties

Melting point/K: n.a.

Boiling point/K: n.a.

$\Delta H_{fusion}/kJ\ mol^{-1}$: n.a.

$\Delta H_{vap}/kJ\ mol^{-1}$: n.a.

Thermodynamic properties (298.15 K, 0.1 MPa)

State	$\Delta_f H^{\ominus}/kJ\ mol^{-1}$	$\Delta_f G^{\ominus}/kJ\ mol^{-1}$	$S^{\ominus}/J\ K^{-1}\ mol^{-1}$	$C_p/J\ K^{-1}\ mol^{-1}$
Solid	0	0	n.a.	n.a.
Gas	n.a.	n.a.	n.a.	n.a.

Density/kg m^{-3}: n.a.

Thermal conductivity/W m^{-1} K^{-1}: 10 (est.) [300 K]

Electrical resistivity/Ω m: n.a.

Mass magnetic susceptibility/kg^{-1} m^3: n.a.

Molar volume/cm^3: n.a.

Coefficient of linear thermal expansion/K^{-1}: n.a.

Lattice structure (cell dimensions/pm), space group

n.a.

X-ray diffraction: mass absorption coefficients $(\mu/\rho)/cm^2\ g^{-1}$:
n.a.

Prepared in 1961 by A. Ghiorso, T. Sikkeland, A. E. Larsh, and
R. M. Latimer at Berkeley, California, USA
[Named after Ernest O. Lawrence]

Lawrencium

Atomic number: 103

Thermal neutron capture cross-section/barns: n.a.

Number of isotopes (including nuclear isomers): 6

Isotope mass range: $255 \rightarrow 260$

Nuclear properties

Key isotopes

Nuclide	Atomic mass	Natural abundance (%)	Half life $T_{1/2}$	Decay mode and energy (MeV)	Nuclear spin I	Nuclear magnetic moment μ	Uses
^{260}Lr		0	$c.$ 3m	$\alpha(8.03)$			

Ground state electron configuration: $[Rn]5f^{14}6d^17s^2$

Term symbol: $^2D_{5/2}$

Electron affinity $(M \rightarrow M^-)$/kJ mol^{-1}: n.a.

Electron shell properties

Ionization energies/kJ mol^{-1}

1. $M \rightarrow M^+$ n.a.
2. $M^+ \rightarrow M^{2+}$
3. $M^{2+} \rightarrow M^{3+}$
4. $M^{3+} \rightarrow M^{4+}$
5. $M^{4+} \rightarrow M^{5+}$
6. $M^{5+} \rightarrow M^{6+}$
7. $M^{6+} \rightarrow M^{7+}$
8. $M^{7+} \rightarrow M^{8+}$
9. $M^{8+} \rightarrow M^{9+}$
10. $M^{9+} \rightarrow M^{10+}$

Principal lines in atomic spectrum

Wavelength/nm	Species	Sensitivity	Application
n.a.			

Abundance: Earth's crust nil; seawater nil

Biological role: None; never encountered, but toxic due to radioactivity

Pb

Atomic number: 82

Relative atomic mass ($^{12}C = 12.0000$): 207.2

Soft, weak, ductile, dull grey metal. Main ore is galena PbS. Tarnishes in moist air but stable to oxygen and water, dissolves in nitric acid. Used in batteries, cables, paints, glass, solder, petrol, radiation shielding etc.

Radii/pm: Pb^{2+} 132; Pb^{4+} 84; atomic 175.0; covalent 154; Pb^{4-} 215

Electronegativity: 2.33 (Pauling); 1.55 (Allred-Rochow)

Effective nuclear charge: 5.65 (Slater); 12.39 (Clementi); 15.33 (Froese–Fischer)

Standard reduction potentials $E^{\ominus}/V$

	IV		II		0		−II
Acid solution	Pb^{4+}	$\xrightarrow{1.69}$	Pb^{2+}	$\xrightarrow{-0.1251}$	Pb	$\xrightarrow{-1.507}$	PbH_2

[Alkaline solutions contain many different forms]

Covalent bonds

	r/pm	E/kJ mol^{-1}
Pb–H	184	180
Pb–C	229	130
Pb–O	192	398
Pb–F	213	314
Pb–Cl	247	244
Pb–Pb	350	100

Oxidation states

Pb^{II} PbO, PbF_2, $PbCl_2$ etc., $PbOH^+$(aq), $Pb(H_2O)_n^{2+}$(aq) salts, complexes.

Pb^{IV} PbO_2, Pb_3O_4 ($=2PbO \cdot PbO_2$) PbF_4, $PbCl_4$, $PbBr_4$, $PbCl_6^{2-}$, $Pb(OH)_6^{2-}$(aq) [Pb^{4+} does not exist in water], organo lead compounds, complexes

Melting point/K: 600.65

Boiling point/K: 2013

ΔH_{fusion}/kJ mol^{-1}: 5.121

ΔH_{vap}/kJ mol^{-1}: 177.8

Thermodynamic properties (298.15 K, 0.1 MPa)

State	$\Delta_f H^{\ominus}$/kJ mol^{-1}	$\Delta_f G^{\ominus}$/kJ mol^{-1}	$S^{\ominus}$/J K^{-1} mol^{-1}	C_p/J K^{-1} mol^{-1}
Solid	0	0	64.81	26.44
Gas	195.0	161.9	175.373	20.786

Density/kg m^{-3}: 11 350 [293 K]; 10 678 [liquid, m.p.]

Thermal conductivity/W m^{-1} K^{-1}: 35.3 [300 K]

Electrical resistivity/Ω m: 20.648×10^{-8} [293 K]

Mass magnetic susceptibility/kg^{-1} m^3: -1.39×10^{-9} (s)

Molar volume/cm^3: 18.26

Coefficient of linear thermal expansion/K^{-1}: 29.1×10^{-9}

Lattice structure (cell dimensions/pm), space group

f.c.c. ($a = 495.00$), Fm3m

X-ray diffraction: mass absorption coefficients (μ/ρ)/cm^2 g^{-1}: CuK$_\alpha$ 232 MoK$_\alpha$ 120

Lead

Atomic number: 82
Thermal neutron capture cross-section/barns: 0.18 ± 0.01
Number of isotopes (including nuclear isomers): 29
Isotope mass range: $194 \rightarrow 214$

Nuclear properties

Key isotopes

Nuclide	Atomic mass	Natural abundance (%)	Half life $T_{1/2}$	Decay mode and energy (MeV)	Nuclear spin I	Nuclear magnetic moment μ	Uses
^{204}Pb	203.9731	1.4	stable				
^{205}Pb		0	3.0×10^7y	EC(c.0.035)			
^{206}Pb	205.9745	24.1	stable		0		
^{207}Pb	206.9759	22.1	stable		1/2	$+0.5783$	NMR
^{208}Pb	207.9766	52.3	stable		0		
^{210}Pb		trace	20.4y	β^- (0.061) 81%; α; γ			tracer
^{214}Pb		trace	10.64h	β^- (0.58); γ			

NMR

^{207}Pb

Relative sensitivity (^{1}H = 1.00)	9.16×10^{-3}
Absolute sensitivity (^{1}H = 1.00)	2.07×10^{-3}
Receptivity (^{13}C = 1.00)	11.8
Magnetogyric ratio/rad T^{-1} s^{-1}	5.5797×10^7
Frequency (^{1}H = 100 MHz; 2.3488 T)/MHz	20.921

Reference: Pb(CH$_3$)$_4$

Ground state electron configuration: [Xe]$4f^{14}5d^{10}6s^26p^2$
Term symbol: $^3P_0$
Electron affinity (M$\rightarrow$M$^-$)/kJ mol^{-1}: 35.2

Electron shell properties

Ionization energies/kJ mol^{-1}

1.	M $\rightarrow$ M$^+$	715.5	6. M$^{5+} \rightarrow$ M^{6+}	(8 100)
2.	M$^+ \rightarrow$ M^{2+}	1450.4	7. M$^{6+} \rightarrow$ M^{7+}	(9 900)
3.	M$^{2+} \rightarrow$ M^{3+}	3081.5	8. M$^{7+} \rightarrow$ M^{8+}	(11 800)
4.	M$^{3+} \rightarrow$ M^{4+}	4083	9. M$^{8+} \rightarrow$ M^{9+}	(13 700)
5.	M$^{4+} \rightarrow$ M^{5+}	6640	10. M$^{9+} \rightarrow$ M^{10+}	(16 700)

Principal lines in atomic spectrum

Wavelength/nm	Species	Sensitivity	Application
216.999	I		AA, AE
261.418	I		AA, AE
283.305*	I		AA
368.346	I	U2	AA, AE
405.781	I	U2	AE

* Lowest energy transition from ground state to nearest empty orbital

Abundance: Earth's crust 13 p.p.m.; seawater 0.003 p.p.m.
Biological role: None; toxic, teratogenic, carcinogenic

Li

Atomic number: 3

Relative atomic mass ($^{12}C = 12.0000$): **6.941**

Chemical properties

Soft, white, silvery metal. Reacts slowly with oxygen and water. Large deposits known of spodumene, $LiAlSi_2O_6$. Used in alloys (with Al and Mg), greases, glass, medicine, and nuclear bombs.

Radii/pm: Li^+ 78; atomic 152; covalent 123

Electronegativity: 0.98 (Pauling); 0.97 (Allred-Rochow)

Effective nuclear charge: 1.30 (Slater); 1.28 (Clementi); 1.55 (Froese–Fischer)

Standard reduction potentials $E^{\ominus}$/V

I		0
Li^+	$\xrightarrow{-3.040}$	Li

Oxidation states

Li^{-1} (s^2)	Li solutions in liquid ammonia
Li^I ([He])	Li_2O, LiOH, LiH, $LiAlH_4$,
	LiF, LiCl etc., $Li(H_2O)_4^+$ (aq),
	Li_2CO_3, salts of Li^+, some
	complexes, $(LiCH_3)_4$

Physical properties

Melting point/K: 453.69

Boiling point/K: 1620

ΔH_{fusion}/kJ mol^{-1}: 4.60

ΔH_{vap}/kJ mol^{-1}: 147.7

Thermodynamic properties (298.15 K, 0.1 MPa)

State	$\Delta_f H^{\ominus}$/kJ mol^{-1}	$\Delta_f G^{\ominus}$/kJ mol^{-1}	$S^{\ominus}$/J K^{-1} mol^{-1}	C_p/J K^{-1} mol^{-1}
Solid	0	0	29.12	24.77
Gas	159.37	126.66	138.77	20.786

Density/kg m^{-3}: 534 [293 K]; 515 [liquid m.p.]

Thermal conductivity/W m^{-1} K^{-1}: 84.7 [300 K]

Electrical resistivity/Ω m: 8.55×10^{-8} [273 K]

Mass magnetic susceptibility/kg^{-1} m^3: $+2.56 \times 10^{-8}$ (s)

Molar volume/cm^3: 13.00

Coefficient of linear thermal expansion/K^{-1}: 56×10^{-6}

Lattice structure (cell dimensions/pm), space group

α-Li b.c.c. ($a = 351.00$), Im3m

β-Li f.c.c. ($a = 437.9$), Fm3m

$T(\alpha \rightarrow \beta) = 77$ K

X-ray diffraction: mass absorption coefficients (μ/ρ)/cm^2 g^{-1}:
CuK$_\alpha$ 0.716   MoK$_\alpha$ 0.217

Discovered in 1817 by J. A. Arfvedson at Stockholm, Sweden.
Isolated by W. T. Brande 1821

[Greek, *lithos* = stone]

Lithium

Atomic number: 3
Thermal neutron capture cross-section/barns: 71 ± 1
Number of isotopes (including nuclear isomers): 5
Isotope mass range: $5 \rightarrow 9$

Key isotopes

Nuclide	Atomic mass	Natural abundance (%)	Half life $T_{1/2}$	Decay mode and energy (MeV)	Nuclear spin I	Nuclear magnetic moment μ	Uses
^{6}Li	6.015 12	7.5	stable		1	+0.822 03	NMR
^{7}Li	7.016 00	92.5	stable		3/2	+3.256 36	NMR

NMR

	$[^6$Li]	^{7}Li
Relative sensitivity (^{1}H = 1.00)	8.50×10^3	0.29
Absolute sensitivity (^{1}H = 1.00)	6.31×10^{-4}	0.27
Receptivity (^{13}C = 1.00)	3.58	1540
Magnetogyric ratio/rad T^{-1} s^{-1}	3.9366×10^7	10.3964×10^7
Quadrupole moment/m$_2$	-8×10^{-32}	-4.5×10^{-30}
Frequency (^{1}H = 100 MHz; 2.3488 T)/MHz	14.716	38.863

Reference: LiCl (aq)

Ground state electron configuration: [He]$2s^1$
Term symbol: $^2S_{1/2}$
Electron affinity (M$\rightarrow$M$^-$)/kJ mol^{-1}: 59.8

Ionization energies/kJ mol^{-1}

1. M $\rightarrow$ M$^+$ 513.3
2. M$^+$ $\rightarrow$ M^{2+} 7298.0
3. M^{2+} $\rightarrow$ M^{3+} 11814.8

Principal lines in atomic spectrum

Wavelength/nm	Species	Sensitivity	Application
323.261	I	U2	AA, AE
460.300	I	U4	AE
610.362	I	U3	AA, AE
670.784*	I	U1	AA, AE

* Lowest energy transition from ground state to nearest empty orbital

Abundance: Earth's crust 18 p.p.m.; seawater 0.2 p.p.m.
Biological role: None; non-toxic; teratogenic; stimulatory; anti-depressant

Lu

Atomic number: 71
Relative atomic mass ($^{12}C = 12.0000$): 174.967

Chemical properties

Hardest, densest, and one of the rarest of the lanthanide (rare earth) metals. Obtained from euxenite, monazite, and bastnaesite. Little used except in research.

Radii/pm: Lu^{3+} 85; atomic 173.4; covalent 156

Electronegativity: 1.27 (Pauling); 1.14 (Allred-Rochow)

Effective nuclear charge: 3.00 (Slater); 8.80 (Clementi); 12.68 (Froese–Fischer)

Standard reduction potentials $E^{\ominus}/V$

	III		0
Acid solution	Lu^{3+}	$\xrightarrow{-2.30}$	Lu
Alkaline solution	$Lu(OH)_3$	$\xrightarrow{-2.83}$	Lu

Oxidation states

Lu^{III} (f^{14})	Lu_2O_3, $Lu(OH)_3$, $[Lu(H_2O)_x]^{3+}$ (aq), Lu^{3+} salts, LuF_3, $LuCl_3$ etc., $LuCl_6^{3-}$, complexes

Physical properties

Melting point/K: 1936
Boiling point/K: 3668
ΔH_{fusion}/kJ mol^{-1}: 19.2
ΔH_{vap}/kJ mol^{-1}: 428

Thermodynamic properties (298.15 K, 0.1 MPa)

State	$\Delta_f H^{\ominus}$/kJ mol^{-1}	$\Delta_f G^{\ominus}$/kJ mol^{-1}	$S^{\ominus}$/J K^{-1} mol^{-1}	C_p/J K^{-1} mol^{-1}
Solid	0	0	50.96	26.86
Gas	427.6	387.8	184.800	20.861

Density/kg m^{-3}: 9840 [298 K]
Thermal conductivity/W m^{-1} K^{-1}: 16.4 [300 K]
Electrical resistivity/Ω m: 79.0×10^{-8} [298 K]
Mass magnetic susceptibility/kg^{-1} m^3: $+1.3 \times 10^{-9}$ (s)
Molar volume/cm^3: 17.78
Coefficient of linear thermal expansion/K^{-1}: 8.12×10^{-6}

Lattice structure (cell dimensions/pm), space group

α-Lu h.c.p. ($a = 350.31$, $c = 555.09$), P6$_3$/mmc
β-Lu b.c.c. ($a = 390$), Im3m

X-ray diffraction: mass absorption coefficients (μ/ρ)/cm^2 g^{-1}:
CuK$_\alpha$ 153   MoK$_\alpha$ 88.2

Discovered in 1907 by G. Urbain at Paris, France and independently by C. James at the University of New Hampshire, USA
[Latin, *Lutetia* = Paris]

Lutetium

Atomic number: 71
Thermal neutron capture cross-section/barns: 75 ± 2
Number of isotopes (including nuclear isomers): 22
Isotope mass range: $167 \rightarrow 180$

Nuclear properties

Key isotopes

Nuclide	Atomic mass	Natural abundance (%)	Half life $T_{1/2}$	Decay mode and energy (MeV)	Nuclear spin I	Nuclear magnetic moment μ	Uses
^{175}Lu	174.9409	97.39	stable		7/2	$+2.203$	NMR
^{176}Lu		2.61	2.2×10^{10}y	β^-(1.02); γ	7	$+3.18$	
^{177}Lu		0	6.74d	β^-(0.497); γ	7/2	$+2.24$	tracer

NMR (not used) ^{175}Lu
Relative sensitivity (^{1}H $= 1.00$) 3.12×10^{-2}
Absolute sensitivity (^{1}H $= 1.00$) 3.03×10^{-2}
Receptivity (^{13}C $= 1.00$)   156
Magnetogyric ratio/rad T^{-1} s^{-1} 3.05×10^7
Quadrupole moment/m^2 5.68×10^{-28}
Frequency (^{1}H $= 100$ MHz; 2.3488 T)/MHz   11.407

Ground state electron configuration: $[Xe]4f^{14}5d^16s^2$
Term symbol: $^2D_{3/2}$
Electron affinity $(M \rightarrow M^-)$/kJ mol^{-1}: $\leqslant 50$

Electron shell properties

Ionization energies/kJ mol^{-1}

1. $M \rightarrow M^+$	523.5	6. $M^{5+} \rightarrow M^{6+}$	
2. $M^+ \rightarrow M^{2+}$	1340	7. $M^{6+} \rightarrow M^{7+}$	
3. $M^{2+} \rightarrow M^{3+}$	2022	8. $M^{7+} \rightarrow M^{8+}$	
4. $M^{3+} \rightarrow M^{4+}$	4360	9. $M^{8+} \rightarrow M^{9+}$	
5. $M^{4+} \rightarrow M^{5+}$		10. $M^{9+} \rightarrow M^{10+}$	

Principal lines in atomic spectrum

Wavelength/nm	Species	Sensitivity	Application
261.542	II		
291.139	II		AE
331.211	I		AA, AE
335.956	I		AE
347.248	II		AE
451.857	I		AA, AE

Abundance: Earth's crust 0.8 p.p.m.; seawater 4×10^{-8} p.p.m.
Biological role: None; low toxicity

<table>
<tr><td>**Mg**</td><td>**Atomic number:** 12
Relative atomic mass ($^{12}C = 12.0000$): **24.3050**</td></tr>
</table>

Chemical properties

Silvery white, lustrous, relatively soft metal. Obtained by electrolysis of fused $MgCl_2$. Burns in air and reacts with hot water. Used as bulk metal and in lightweight alloys for engines, also as a sacrificial electrode to protect other metals.

Radii/pm: Mg^{2+} 78; atomic 160; covalent 136
Electronegativity: 1.31 (Pauling); 1.23 (Allred-Rochow)
Effective nuclear charge: 2.85 (Slater); 3.31 (Clementi); 4.15 (Froese–Fischer)

Standard reduction potentials $E^{\ominus}/V$

	II		I		0
Acid solution	Mg^{2+}	$\underset{-2.356}{\overset{-2.054}{\rule{2cm}{0.4pt}}}$	Mg^{+}	$\overset{-2.657}{\rule{2cm}{0.4pt}}$	Mg
Alkaline solution	$Mg(OH)_2$		$\overset{-2.687}{\rule{4cm}{0.4pt}}$		Mg

Oxidation states

Mg^{II}	MgO, MgO_2, $Mg(OH)_2$, $[Mg(H_2O)_6]^{2+}$(aq), MgH_2, $MgCO_3$, Mg^{2+} salts, MgF_2, $MgCl_2$, etc., CH_3MgI, complexes

Physical properties

Melting point/K: 922.0
Boiling point/K: 1363
ΔH_{fusion}/kJ mol^{-1}: 9.04
ΔH_{vap}/kJ mol^{-1}: 127.6

Thermodynamic properties (298.15 K, 0.1 MPa)

State	$\Delta_f H^{\ominus}$/kJ mol^{-1}	$\Delta_f G^{\ominus}$/kJ mol^{-1}	$S^{\ominus}$/J K^{-1} mol^{-1}	C_p/J K^{-1} mol^{-1}
Solid	0	0	32.68	24.89
Gas	147.70	113.10	148.650	20.786

Density/kg m^{-3}: 1738 [293 K]; 1585 [liquid at m.p.]
Thermal conductivity/W m^{-1} K^{-1}: 156 [300 K]
Electrical resistivity/Ω m: 4.45×10^{-8} [293 K]
Mass magnetic susceptibility/kg^{-1} m^3: $+6.8 \times 10^{-9}$ (s)
Molar volume/cm^3: 13.98
Coefficient of linear thermal expansion/K^{-1}: 26.1×10^{-6}

Lattice structure (cell dimensions/pm), space group

h.c.p. ($a = 320.94$; $c = 521.03$), P6$_3$/mmc

X-ray diffraction: mass absorption coefficients (μ/ρ)/cm^2 g^{-1}:
CuK$_{\alpha}$ 38.6   MoK$_{\alpha}$ 4.11

Magnesium

Atomic number: 12
Thermal neutron capture cross-section/barns: 0.064 ± 0.002
Number of isotopes (including nuclear isomers): 8
Isotope mass range: $20 \rightarrow 28$

Nuclear properties

Key isotopes

Nuclide	Atomic mass	Natural abundance (%)	Half life $T_{1/2}$	Decay mode and energy (MeV)	Nuclear spin I	Nuclear magnetic moment μ	Uses
^{24}Mg	23.98504	78.99	stable		0		
^{25}Mg	24.98584	10.00	stable		5/2	-0.8564	NMR
^{26}Mg	25.98259	11.01	stable		0		
^{28}Mg*		0	21.2h	$\beta^-(1.84); \gamma$			

* longest lived radioactive isotope

NMR ^{25}Mg

Relative sensitivity (^{1}H $= 1.00$) 2.67×10^{-3}
Absolute sensitivity (^{1}H $= 1.00$) 2.71×10^{-4}
Receptivity (^{13}C $= 1.00$) 1.54
Magnetogyric ratio/rad T^{-1} s^{-1} 1.6375×10^7
Quadrupole moment/m$_2$ 0.22×10^{-28}
Frequency (^{1}H $= 100$ MHz; 2.3488 T)/MHz   6.1195
Reference: $MgCl_2$(aq)

Ground state electron configuration: [Ne]$3s^2$
Term symbol: $^1S_0$
Electron affinity $(M \rightarrow M^-)$/kJ mol^{-1}: -21

Electron shell properties

Ionization energies/kJ mol^{-1}

1. $M \rightarrow M^+$	737.7	6. $M^{5+} \rightarrow M^{6+}$	17 995
2. $M^+ \rightarrow M^{2+}$	1 450.7	7. $M^{6+} \rightarrow M^{7+}$	21 703
3. $M^{2+} \rightarrow M^{3+}$	7 732.6	8. $M^{7+} \rightarrow M^{8+}$	25 656
4. $M^{3+} \rightarrow M^{4+}$	10 540	9. $M^{8+} \rightarrow M^{9+}$	31 642
5. $M^{4+} \rightarrow M^{5+}$	13 630	10. $M^{9+} \rightarrow M^{10+}$	35 461

Principal lines in atomic spectrum

Wavelength/nm	Species	Sensitivity	Application
202.582	I		AA
279.553	II	V1	AA, AE
285.213	I	U1	AA, AE
383.230	I	U3	AA
383.826	I	U2	AE
457.110*			

* Lowest energy transition from ground state to nearest empty orbital

Abundance: Earth's crust 27 640 p.p.m.; seawater 1300 p.p.m.
Biological role: Essential; non-toxic

Mn

Atomic number: 25

Relative atomic mass ($^{12}C = 12.0000$): 54.93805

Chemical properties

Hard, brittle, silvery metal. Occurs as pyrolusite MnO_2. Reactive when impure and will burn in oxygen. Surface oxidation occurs in air; will react with water, dissolves in dilute acids. Most Mn is used in steel production.

Radii/pm: Mn^{2+} 91; Mn^{3+} 70; Mn^{4+} 52; atomic 124; covalent 117

Electronegativity: 1.55 (Pauling); 1.60 (Allred-Rochow)

Effective nuclear charge: 3.60 (Slater); 5.23 (Clementi); 7.17 (Froese–Fischer)

Standard reduction potentials $E^{\ominus}/V$

Oxidation states

Mn^{-III}	(d^{10})	$[Mn(NO)_3(CO)]$
Mn^{-II}	(d^9)	some complexes known
Mn^{-I}	(d^8)	$[Mn(CO)_5]^-$
Mn^0	(d^7)	$Mn_2(CO)_{10}$
Mn^I	(d^6)	$[Mn(CN)_6]^-$
Mn^{II}	(d^5)	MnO, $Mn_3O_4(= Mn^{II}Mn_2^{III}O_4)$, $[Mn(H_2O)_6]^{2+}$(aq), MnF_2, $MnCl_2$ etc., salts, complexes
Mn^{III}	(d^4)	Mn_2O_3 $[Mn(H_2O)_6]^{3+}$(aq) unstable; MnF_3, $MnCl_5^{2-}$
Mn^{IV}	(d^3)	MnO_2, MnF_4, MnF_6^{2-}
Mn^V	(d^2)	MnO_4^{3-}
Mn^{VI}	(d^1)	MnO_4^{2-}
Mn^{VII}	$(d^0, [Ar])$	Mn_2O_7, MnO_4^-

Physical properties

Melting point/K: 1517

Boiling point/K: 2235

ΔH_{fusion}/kJ mol^{-1}: 14.4 ΔH_{vap}/kJ mol^{-1}: 220.5

Thermodynamic properties (298.15 K, 0.1 MPa)

State	$\Delta_f H^{\ominus}$/kJ mol^{-1}	$\Delta_f G^{\ominus}$/kJ mol^{-1}	$S^{\ominus}$/J K^{-1} mol^{-1}	C_p/J K^{-1} mol^{-1}
Solid	0	0	32.01	26.32
Gas	280.7	238.5	173.70	20.79

Density/kg m^{-3}: 7440 (α) [293 K]; 6430 [liquid at m.p.]

Thermal conductivity/W m^{-1} K^{-1}: 7.82 [300 K]

Electrical resistivity/Ω m: 185.0×10^{-8} [298 K]

Mass magnetic susceptibility/kg^{-1} m^3: $+1.21 \times 10^{-7}$ (s)

Molar volume/cm^3: 7.38

Coefficient of linear thermal expansion/K^{-1}: 22×10^{-6}

Lattice structure (cell dimensions/pm), space group

α-Mn b.c.c. ($a = 891.39$), I$\bar{4}$3m	γ-Mn f.c.c. ($a = 386.3$), Fm3m
β-Mn b.c.c. ($a = 631.45$), P4$_1$32	δ-Mn b.c.c. ($a = 308.1$), Im3m

$T(\alpha \rightarrow \beta) = 973$ K; $T(\beta \rightarrow \gamma) = 1352$ K; $T(\gamma \rightarrow \delta) = 1413$ K

X-ray diffraction: mass absorption coefficients (μ/ρ)/cm^2 g^{-1}:
CuK$_\alpha$ 285   MoK$_\alpha$ 34.7

Manganese

Atomic number: 25
Thermal neutron capture cross-section/barns: 13.3 ± 0.1
Number of isotopes (including nuclear isomers): 11
Isotope mass range: $50 \rightarrow 58$

Nuclear properties

Key isotopes

Nuclide	Atomic mass	Natural abundance (%)	Half life $T_{1/2}$	Decay mode and energy (MeV)	Nuclear spin I	Nuclear magnetic moment μ	Uses
^{53}Mn	52.9413	0	2×10^6y	EC(0.598); no γ	7/2	± 5.02	
^{54}Mn	53.9402	0	303d	EC(1.379); γ	3	$+3.278$	tracer
^{55}Mn	54.9381	100	stable		5/2	$+3.449$	NMR
^{56}Mn		0	2.576h	β^-(3.702); γ	3	$+3.223$	tracer

NMR

^{55}Mn

Relative sensitivity (^{1}H = 1.00)	0.18
Absolute sensitivity (^{1}H = 1.00)	0.18
Receptivity (^{13}C = 1.00)	994
Magnetogyric ratio/rad T^{-1} s^{-1}	6.6195×10^7
Quadrupole moment/m^2	0.55×10^{-28}
Frequency (^{1}H = 100 MHz; 2.3488 T)/MHz	24.664

Reference: $KMnO_4$(aq)

Ground state electron configuration: $[Ar]3d^5 4s^2$
Term symbol: $^6S_{5/2}$
Electron affinity (M $\rightarrow$ M$^-$)/kJ mol^{-1}: -94

Electron shell properties

Ionization energies/kJ mol^{-1}

1. M $\rightarrow$ M$^+$	717.4	6. M$^{5+} \rightarrow$ M^{6+}	9 200
2. M$^+ \rightarrow$ M^{2+}	1509.0	7. M$^{6+} \rightarrow$ M^{7+}	11 508
3. M$^{2+} \rightarrow$ M^{3+}	3248.4	8. M$^{7+} \rightarrow$ M^{8+}	18 956
4. M$^{3+} \rightarrow$ M^{4+}	4940	9. M$^{8+} \rightarrow$ M^{9+}	21 400
5. M$^{4+} \rightarrow$ M^{5+}	6990	10. M$^{9+} \rightarrow$ M^{10+}	23 960

Principal lines in atomic spectrum

Wavelength/nm	Species	Sensitivity	Application
279.482	I		AA
279.827	I		AA, AE
403.076	I	U2	AA, AE
403.307	I	U2	AE
539.467*	I		

* Lowest energy transition from ground state to nearest empty orbital

Abundance: Earth's crust 1060 p.p.m.; seawater 0.002 p.p.m.
Biological role: Essential; non-toxic, suspected carcinogenic

<table>
<tr><td>**Md**</td><td>Atomic number: **101**</td></tr>
<tr><td></td><td>Relative atomic mass ($^{12}C = 12.0000$): **(258)**</td></tr>
</table>

Chemical properties

Radioactive metal of which only a few atoms have been made by bombarding ^{253}Es with α particles (^{4}He)

Radii/pm: Md^{2+} 114; Md^{3+} 96; Md^{4+} 84

Electronegativity: 1.3 (Pauling); n.a. (Allred-Rochow)

Effective nuclear charge: 1.65 (Slater)

Standard reduction potentials $E^{\ominus}/V$

	III	II	0

Acid solution $\quad\quad Md^{3+} \xrightarrow{-0.15} Md^{2+} \xrightarrow{-2.4} Md$

(−1.7)

Oxidation states

Md^{II} (f^{13})   stable ?
Md^{III} (f^{12}) $[Md(H_2O)_x]^{3+}$(aq)

Physical properties

Melting point/K: n.a.
Boiling point/K: n.a.
ΔH_{fusion}/kJ mol^{-1}: n.a.
ΔH_{vap}/kJ mol^{-1}: n.a.

Thermodynamic properties (298.15 K, 0.1 MPa)

State	$\Delta_f H^{\ominus}$/kJ mol^{-1}	$\Delta_f G^{\ominus}$/kJ mol^{-1}	$S^{\ominus}$/J K^{-1} mol^{-1}	C_p/J K^{-1} mol^{-1}
Solid	0	0	n.a.	n.a.
Gas	n.a.	n.a.	n.a.	n.a.

Density/kg m^{-3}: n.a.
Thermal conductivity/W m^{-1} K^{-1}: 10 (est.) [300 K]
Electrical resistivity/Ω m: n.a.
Mass magnetic susceptibility/kg^{-1} m^3: n.a.
Molar volume/cm^3: n.a.
Coefficient of linear thermal expansion/K^{-1}: n.a.

Lattice structure (cell dimensions/pm), space group

n.a.

X-ray diffraction: mass absorption coefficients (μ/ρ)/cm^2 g^{-1}:
n.a.

Prepared in 1955 by A. Ghiorso, B. G. Harvey, G. R. Choppin, S. G. Thompson, and G. T. Seaborg at Berkeley, California, USA
[Named after Dmitri Mendeleyev]

Mendelevium

Atomic number: 101
Thermal neutron capture cross-section/barns: n.a.
Number of isotopes (including nuclear isomers): 10
Isotope mass range: $248 \rightarrow 258$

Key isotopes

Nuclide	Atomic mass	Natural abundance (%)	Half life $T_{1/2}$	Decay mode and energy (MeV)	Nuclear spin I	Nuclear magnetic moment μ	Uses
^{258}Md		0	54d	EC; $\alpha(6.79)$; SF			

Ground state electron configuration: $[Rn]5f^{13}7s^2$
Term symbol: $^2F_{7/2}$
Electron affinity $(M \rightarrow M^-)/kJ\ mol^{-1}$: n.a.

Ionization energies/kJ mol^{-1}

1. $M \rightarrow M^+$	635		6. $M^{5+} \rightarrow M^{6+}$	
2. $M^+ \rightarrow M^{2+}$			7. $M^{6+} \rightarrow M^{7+}$	
3. $M^{2+} \rightarrow M^{3+}$			8. $M^{7+} \rightarrow M^{8+}$	
4. $M^{3+} \rightarrow M^{4+}$			9. $M^{8+} \rightarrow M^{9+}$	
5. $M^{4+} \rightarrow M^{5+}$			10. $M^{9+} \rightarrow M^{10+}$	

Principal lines in atomic spectrum

Wavelength/nm	Species	Sensitivity	Application
n.a.			

Abundance: Earth's crust nil; seawater nil
Biological role: None; toxic due to radiation

Hg

Atomic number: 80
Relative atomic mass ($^{12}C = 12.0000$): **200.59**

Chemical properties

Liquid silvery metal obtained by roasting cinnabar, HgS. Stable with air and water, unreactive to acids (except conc HNO_3) and alkalis. Used in chlorine and NaOH manufacture, street lights, fungicides, electrical apparatus, etc.

Radii/pm: Hg^+ 127; Hg^{2+} 112; atomic 160; covalent 144
Electronegativity: 2.00 (Pauling); 1.44 (Allred-Rochow)
Effective nuclear charge: 4.35 (Slater); 11.15 (Clementi); 16.22 (Froese–Fischer)

Standard reduction potentials $E^\ominus/V$

	II	I	0
		0.8535	
Acid solution	Hg^{2+} $\xrightarrow{-0.9110}$	Hg_2^{2+} $\xrightarrow{0.7960}$	Hg
Alkaline solution	HgO $\xrightarrow{\hspace{1cm} 0.0977 \hspace{1cm}}$		Hg

Oxidation states

Hg^I (d^{10} s^1)	Hg_2F_2, Hg_2Cl_2 etc., Hg_2^{2+} salts, most insoluble, $Hg_2(NO_3)_2 \cdot 2H_2O$ soluble.
Hg^{II} (d^{10})	HgO, HgS, HgF_2, HgCl_2, etc., Hg^{2+} salts, $[Hg(H_2O)_6]^{2+}$ (aq) HgN(OH), complexes eg $[Hg(SCN)_4]^{2-}$, $Hg(CH_3)_2$, etc.

Physical properties

Melting point/K: 234.28
Boiling point/K: 629.73
ΔH_{fusion}/kJ mol^{-1}: 2.331
ΔH_{vap}/kJ mol^{-1}: 59.11

Thermodynamic properties (298.15 K, 0.1 MPa)

State	$\Delta_f H^\ominus$/kJ mol^{-1}	$\Delta_f G^\ominus$/kJ mol^{-1}	$S^\ominus$/J K^{-1} mol^{-1}	C_p/J K^{-1} mol^{-1}
Solid	0	0	76.02	27.983
Gas	61.317	31.820	174.96	20.786

Density/kg m^{-3}: 13 546 [293 K]
Thermal conductivity/W m^{-1} K^{-1}: 8.34 [300 K]
Electrical resistivity/Ω m: 94.1×10^{-8} [273 K]
Mass magnetic susceptibility/kg^{-1} m^3: -2.095×10^{-9} (l)
Molar volume/cm^3: 14.81
Coefficient of cubical thermal expansion/K^{-1}: 18.1×10^{-5}

Lattice structure (cell dimensions/pm), space group

α-Hg rhombohedral ($a = 299.25$, $\alpha = 70°$ 44.6'), $R\bar{3}m$
β-Hg tetragonal ($a = 399.5$, $c = 282.5$), I4/mmm
$\alpha \rightarrow \beta$ high pressure

X-ray diffraction: mass absorption coefficients (μ/ρ)/cm^2 g^{-1}:
CuK$_\alpha$ 216   MoK$_\alpha$ 117

Known to ancient civilizations

[Named after planet Mercury. Latin, *hydragyrum* = liquid silver]

Mercury

Atomic number: 80
Thermal neutron capture cross-section/barns: 375 ± 5
Number of isotopes (including nuclear isomers): 26
Isotope mass range: $185 \rightarrow 206$

Nuclear properties

Key isotopes

Nuclide	Atomic mass	Natural abundance (%)	Half life $T_{1/2}$	Decay mode and energy (MeV)	Nuclear spin I	Nuclear magnetic moment μ	Uses
^{196}Hg	195.9658	0.2	stable				
^{197}Hg		0	65h	EC(0.42); γ	1/2	+0.5241	tracer, medical
^{198}Hg	197.9668	10.1	stable		0		
^{199}Hg	198.9683	17.0	stable		1/2	+0.502 71	NMR
^{200}Hg	199.9682	23.1	stable		0		
^{201}Hg	200.9703	13.2	stable		3/2	−0.555 71	NMR
^{202}Hg	201.9706	29.6	stable		0		
^{203}Hg		0	46.9	β^-(0.492); γ	5/2	+0.86	tracer, medical
^{204}Hg	203.9735	6.8	stable	0			

NMR

	^{199}Hg	$[^{201}$Hg]
Relative sensitivity (^{1}H $= 1.00$)	5.67×10^{-3}	1.44×10^{-3}
Absolute sensitivity (^{1}H $= 1.00$)	9.54×10^{-4}	1.90×10^{-4}
Receptivity (^{13}C $= 1.00$)	5.42	1.08
Magnetogyric ratio/rad T^{-1} s^{-1}	4.7912×10^7	-1.7686×10^7
Quadrupole moment/m^2	—	0.5×10^{-28}
Frequency (^{1}H $= 100$ MHz; 2.3488 T)/MHz	17.827	6.599

Reference: $Hg(CH_3)_2$

Ground state electron configuration: $[Xe]4f^{14}5d^{10}6s^2$
Term symbol: $^1S_0$
Electron affinity $(M \rightarrow M^-)$/kJ mol^{-1}: -18

Electron shell properties

Ionization energies/kJ mol^{-1}

1. $M \rightarrow M^+$ 1007.0	6. $M^{5+} \rightarrow M^{6+}$ (7 400)	
2. $M^+ \rightarrow M^{2+}$ 1809.7	7. $M^{6+} \rightarrow M^{7+}$ (9 100)	
3. $M^{2+} \rightarrow M^{3+}$ 3300	8. $M^{7+} \rightarrow M^{8+}$ (11 600)	
4. $M^{3+} \rightarrow M^{4+}$ (4400)	9. $M^{8+} \rightarrow M^{9+}$ (13 400)	
5. $M^{4+} \rightarrow M^{5+}$ (5900)	10. $M^{9+} \rightarrow M^{10+}$ (15 300)	

Principal lines in atomic spectrum

Wavelength/nm	Species	Sensitivity	Application
253.652*	I	U2	AA, AE
365.015	I	U3	AE
365.483	I	U4	AE
366.328	I	U5	AE
435.833	I		AE

* Lowest energy transition from ground state to nearest empty orbital

Abundance: Earth's crust 0.08 p.p.m.; seawater 3×10^{-5} p.p.m.
Biological role: None; toxic, methyl mercury very toxic; teratogenic

Mo

Atomic number: 42

Relative atomic mass ($^{12}C = 12.0000$): 95.94

Chemical properties

Metal is lustrous, silvery, and fairly soft when pure. Usually obtained as grey powder. Key ore is molybdenite, MoS_2, but is also obtained as by-product of copper production. Used in alloys, electrodes, and catalysts.

Radii/pm: Mo^{2+} 92; Mo^{6+} 6.2; atomic 136.2; covalent 129

Electronegativity: 2.16 (Pauling); 1.30 (Allred-Rochow)

Effective nuclear charge: 3.45 (Slater); 6.98 (Clementi); 9.95 (Froese–Fischer)

Standard reduction potentials $E^{\ominus}/V$

Oxidation states

Mo^{-II}	(d^8)	rare $[Mo(CO)_5]^{2-}$
Mo^0	(d^6)	rare $Mo(CO)_6$
Mo^I	(d^5)	rare $[Mo(C_6H_6)_2]^+$
Mo^{II}	(d^4)	Mo_6Cl_{12}, $[Mo_2Cl_8]^{4-}$, Mo_2^{4+}(aq)
Mo^{III}	(d^3)	MoF_3, $MoCl_3$ etc., $[Mo(H_2O)_6]^{3+}$ (aq)
Mo^{IV}	(d^2)	MoO_2, MoS_2, MoF_4, $MoCl_4$, etc.
Mo^V	(d^1)	Mo_2O_5, MoF_5, $MoCl_5$
$\mathbf{Mo^{VI}}$	$(d^0, [Kr])$	MoO_3, MoO_4^{2-}(aq), MoF_6, MoF_8^{2-}, $MoOF_4$

Physical properties

Melting point/K: 2890

Boiling point/K: 4885

ΔH_{fusion}/kJ mol^{-1}: 27.6   **ΔH_{vap}/kJ mol^{-1}:** 589.9

Thermodynamic properties (298.15 K, 0.1 MPa)

State	$\Delta_f H^{\ominus}$/kJ mol^{-1}	$\Delta_f G^{\ominus}$/kJ mol^{-1}	$S^{\ominus}$/J K^{-1} mol^{-1}	C_p/J K^{-1} mol^{-1}
Solid	0	0	28.66	24.06
Gas	658.1	612.5	181.950	20.786

Density/kg m^{-3}: 10 220 [293 K]; 9330 [liquid at m.p.]

Thermal conductivity/W m^{-1} K^{-1}: 138 [300 K]

Electrical resistivity/Ω m: 5.2×10^{-8} [273 K]

Mass magnetic susceptibility/kg^{-1} m^3: $+1.2 \times 10^{-8}$ (s)

Molar volume/cm^3: 9.39

Coefficient of linear thermal expansion/K^{-1}: 5.43×10^{-6}

Lattice structure (cell dimensions/pm), space group

b.c.c. ($a = 314.700$), Im3m

X-ray diffraction: mass absorption coefficients (μ/ρ)/cm^2 g^{-1}:
CuK$_\alpha$ 162   MoK$_\alpha$ 18.4

Molybdenum

Atomic number: 42

Thermal neutron capture cross-section/barns: 2.65 ± 0.05

Number of isotopes (including nuclear isomers): 20

Isotope mass range: $88 \rightarrow 105$

Key isotopes

Nuclide	Atomic mass	Natural abundance (%)	Half life $T_{1/2}$	Decay mode and energy (MeV)	Nuclear spin I	Nuclear magnetic moment μ	Uses
^{92}Mo	91.9063	14.84	stable		0		
^{94}Mo	93.9047	9.25	stable		0		
^{95}Mo	94.90584	15.92	stable		5/2	-0.9135	NMR
^{96}Mo	95.9046	16.68	stable		0		
^{97}Mo	96.9058	9.55	stable		5/2	-0.9327	NMR
^{98}Mo	97.9055	24.13	stable		0		
^{99}Mo		0	66.69h	β^- (1.37); γ			tracer
^{100}Mo	99.9076	9.63	stable		0		

NMR

	^{95}Mo	[^{97}Mo]
Relative sensitivity (^{1}H = 1.00)	3.23×10^{-3}	3.43×10^{-3}
Absolute sensitivity (^{1}H = 1.00)	5.07×10^{-4}	3.24×10^{-4}
Receptivity (^{13}C = 1.00)	2.88	1.84
Magnetogyric ratio/rad T^{-1} s^{-1}	1.7433×10^7	-1.7799×10^7
Quadrupole moment/m^2	0.12×10^{-28}	1.1×10^{-28}
Frequency (^{1}H = 100 MHz; 2.3488 T)/MHz	6.514	6.652

Reference: MoO_4^{2-} (aq)

Ground state electron configuration: [Kr]$4d^5 5s^1$

Term symbol: $^7S_3$

Electron affinity (M$\rightarrow$M$^-$)/kJ mol^{-1}: 114

Ionization energies/kJ mol^{-1}

1.	M $\rightarrow$ M$^+$	685.0	6. M^{5+} $\rightarrow$ M^{6+}	6 560
2.	M$^+$ $\rightarrow$ M^{2+}	1558	7. M^{6+} $\rightarrow$ M^{7+}	12 230
3.	M^{2+} $\rightarrow$ M^{3+}	2621	8. M^{7+} $\rightarrow$ M^{8+}	14 800
4.	M^{3+} $\rightarrow$ M^{4+}	4480	9. M^{8+} $\rightarrow$ M^{9+}	(16 800)
5.	M^{4+} $\rightarrow$ M^{5+}	5900	10. M^{9+} $\rightarrow$ M^{10+}	(19 700)

Principal lines in atomic spectrum

Wavelength/nm	Species	Sensitivity	Application
202.030	II		
313.259	I		AA
317.035	I		AA
379.825	I	U1	AA, AE
386.411	I	U2	AE
390.296	I	U3	AE

Abundance: Earth's crust 1.2 p.p.m.; seawater 0.01 p.p.m.

Biological role: Essential; moderately toxic; teratogenic

<table>
<tr><td>**Nd**</td><td>Atomic number: **60**
Relative atomic mass ($^{12}C = 12.0000$): **144.24**</td></tr>
</table>

Chemical properties

Silvery-white metal of the lanthanide (rare earth) group. Obtained from monazite and bastnaesite. Tarnishes in air, reacts slowly with cold water, rapidly with hot. Used in alloys, flints, glazes, and glass.

Radii/pm: Nd^{3+} 104; atomic 182.1; covalent 164

Electronegativity: 1.14 (Pauling); 1.07 (Allred-Rochow)

Effective nuclear charge: 2.85 (Slater); 9.31 (Clementi); 10.83 (Froese–Fischer)

Standard reduction potentials $E^{\ominus}/V$

	IV		III		II		0
Acid solution	Nd^{4+}	—4.9—	Nd^{3+}	—(−2.6)—	Nd^{2+}	—(−2.2)—	Nd
Alkaline solution	$[NdO_2]$	—2.5—	$Nd(OH)_3$		—(−2.78)—		Nd

(Acid solution III→0: −2.32)

Oxidation states

Nd^{II} (f^4)	NdO, $NdCl_2$, NdI_2
Nd^{III} (f^3)	Nd_2O_3, $Nd(OH)_3$, $[Nd(H_2O)_x]^{3+}$ (aq), NdF_3, $NdCl_3$, etc., Nd^{3+} salts, complexes
Nd^{IV} (f^2)	Cs_3NdF_7

Physical properties

Melting point/K: 1294
Boiling point/K: 3341
ΔH_{fusion}/kJ mol^{-1}: 7.113
ΔH_{vap}/kJ mol^{-1}: 328

Thermodynamic properties (298.15 K, 0.1 MPa)

State	$\Delta_f H^{\ominus}$/kJ mol^{-1}	$\Delta_f G^{\ominus}$/kJ mol^{-1}	$S^{\ominus}$/J K^{-1} mol^{-1}	C_p/J K^{-1} mol^{-1}
Solid	0	0	71.5	27.45
Gas	327.6	292.4	189.406	22.092

Density/kg m^{-3}: 7007 [293 K]
Thermal conductivity/W m^{-1} K^{-1}: 16.5 [300 K]
Electrical resistivity/Ω m: 64.0×10^{-8} [293 K]
Mass magnetic susceptibility/kg^{-1} m^3: $+4.902 \times 10^{-7}$ (s)
Molar volume/cm^3: 20.59
Coefficient of linear thermal expansion/K^{-1}: 6.7×10^{-6}

Lattice structure (cell dimensions/pm), space group

α-Nd hexagonal ($a = 365.79$, $c = 1179.92$), P6$_3$/mmc
β-Nd b.c.c. ($a = 413$), Im3m

$T(\alpha \rightarrow \beta) = 1135$ K

High pressure form: f.c.c. ($a = 480$), Fm3m

X-ray diffraction: mass absorption coefficients (μ/ρ)/cm^2 g^{-1}:
CuK$_\alpha$ 374   MoK$_\alpha$ 53.2

Neodymium

Atomic number: 60

Thermal neutron capture cross-section/barns: 49 ± 20

Number of isotopes (including nuclear isomers): 16

Isotope mass range: $138 \rightarrow 151$

Key isotopes

Nuclide	Atomic mass	Natural abundance (%)	Half life $T_{1/2}$	Decay mode and energy (MeV)	Nuclear spin I	Nuclear magnetic moment μ	Uses
^{142}Nd	141.9075	27.16	stable				
^{143}Nd	142.9096	12.18	stable		7/2	-1.063	NMR
^{144}Nd	143.9099	23.80	2.4×10^{15}y	$\alpha(1.83)$			
^{145}Nd	144.9122	8.29	stable		7/2	-0.654	NMR
^{146}Nd	145.9172	17.19	stable				
^{147}Nd		0	11.06d	$\beta^-(0.91)$; γ	5/2	± 0.55	tracer
^{148}Nd	147.9165	5.75	stable				
^{150}Nd	149.9207	5.63	stable				

NMR

	^{143}Nd	^{145}Nd
Relative sensitivity (^{1}H = 1.00)	3.38×10^{-3}	7.86×10^{-4}
Absolute sensitivity (^{1}H = 1.00)	4.11×10^{-4}	6.52×10^{-5}
Receptivity (^{13}C = 1.00)	2.43	0.393
Magnetogyric ratio/rad T^{-1} s^{-1}	-1.474×10^7	-0.913×10^7
Quadrupole moment/m^2	-0.48×10^{-28}	-0.25×10^{-28}
Frequency (^{1}H = 100 MHz; 2.3488 T)/MHz	5.437	3.346

Ground state electron configuration: $[Xe]4f^4 6s^2$

Term symbol: $^5I_4$

Electron affinity $(M \rightarrow M^-)$/kJ mol^{-1}: $\leqslant 50$

Ionization energies/kJ mol^{-1}

1. $M \rightarrow M^+$	529.6	6. $M^{5+} \rightarrow M^{6+}$	
2. $M^+ \rightarrow M^{2+}$	1035	7. $M^{6+} \rightarrow M^{7+}$	
3. $M^{2+} \rightarrow M^{3+}$	2130	8. $M^{7+} \rightarrow M^{8+}$	
4. $M^{3+} \rightarrow M^{4+}$	3899	9. $M^{8+} \rightarrow M^{9+}$	
5. $M^{4+} \rightarrow M^{5+}$		10. $M^{9+} \rightarrow M^{10+}$	

Principal lines in atomic spectrum

Wavelength/nm	Species	Sensitivity	Application
395.115	II		AE
417.732	II		AE
430.357	II		AE
463.424	I		AA
492.453	I		AE

Abundance: Earth's crust 40 p.p.m.; seawater 3×10^{-6} p.p.m.

Biological role: None; moderately toxic, eye irritant

Atomic number: **10**

Relative atomic mass (^{12}C = 12.0000): **20.1797**

Chemical properties

Colourless, odourless gas, obtained from liquid air. Chemically inert towards everything including fluorine gas. Used in ornamental lighting ('neon' signs)

Radii/pm: van der Waals 160

Electronegativity: n.a. (Pauling); n.a. (Allred-Rochow)

Effective nuclear charge: 5.85 (Slater); 5.76 (Clementi); 5.18 (Froese–Fischer)

Standard reduction potentials $E^{\ominus}$/V

n.a.

Covalent bonds	r/pm	E/kJ mol^{-1}	Oxidation states
n.a.			n.a.

Physical properties

Melting point/K: 24.48
Boiling point/K: 27.10
ΔH_{fusion}/kJ mol^{-1}: 0.324
ΔH_{vap}/kJ mol^{-1}: 1.736

Thermodynamic properties (298.15 K, 0.1 MPa)

State	$\Delta_f H^{\ominus}$/kJ mol^{-1}	$\Delta_f G^{\ominus}$/kJ mol^{-1}	$S^{\ominus}$/J K^{-1} mol^{-1}	C_p/J K^{-1} mol^{-1}
Gas	0	0	146.328	20.786

Density/kg m^{-3}: 1444 [solid, m.p.]; 1207.3 [liquid, b.p.]; 0.899 94 [gas, 273 K]

Thermal conductivity/W m^{-1} K^{-1}: 0.0493 [300 K]$_g$

Mass magnetic susceptibility/kg^{-1} m^3: -4.2×10^{-9} (g);

Molar volume/cm^3: 13.97 [24 K]

Lattice structure (cell dimensions/pm), space group

f.c.c. (a = 445.462), Fm3m
h.c.p. (a = 314.5, c = 514), P6$_3$/mmc [3 K]

X-ray diffraction: mass absorption coefficients (μ/ρ)/cm^2 g^{-1}:
CuK$_\alpha$ 22.9   MoK$_\alpha$ 2.47

Neon

Atomic number: 10
Thermal neutron capture cross-section/barns: 0.038
Number of isotopes (including nuclear isomers): 8
Isotope mass range: $17 \rightarrow 24$

Nuclear properties

Key isotopes

Nuclide	Atomic mass	Natural abundance (%)	Half life $T_{1/2}$	Decay mode and energy (MeV)	Nuclear spin I	Nuclear magnetic moment μ	Uses
^{20}Ne	19.992 44	90.51	stable		0		
^{21}Ne	20.993 95	0.27	stable		3/2	$-0.661\,76$	NMR
^{22}Ne	21.991 38	9.22	stable		0		

NMR (no known compounds) ^{21}Ne

Relative sensitivity (^{1}H = 1.00) 2.50×10^{-3}
Absolute sensitivity (^{1}H = 1.00) 6.43×10^{-6}
Receptivity (^{13}C = 1.00) 0.0359
Magnetogyric ratio/rad T^{-1} s^{-1} -2.1118×10^7
Quadrupole moment/m^2 9×10^{-30}
Frequency (^{1}H = 100 MHz; 2.3488 T)/MHz   7.894

Ground state electron configuration: $[He]2s^2 2p^6$
Term symbol: $^1S_0$
Electron affinity $(M \rightarrow M^-)$/kJ mol^{-1}: -29 (calc.)

Electron shell properties

Ionization energies/kJ mol^{-1}

1. M $\rightarrow$ M$^+$	2 080.6	6. M$^{5+} \rightarrow$ M^{6+}	15 238
2. M$^+ \rightarrow$ M^{2+}	3 952.2	7. M$^{6+} \rightarrow$ M^{7+}	19 998
3. M$^{2+} \rightarrow$ M^{3+}	6 122	8. M$^{7+} \rightarrow$ M^{8+}	23 069
4. M$^{3+} \rightarrow$ M^{4+}	9 370	9. M$^{8+} \rightarrow$ M^{9+}	115 377
5. M$^{4+} \rightarrow$ M^{5+}	12 177	10. M$^{9+} \rightarrow$ M^{10+}	131 429

Principal lines in atomic spectrum

Wavelength/nm	Species	Sensitivity	Application
74.370*	I		
540.056	I		AE
585.249	I		AE
640.225	I		AE
865.438	I		

* Lowest energy transition from ground state to nearest empty orbital

Abundance: Atmosphere 18 p.p.m. (volume); Earth's crust 0.04 p.p.m.; seawater 0.45 p.p.m.
Biological role: None; non-toxic

Np	Atomic number: 93
	Relative atomic mass ($^{12}C = 12.0000$): 237.0482

Chemical properties

Radioactive silvery metal obtained in kg quantities as ^{237}Np from uranium fuel elements. Attacked by oxygen, steam, and acids, but not alkalis.

Radii/pm: Np^{3+} 110; Np^{4+} 95; Np^{5+} 88; Np^{6+} 82; atomic 131

Electronegativity: 1.36 (Pauling); 1.22 (Allred-Rochow)

Effective nuclear charge: 1.80 (Slater)

Standard reduction potentials $E^{\ominus}/V$

	VII		VI		V		IV		III		0
Acid solution	NpO_3^+	$\xrightarrow{2.04}$	NpO_2^{2+}	$\xrightarrow{1.24}$	NpO_2^+	$\xrightarrow{0.66}$	Np^{4+}	$\xrightarrow{0.18}$	Np^{3+}	$\xrightarrow{-1.79}$	Np

$NpO_2^{2+} \xrightarrow{0.95} NpO_2^+$

$Np^{4+} \xrightarrow{-1.30} Np$

Alkaline solution $NpO_5^{3-} \xrightarrow{0.58} NpO_2(OH)_2 \xrightarrow{0.6} NpO_2OH \xrightarrow{0.3} NpO_2 \xrightarrow{-2.1} Np(OH)_3 \xrightarrow{-2.2} Np$

Oxidation states

Np^{II}	(f^4d^1)	NpO
Np^{III}	(f^4)	NpF_3, $NpCl_3$, etc., $[NpCl_6]^{3-}$, $(Np(H_2O)_x)^{3+}$ (aq),
Np^{IV}	(f^3)	NpO_2, $[Np(H_2O)_x]^{4+}$ (aq), NpF_4, $NpCl_4$, $NpBr_4$,
		$[NpCl_6]^{2-}$, complexes
Np^V	(f^2)	Np_2O_5, NpF_5, $CsNpF_6$, Na_3NpF_8, NpO_2^+ (aq)
Np^{VI}	(f^1)	$NpO_3 \cdot H_2O$, NpO_2^{2+} (aq), NpF_6
Np^{VII}	([Rn])	Li_5NpO_6

Physical properties

Melting point/K: 913

Boiling point/K: 4175

ΔH_{fusion}/kJ mol^{-1}: 9.46

ΔH_{vap}/kJ mol^{-1}: 336.6

Thermodynamic properties (298.15 K, 0.1 MPa)

State	$\Delta_f H^{\ominus}$/kJ mol^{-1}	$\Delta_f G^{\ominus}$/kJ mol^{-1}	$S^{\ominus}$/J K^{-1} mol^{-1}	C_p/J K^{-1} mol^{-1}
Solid	0	0	n.a.	n.a.
Gas	n.a.	n.a.	n.a.	n.a.

Density/kg m^{-3}: 20 250 [293 K]

Thermal conductivity/W m^{-1} K^{-1}: 6.3 [300 K]

Electrical resistivity/Ω m: 122×10^{-8} [293 K]

Mass magnetic susceptibility/kg^{-1} m^3: n.a.

Molar volume/cm^3: 11.71

Coefficient of linear thermal expansion/K^{-1}: 27.5×10^{-6}

Lattice structure (cell dimensions/pm), space group

α-Np orthorhombic ($a = 472.3$, $b = 488.7$, $c = 666.3$), Pmcn
β-Np tetragonal ($a = 489.7$, $c = 338.8$), P42$_1$2
γ-Np cubic ($a = 352$), Im3m

$T(\alpha \rightarrow \beta) = 551$ K
$T(\beta \rightarrow \gamma) = 850$ K

X-ray diffraction: mass absorption coefficients (μ/ρ)/cm^2 g^{-1}:
n.a.

Neptunium

Atomic number: 93

Thermal neutron capture cross-section/barns: 170 ± 5 (^{237}Np)

Number of isotopes (including nuclear isomers): 15

Isotope mass range: $229 \rightarrow 241$

Nuclear properties

Key isotopes

Nuclide	Atomic mass	Natural abundance (%)	Half life $T_{1/2}$	Decay mode and energy (MeV)	Nuclear spin I	Nuclear magnetic moment μ	Uses
^{237}Np	237.0480	0	2.14×10^6y	$\alpha(4.956)$; γ	5/2	$+2.4$	tracer

NMR

^{237}Np

Relative sensitivity (^{1}H $= 1.00$)	—
Absolute sensitivity (^{1}H $= 1.00$)	—
Receptivity (^{13}C $= 1.00$)	—
Magnetogyric ratio/rad T^{-1} s^{-1}	3.1×10^7
Quadrupole moment/m^2	4.2×10^{-28}
Frequency (^{1}H $= 100$ MHz; 2.3488 T)/MHz	11.25

Ground state electron configuration: [Rn]5f^{4}6d^{1}7s^2

Term symbol: $^6L_{11/2}$

Electron affinity (M$\rightarrow$M$^-$)/kJ mol^{-1}: n.a.

Electron shell properties

Ionization energies/kJ mol^{-1}

1. M $\rightarrow$M$^+$	597	6. M$^{5+}\rightarrow$M^{6+}
2. M$^+\rightarrow$M^{2+}		7. M$^{6+}\rightarrow$M^{7+}
3. M$^{2+}\rightarrow$M^{3+}		8. M$^{7+}\rightarrow$M^{8+}
4. M$^{3+}\rightarrow$M^{4+}		9. M$^{8+}\rightarrow$M^{9+}
5. M$^{4+}\rightarrow$M^{5+}		10. M$^{9+}\rightarrow$M^{10+}

Principal lines in atomic spectrum

Wavelength/nm	Species	Wavelength	Species
901.618	I	1214.818	I
1009.199	I	1237.742	I
1081.745	I	1240.799	I
1169.515	I	1383.433	I
1177.664	I		

Abundance: Earth's crust: trace amounts in uranium minerals; seawater nil

Biological role: None; toxic due to radioactivity

Ni

Atomic number: 28
Relative atomic mass ($^{12}C = 12.0000$): 58.69

Chemical properties

Silvery-white metal, lustrous, malleable, and ductile. Obtained from garnierite $(Ni,Mg)_6Si_4O_{10}(OH)_2$ and pentlandite $(Ni,Fe)_9S_8$ ores. Resists corrosion, soluble in acids, except concentrated HNO_3, unaffected by alkalis. Used in alloys, coins, metal plating, and catalysts.

Radii/pm: Ni^{2+} 78; Ni^{3+} 62; atomic 124.6; covalent 115
Electronegativity: 1.91 (Pauling); 1.75 (Allred-Rochow)
Effective nuclear charge: 4.05 (Slater); 5.71 (Clementi); 7.86 (Froese–Fischer)

Standard reduction potentials $E^\ominus$/V

	VI		IV		II		0
			> 1.6				
Acid solution	NiO_4^{2-}	> 1.8	NiO_2	1.593	Ni^{2+}	−0.257	Ni
Alkaline solution	NiO_4^{2-}	> 0.4	NiO_2	0.490	$Ni(OH)_2$	−0.72	Ni

Oxidation states

Ni^{-I}	$(d^{10}s^1)$	$[Ni_2(CO)_6]^{2-}$
Ni^0	(d^{10})	$Ni(CO)_4$, $K_4[Ni(CN)_4]$
Ni^I	(d^9)	$K_2[Ni(CN)_4]$
Ni^{II}	(d^8)	NiO, $Ni(OH)_2$, $[Ni(H_2O)_6]^{2+}$ (aq), NiF_2, $NiCl_2$, etc., salts, NiAs, $[NiCl_4]^{2-}$, complexes, $[Ni(C_5H_5)_2]$
Ni^{III}	(d^7)	$NiO(OH)$, NiF_3 ?, NiF_6^{3-}
Ni^{IV}	(d^6)	NiO_2 ?, NiF_6^{2-}
Ni^{VI}	(d^4)	K_2NiO_4 ?

Physical properties

Melting point/K: 1726
Boiling point/K: 3005
ΔH_{fusion}/kJ mol^{-1}: 17.6
ΔH_{vap}/kJ mol^{-1}: 374.8

Thermodynamic properties (298.15 K, 0.1 MPa)

State	$\Delta_f H^\ominus$/kJ mol^{-1}	$\Delta_f G^\ominus$/kJ mol^{-1}	$S^\ominus$/J K^{-1} mol^{-1}	C_p/J K^{-1} mol^{-1}
Solid	0	0	29.87	26.07
Gas	429.7	384.5	182.193	23.359

Density/kg m^{-3}: 8902 [298 K]; 7780 [liquid at m.p.]
Thermal conductivity/W m^{-1} K^{-1}: 90.7 [300 K]
Electrical resistivity/Ω m: 6.84×10^{-8} [293 K]
Mass magnetic susceptibility/kg^{-1} m^3: ferromagnetic
Molar volume/cm^3: 6.59
Coefficient of linear thermal expansion/K^{-1}: 13.3×10^{-6}

Lattice structure (cell dimensions/pm), space group

f.c.c. ($a = 352.38$), Fm3m
'Hexagonal' nickel* ($a = 266$, $c = 432$), P6$_3$/mmc

*impure form of nickel

X-ray diffraction: mass absorption coefficients (μ/ρ)/cm^2 g^{-1}:
CuK$_\alpha$ 45.7   MoK$_\alpha$ 46.6

Nickel

Atomic number: 28
Thermal neutron capture cross-section/barns: 4.51 ± 0.1
Number of isotopes (including nuclear isomers): 9
Isotope mass range: $56 \rightarrow 64$

Nuclear properties

Key isotopes

Nuclide	Atomic mass	Natural abundance (%)	Half life $T_{1/2}$	Decay mode and energy (MeV)	Nuclear spin I	Nuclear magnetic moment μ	Uses
^{58}Ni	57.9353	68.27	stable				
^{59}Ni	58.934	0	8×10^4y	EC(1.072); no γ			
^{60}Ni	59.9332	26.10	stable				
^{61}Ni	60.9310	1.13	stable		3/2	-0.7498	NMR
^{62}Ni	61.9283	3.59	stable				
^{63}Ni	62.930	0	92y	β^- (0.067); no γ			tracer
^{64}Ni	63.9280	0.91	stable				

NMR

^{61}Ni

Relative sensitivity (^{1}H = 1.00) 3.57×10^{-3}
Absolute sensitivity (^{1}H = 1.00) 4.25×10^{-5}
Receptivity (^{13}C = 1.00)   0.242
Magnetogyric ratio/rad T^{-1} s^{-1} -2.3948×10^7
Quadrupole moment/m^2 0.16×10^{-28}
Frequency (^{1}H = 100 MHz; 2.3488 T)/MHz   8.936

Ground state electron configuration: [Ar]$3d^8 4s^2$
Term symbol: $^3F_4$
Electron affinity (M$\rightarrow$M$^-$)/kJ mol^{-1}: 156

Electron shell properties

Ionization energies/kJ mol^{-1}

1. M $\rightarrow$ M$^+$	736.7	6. M$^{5+}\rightarrow$M^{6+}	10 400
2. M$^+ \rightarrow$ M^{2+}	1753.0	7. M$^{6+}\rightarrow$M^{7+}	12 800
3. M$^{2+}\rightarrow$M^{3+}	3393	8. M$^{7+}\rightarrow$M^{8+}	15 600
4. M$^{3+}\rightarrow$M^{4+}	5300	9. M$^{8+}\rightarrow$M^{9+}	18 600
5. M$^{4+}\rightarrow$M^{5+}	7280	10. M$^{9+}\rightarrow$M^{10+}	21 660

Principal lines in atomic spectrum

Wavelength/nm	Species	Sensitivity	Application
231.096	I		AA
232.003	I		AA
336.957*	I		
341.477	I	U1	AA, AE
349.296	I	U2	AE
352.454	I		AE

* Lowest energy transition from ground state to nearest empty orbital

Abundance: Earth's crust 99 p.p.m.; seawater 0.0005 p.p.m.
Biological role: Uncertain; carcinogenic; stimulatory

Atomic number: 41

Relative atomic mass ($^{12}C = 12.0000$): **92.90638**

Chemical properties

Shiny silvery metal, soft when pure. Found as columbite, $(Fe,Mn)Nb_2O_6$, but obtained as by-product of tin extraction. Resists corrosion due to oxide film, attacked by hot, concentrated acids but resists fused alkalis. Used in stainless steels.

Radii/pm: Nb^{4+} 74; Nb^{5+} 69; atomic 142.9; covalent 134

Electronegativity: 1.6 (Pauling); 1.23 (Allred-Rochow)

Effective nuclear charge: 3.30 (Slater); 6.70 (Clementi); 9.60 (Froese–Fischer)

Standard reduction potentials $E^{\ominus}$/V

	V		III		0
Acid solution	Nb_2O_5	$\xrightarrow{-0.1}$	Nb^{3+}	$\xrightarrow{-1.1}$	Nb
			$\xrightarrow{-0.65}$		

Oxidation states

Nb^{-III}	(d^8)	$[Nb(CO)_5]^{3-}$
Nb^{-I}	(d^6)	$[Nb(CO)_5]^-$
Nb^{I}	(d^4)	$[(C_5H_5)Nb(CO)_4]$
Nb^{II}	(d^3)	NbO
Nb^{III}	(d^2)	$LiNbO_2$, $NbCl_3$, $NbBr_3$, NbI_3, $[Nb(CN)_8]^{5-}$
Nb^{IV}	(d^1)	NbO_2, NbF_4, $NbCl_4$ etc., $NbOCl_2$
Nb^V	(d^0, f^{14})	Nb_2O_5, $[HNb_6O_{19}]^-$ (aq), NbF_5, $NbCl_5$, etc., NbO_2F, $NbOCl_3$

Physical properties

Melting point/K: 2741

Boiling point/K: 5015

ΔH_{fusion}/kJ mol^{-1}: 27.2

ΔH_{vap}/kJ mol^{-1}: 680.19

Thermodynamic properties (298.15 K, 0.1 MPa)

State	$\Delta_f H^{\ominus}$/kJ mol^{-1}	$\Delta_f G^{\ominus}$/kJ mol^{-1}	$S^{\ominus}$/J K^{-1} mol^{-1}	C_p/J K^{-1} mol^{-1}
Solid	0	0	36.40	24.60
Gas	725.9	681.1	186.256	30.158

Density/kg m^{-3}: 8570 [293 K]; 7830 [liquid at m.p.]

Thermal conductivity/W m^{-1} K^{-1}: 53.7 [300 K]

Electrical resistivity/Ω m: 12.5×10^{-8} [273 K]

Mass magnetic susceptibility/kg^{-1} m^3: $+2.76 \times 10^{-8}$ (s)

Molar volume/cm^3: 10.84

Coefficient of linear thermal expansion/K^{-1}: 7.07×10^{-6}

Lattice structure (cell dimensions/pm), space group

b.c.c. ($a = 329.86$), Im3m

X-ray diffraction: mass absorption coefficients (μ/ρ)/cm^2 g^{-1}: CuK_α 153 MoK_α 17.1

Niobium

Atomic number: 41
Thermal neutron capture cross-section/barns: 1.15 ± 0.05
Number of isotopes (including nuclear isomers): 24
Isotope mass range: $88 \rightarrow 101$

Key isotopes

Nuclide	Atomic mass	Natural abundance (%)	Half life $T_{1/2}$	Decay mode and energy (MeV)	Nuclear spin I	Nuclear magnetic moment μ	Uses
^{93}Nb	92.9060	100	stable		9/2	$+6.167$	NMR
^{94}Nb	93.9057	0	2×10^4y	β^- (2.06); γ			

NMR

	^{93}Nb
Relative sensitivity (^{1}H = 1.00)	0.48
Absolute sensitivity (^{1}H = 1.00)	0.48
Receptivity (^{13}C = 1.00)	2740
Magnetogyric ratio/rad T^{-1} s^{-1}	6.5476×10^7
Quadrupole moment/m^2	-0.2×10^{-28}
Frequency (^{1}H = 100 MHz; 2.3488 T)/MHz	24.442

Reference: NbF$_6^-$ (conc. HF)

Ground state electron configuration: [Kr]4d^{4}5s^1
Term symbol: ^{6}D$_{1/2}$
Electron affinity (M→M$^-$)/kJ mol^{-1}: 109

Ionization energies/kJ mol^{-1}

1. M $\rightarrow$ M$^+$	664	6. M^{5+} $\rightarrow$ M^{6+}	9 899
2. M$^+$ $\rightarrow$ M^{2+}	1382	7. M^{6+} $\rightarrow$ M^{7+}	12 100
3. M^{2+} $\rightarrow$ M^{3+}	2416	8. M^{7+} $\rightarrow$ M^{8+}	
4. M^{3+} $\rightarrow$ M^{4+}	3695	9. M^{8+} $\rightarrow$ M^{9+}	
5. M^{4+} $\rightarrow$ M^{5+}	4877	10. M^{9+} $\rightarrow$ M^{10+}	

Principal lines in atomic spectrum

Wavelength/nm	Species	Sensitivity	Application
334.370	I		AA
358.027	I		AA
405.894	I	U1	AA, AE
407.973	I	U2	AA, AE
410.092	I	U3	AE
525.162*	I		

* Lowest energy transition from ground state to nearest empty orbital

Abundance: Earth's crust 20 p.p.m.; seawater 1×10^{-5} p.p.m.
Biological role: None; moderately toxic

<table>
<tr><td>

N

</td><td>

Atomic number: 7

Relative atomic mass ($^{12}C = 12.0000$): **14.00674**

</td></tr>
</table>

<table>
<tr><td>

Chemical properties

</td><td>

Colourless, odourless gas (N_2) obtained from liquid air. Generally unreactive at room temperature. Extensive inorganic and organic chemistry. Used in fertilizers, acids (HNO_3), explosives, plastics, dyes, etc.

</td></tr>
</table>

Radii/pm: N^{3+} 16; atomic 71; covalent 70 (single bond); van der Waals 154; N^{3-} 171

Electronegativity: 3.04 (Pauling); 3.07 (Allred–Rochow)

Effective nuclear charge: 3.90 (Slater); 3.83 (Clementi); 3.46 (Froese–Fischer)

Standard reduction potentials $E^{\ominus}$/V

Covalent bonds r/pm E/kJ mol^{-1} Oxidation states

	r/pm	E/kJ mol^{-1}
N—H	101	390
N—N	147	160
N=N	125	415
N≡N	110	946
N—Cl	195	193

N^{-III}	NH_3, NH_4^+ (aq)		
N^{-II}	N_2H_4, $N_2H_5^+$ (aq)	N^{III}	HNO_2, NO_2^- (aq),
N^{-I}	NH_2OH		NF_3
N^0	N_2	N^{IV}	$N_2O_4 \rightleftharpoons 2NO_2$
N^{II}	NO	N^V	HNO_3, NO_3^- (aq)

<table>
<tr><td>

Physical properties

</td><td>

Melting point/K: 63.29 ΔH_{fusion}/kJ mol^{-1}: 0.720

Boiling point/K: 77.4 ΔH_{vap}/kJ mol^{-1}: 5.577

</td></tr>
</table>

Thermodynamic properties (298.15 K, 0.1 MPa)

State	$\Delta_f H^{\ominus}$/kJ mol^{-1}	$\Delta_f G^{\ominus}$/kJ mol^{-1}	$S^{\ominus}$/J K^{-1} mol^{-1}	C_p/J K^{-1} mol^{-1}
Gas (N_2)	0	0	191.61	29.125
Gas (atoms)	472.704	455.563	153.298	20.786

Density/kg m^{-3}: 1026 [solid, 21 K]; 880 [liquid, b.p.]; 1.2506 [gas, 273 K]

Thermal conductivity/W m^{-1} K^{-1}: 0.025 98 [300 K]$_g$

Mass magnetic susceptibility/kg^{-1} m^3: -5.4×10^{-9} (g)

Molar volume/cm^3: 13.65 [21 K]

Lattice structure (cell dimensions/pm), space group

α-N_2 cubic ($a = 564.4$), P2$_1$3
β-N_2 h.c.p. ($a = 404.2$, $c = 660.1$), P6$_3$/mmc
$T(\alpha \rightarrow \beta) = 35$ K

X-ray diffraction: mass absorption coefficients (μ/ρ)/cm^2 g^{-1}: CuK$_\alpha$ 7.52   MoK$_\alpha$ 0.916

Nitrogen

Atomic number: 7
Thermal neutron capture cross-section/barns: 0.075 ± 0.008
Number of isotopes (including nuclear isomers): 8
Isotope mass range: $12 \rightarrow 18$

Nuclear properties

Key isotopes

Nuclide	Atomic mass	Natural abundance (%)	Half life $T_{1/2}$	Decay mode and energy (MeV)	Nuclear spin I	Nuclear magnetic moment μ	Uses
^{14}N	14.003 07	99.63	stable		1	+0.4036	NMR
^{15}N	15.000 11	0.37	stable		1/2	−0.2831	NMR

No long-lived radioactive isotopes: ^{13}N has $T_{1/2} = 9.97$ m

NMR

	$[^{14}N]$	^{15}N
Relative sensitivity (^{1}H = 1.00)	1.01×10^{-3}	1.04×10^{-3}
Absolute sensitivity (^{1}H = 1.00)	1.01×10^{-3}	3.85×10^{-6}
Receptivity (^{13}C = 1.00)	5.69	0.0219
Magnetogyric ratio/rad T^{-1} s^{-1}	1.9331×10^7	-2.7116×10^7
Quadrupole moment/m^2	1.6×10^{-30}	—
Frequency (^{1}H = 100 MHz; 2.3488 T)/MHz	7.224	10.133

Reference: $MeNO_2$ or NO_3^-

Ground state electron configuration: $[He]2s^2 2p^3$
Term symbol: $^4S_{3/2}$
Electron affinity ($M \rightarrow M^-$)/kJ mol^{-1}: -7

Electron shell properties

Ionization energies/kJ mol^{-1}

1. $M \rightarrow M^+$ 1402.3	6. $M^{5+} \rightarrow M^{6+}$ 53 265.6
2. $M^+ \rightarrow M^{2+}$ 2856.1	7. $M^{6+} \rightarrow M^{7+}$ 64 358.7
3. $M^{2+} \rightarrow M^{3+}$ 4578.0	
4. $M^{3+} \rightarrow M^{4+}$ 7474.9	
5. $M^{4+} \rightarrow M^{5+}$ 9440.0	

Principal lines in atomic spectrum

Wavelength/nm	Species	Sensitivity	Application
120.002* (vac)	I		
399.500	II		
409.995	I	U3	AE
410.998	I	U2	AE
567.956	II	V2	AE

* Lowest energy transition from ground state to nearest empty orbital

Abundance: Atmosphere 780 900 p.p.m. (volume); Earth's crust 19 p.p.m.; seawater 0.8 p.p.m.
Biological role: Basis of life (part of DNA); nitrogen cycle in nature

No	Atomic number: 102
	Relative atomic mass ($^{12}C=12.0000$): **(259)**

Chemical properties

Radioactive metal, only a few atoms of which have ever been made. Obtained by bombardment of ^{246}Cm with carbon nuclei.

Radii/pm: No^{2+} 113; No^{3+} 95; No^{4+} 83

Electronegativity: 1.3 (Pauling); n.a. (Allred-Rochow)

Effective nuclear charge: 1.65 (Slater)

Standard reduction potentials $E^{\ominus}/V$

	III		II		0
		-1.2			
Acid solution	No^{3+}	$\underline{1.4}$	No^{2+}	$\underline{-2.5}$	No

Oxidation states

No^{II} (f^{14})	$[No(H_2O)_x]^{2+}$ (aq)
No^{III} (f^{13})	$[No(H_2O)_x]^{3+}$ (aq)

Physical properties

Melting point/K: n.a.

Boiling point/K: n.a.

ΔH_{fusion}/kJ mol^{-1}: n.a.

ΔH_{vap}/kJ mol^{-1}: n.a.

Thermodynamic properties (298.15 K, 0.1 MPa)

State	$\Delta_f H^{\ominus}$/kJ mol^{-1}	$\Delta_f G^{\ominus}$/kJ mol^{-1}	$S^{\ominus}$/J K^{-1} mol^{-1}	C_p/J K^{-1} mol^{-1}
Solid	0	0	n.a.	n.a.
Gas	n.a.	n.a.	n.a.	n.a.

Density/kg m^{-3}: n.a.

Thermal conductivity/W m^{-1} K^{-1}: 10 (est.) [300 K]

Electrical resistivity/Ω m: n.a.

Mass magnetic susceptibility/kg^{-1} m^3: n.a.

Molar volume/cm^3: n.a.

Coefficient of linear thermal expansion/K^{-1}: n.a.

Lattice structure (cell dimensions/pm), space group

n.a.

X-ray diffraction: mass absorption coefficients (μ/ρ)/cm^2 g^{-1}:
n.a.

130

Conclusively identified in 1958 by A. Ghiorso, T. Sikkeland, J. R. Walton, and G. T. Seaborg at Berkeley, California, USA
[Named after Alfred Nobel]

Nobelium

Atomic number: 102

Thermal neutron capture cross-section/barns: n.a.

Number of isotopes (including nuclear isomers): 9

Isotope mass range: $251 \rightarrow 259$

Nuclear properties

Key isotopes

Nuclide	Atomic mass	Natural abundance (%)	Half life $T_{1/2}$	Decay mode and energy (MeV)	Nuclear spin I	Nuclear magnetic moment μ	Uses
^{259}No		0	58m	$\alpha(7.50)$; SF			

Ground state electron configuration: $[Rn]5f^{14}7s^2$

Term symbol: $^1S_0$

Electron affinity $(M \rightarrow M^-)$/kJ mol^{-1}: n.a.

Electron shell properties

Ionization energies/kJ mol^{-1}

1. $M \rightarrow M^+$	642	6. $M^{5+} \rightarrow M^{6+}$
2. $M^+ \rightarrow M^{2+}$		7. $M^{6+} \rightarrow M^{7+}$
3. $M^{2+} \rightarrow M^{3+}$		8. $M^{7+} \rightarrow M^{8+}$
4. $M^{3+} \rightarrow M^{4+}$		9. $M^{8+} \rightarrow M^{9+}$
5. $M^{4+} \rightarrow M^{5+}$		10. $M^{9+} \rightarrow M^{10+}$

Principal lines in atomic spectrum

Wavelength/nm	Species	Sensitivity	Application
n.a.			

Abundance: Earth's crust nil; seawater nil

Biological role: None; toxic due to radioactivity

Os

Atomic number: 76
Relative atomic mass ($^{12}C = 12.0000$): 190.2

Chemical properties

Lustrous silvery metal of platinum group. Found in free state but obtained from wastes of nickel refining. Unaffected by air, water, and acids, but dissolves in molten alkalis. Smells, due to formation of volatile OsO_4. Used in alloys and catalysts.

Radii/pm: Os^{2+} 89; Os^{3+} 81; Os $^{4+}$ 67; atomic 135; covalent 126

Electronegativity: 2.2 (Pauling); 1.52 (Allred-Rochow)

Effective nuclear charge: 3.75 (Slater); 10.32 (Clementi); 14.90 (Froese–Fischer)

Standard reduction potentials $E^{\ominus}/V$

VIII	IV	III	II	0

OsO_4 —— 1.005 —— OsO_2 —— 0.687 —————————————— Os

(0.85 over VIII–IV span)

$OsCl_6^{2-}$ —— 0.45 —— $OsCl_6^{3-}$

$Os(CN)_4(OH)_2^{3-}$ —— 0.634 —— $Os(CN)_4(OH)_2^{4-}$

Oxidation states

Os^{-II} (d^{10})	$[Os(CO)_4]^{2-}$
Os^{0} (d^{8})	$Os(CO)_5$, $Os_2(CO)_9$
Os^{I} (d^{7})	OsI
Os^{II} (d^{6})	$OsCl_2$, OsI_2
Os^{III} (d^{5})	$OsCl_3$, $OsBr_3$, OsI_3, complexes
Os^{IV} (d^{4})	OsO_2, OsO_2 (aq), OsF_4, $OsCl_4$, $OsBr_4$, $OsCl_6^{2-}$, complexes
Os^{V} (d^{3})	OsF_5, $OsCl_5$
Os^{VI} (d^{2})	OsO_3?, OsF_6
Os^{VII} (d^{1})	OsF_7
Os^{VIII} (d^{0}, f^{14})	OsO_4, $[OsO_4(OH)_2]^{2-}$ (aq)

Physical properties

Melting point/K: 3327
Boiling point/K: 5300
ΔH_{fusion}/kJ mol^{-1}: 29.3
ΔH_{vap}/kJ mol^{-1}: 738.06

Thermodynamic properties (298.15 K, 0.1 MPa)

State	$\Delta_f H^{\ominus}$/kJ mol^{-1}	$\Delta_f G^{\ominus}$/kJ mol^{-1}	$S^{\ominus}$/J K^{-1} mol^{-1}	C_p/J K^{-1} mol^{-1}
Solid	0	0	32.6	24.7
Gas	791	745	192.573	20.786

Density/kg m^{-3}: 22 590 [293 K]; 20 100 [liquid at m.p.]
Thermal conductivity/W m^{-1} K^{-1}: 87.6 [300 K]
Electrical resistivity/Ω m: 8.12×10^{-8} [273 K]
Mass magnetic susceptibility/kg^{-1} m^{3}: $+6.5 \times 10^{-10}$ (s)
Molar volume/cm^{3}: 8.43
Coefficient of linear thermal expansion/K^{-1}:
4.3×10^{-6} (a axis); 6.1×10^{-6} (b axis); 6.8×10^{-6} (c axis)

Lattice structure (cell dimensions/pm), space group

h.c.p. ($a = 273.43$; $c = 432.00$), $P6_3/mmc$

X-ray diffraction: mass absorption coefficients (μ/ρ)/cm^2 g^{-1}:
CuK$_{\alpha}$ 186   MoK$_{\alpha}$ 106

Osmium

Atomic number: 76
Thermal neutron capture cross-section/barns: 15.3 ± 0.7
Number of isotopes (including nuclear isomers): 19
Isotope mass range: $181 \rightarrow 195$

Nuclear properties

Key isotopes

Nuclide	Atomic mass	Natural abundance (%)	Half life $T_{1/2}$	Decay mode and energy (MeV)	Nuclear spin I	Nuclear magnetic moment μ	Uses
^{184}Os	183.9526	0.020	stable				
^{185}Os		0	96.6d	EC(0.982); γ			tracer
^{186}Os	185.9539	1.58	stable				
^{187}Os	186.9560	1.6	stable		1/2	+0.0643	NMR
^{188}Os	187.9560	13.3	stable				
^{189}Os	188.9586	16.1	stable		3/2	+0.6565	NMR
^{190}Os	189.9586	26.4	stable				
^{191}Os		0	15.0d	β^-(0.310); γ			tracer
^{192}Os		41.0	stable				

NMR (Only OsO$_4$ studied)	^{187}Os	^{189}Os
Relative sensitivity (^{1}H = 1.00)	1.22×10^{-5}	2.34×10^{-3}
Absolute sensitivity (^{1}H = 1.00)	2.00×10^{-7}	3.76×10^{-4}
Receptivity (^{13}C = 1.00)	1.14×10^{-3}	2.13
Magnetogyric ratio/rad T^{-1} s^{-1}	0.6105×10^7	2.0773×10^7
Quadrupole moment/rad T^{-1} s^{-1}	—	0.8×10^{-28}
Frequency (^{1}H = 100 MHz; 2.3488 T)/MHz	2.282	7.758

Reference: OsO$_4$

Ground state electron configuration: [Xe]4f^{14}5d^{6}6s^2
Term symbol: $^5D_4$
Electron affinity (M$\rightarrow$M$^-$)/kJ mol^{-1}: 139

Electron shell properties

Ionization energies/kJ mol^{-1}

1. M $\rightarrow$M$^+$	840	6. M$^{5+}\rightarrow$M^{6+}	(6600)
2. M$^+\rightarrow$M^{2+}	(1600)	7. M$^{6+}\rightarrow$M^{7+}	(8100)
3. M$^{2+}\rightarrow$M^{3+}	(2400)	8. M$^{7+}\rightarrow$M^{8+}	(9500)
4. M$^{3+}\rightarrow$M^{4+}	(3900)	9. M$^{8+}\rightarrow$M^{9+}	
5. M$^{4+}\rightarrow$M^{5+}	(5200)	10. M$^{9+}\rightarrow$M^{10+}	

Principal lines in atomic spectrum

Wavelength/nm	Species	Sensitivity	Application
201.814	I		
202.026	I		
290.906	I	U2	AA, AE
305.866	I		AA, AE
442.047*	I		AE

* Lowest energy transition from ground state to nearest empty orbital

Abundance: Earth's crust 0.005 p.p.m.; seawater n.a. ($<1 \times 10^{-10}$ p.p.m.?)
Biological role: None; very toxic

O

Atomic number: 8
Relative atomic mass ($^{12}C = 12.0000$): 15.9994

Chemical properties

Colourless, odourless gas. Obtained from liquid air. Very reactive and forms oxides of all elements except He, Ne, Ar, and Kr. Moderately soluble in water (30.8 cm^3 per dm^3) at 293 K. Used in steel making, metal cutting, chemical industry.

Radii/pm: O^+ 22; O^{2-} 132; covalent (single bonds) 66; van der Waals 140

Electronegativity: 3.44 (Pauling); 3.50 (Allred-Rochow)

Effective nuclear charge: 4.55 (Slater); 4.45 (Clementi); 4.04 (Froese–Fischer)

Standard reduction potentials $E^{\ominus}/V$

	0	$-I$	$-II$

Acid solutions

$$O_2 \xrightarrow{0.695} H_2O_2 \xrightarrow{1.763} H_2O \qquad O_3 \xrightarrow{1.246} O_2$$

(overall $O_2 \xrightarrow{1.229} H_2O$)

Alkaline solutions

$$O_2 \xrightarrow{-0.0649} HO_2^- \xrightarrow{0.867} OH^- \qquad O_3 \xrightarrow{2.075} O_2$$

(overall $O_2 \xrightarrow{0.401} OH^-$)

Covalent bonds r/pm E/kJ mol^{-1}

bond	r/pm	E/kJ mol^{-1}
O—O	148	146
O=O (O$_2$)	120.8	498
N—O	146	200
N=O	115	678
N≡O (NO)	106	1063

For other covalent bonds to oxygen see other elements

Oxidation states

O^{-II}	H_2O, H_3O^+, OH^-, oxides, etc.
O^{-I}	H_2O_2, peroxides
O^0	O_2, O_3
O^I	O_2F_2
O^{II}	OF_2

Physical properties

Melting point/K: 54.8
Boiling point/K: 90.188
ΔH_{fusion}/kJ mol^{-1}: 0.444
ΔH_{vap}/kJ mol^{-1}: 6.82

Thermodynamic properties (298.15 K, 0.1 MPa)

State	$\Delta_f H^{\ominus}$/kJ mol^{-1}	$\Delta_f G^{\ominus}$/kJ mol^{-1}	$S^{\ominus}$/J K^{-1} mol^{-1}	C_p/J K^{-1} mol^{-1}
Gas (O$_2$)	0	0	205.138	29.355
Gas (atoms)	249.170	231.731	161.055	21.912

Density/kg m^{-3}: 2000 [solid, m.p.]; 1140 [liquid, b.p.]; 1.429 [gas, 273 K]
Thermal conductivity/W m^{-1} K^{-1}: 0.2674 [300 K]
Mass magnetic susceptibility/kg^{-1} m^3: $+1.355 \times 10^{-6}$ (g)
Molar volume/cm^3: 8.00 [54 K]

Lattice structure (cell dimensions/pm), space group

α-O$_2$ rhombic ($a = 540.3$; $b = 342.9$; $c = 508.6$; $\beta = 132.53°$), C2/m
β-O$_2$ rhombohedral ($a = 330.7$; $c = 1125.6$), R$\bar{3}$m
γ-O$_2$ cubic ($a = 683$), Pm3n
$T(\alpha \rightarrow \beta) = 23.8$ K
$T(\beta \rightarrow \gamma) = 43.8$ K

X-ray diffraction: mass absorption coefficients (μ/ρ)/cm^2 g^{-1}:
CuK$_\alpha$ 11.5 MoK$_\alpha$ 1.31

Oxygen

Atomic number: 8

Thermal neutron capture cross-section/barns: 0.178×10^{-3} (^{16}O)

Number of isotopes (including nuclear isomers): 8

Isotope mass range: $13 \rightarrow 20$

Key isotopes

Nuclide	Atomic mass	Natural abundance (%)	Half life $T_{1/2}$	Decay mode and energy (MeV)	Nuclear spin I	Nuclear magnetic moment μ	Uses
^{16}O	15.99491	99.762	stable		0		
^{17}O	16.9991	0.038	stable		5/2	−1.8937	NMR
^{18}O	17.9992	0.200	stable		0		

No long-lived radioactive isotopes: ^{15}O has $T_{1/2} = 124$ s

NMR

^{17}O

Relative sensitivity (^{1}H = 1.00)	2.91×10^{-2}
Absolute sensitivity (^{1}H = 1.00)	1.08×10^{-5}
Receptivity (^{13}C = 1.00)	0.061
Magnetogyric ratio/rad T^{-1} s^{-1}	-3.6264×10^7
Quadrupole moment/m^2	-2.6×10^{-30}
Frequency (^{1}H = 100 MHz; 2.3488 T)/MHz	13.557
Reference: H$_2$O	

Ground state electron configuration: [He]$2s^2 2p^4$

Term symbol: ^{3}P$_2$

Electron affinity (M$\rightarrow$M$^-$)/kJ mol^{-1}: 141

Ionization energies/kJ mol^{-1}

1. M $\rightarrow$M$^+$	1313.9	6. M$^{5+}\rightarrow$M^{6+}	13326.2
2. M$^+\rightarrow$M^{2+}	3388.2	7. M$^{6+}\rightarrow$M^{7+}	71333.3
3. M$^{2+}\rightarrow$M^{3+}	5300.3	8. M$^{7+}\rightarrow$M^{8+}	84076.3
4. M$^{3+}\rightarrow$M^{4+}	7469.1		
5. M$^{4+}\rightarrow$M^{5+}	10989.3		

Principal lines in atomic spectrum

Wavelength/nm	Species	Sensitivity	Application
135.852*			
777.193	I	U2	AE
777.414	I	U1	AE
777.543	I	U3	AE
844.636			

* Lowest energy transition from ground state to nearest empty orbital

Abundance: Atmosphere 209 500 p.p.m. (volume); Earth's crust 455 000 p.p.m.; seawater, constituent

Biological role: Basis of life (part of DNA)

Pd	Atomic number: **46**
	Relative atomic mass ($^{12}C = 12.0000$): **106.42**

Chemical properties

Silvery-white metal, lustrous, malleable, ductile. Extracted from Cu–Zn ores. Resists corrosion, dissolves in oxidising acids and fused alkalis. Readily absorbs hydrogen gas. Main use is as catalyst.

Radii/pm: Pd^{2+} 86; Pd^{4+} 64; atomic 137.6; covalent 128

Electronegativity: 2.20 (Pauling); 1.35 (Allred-Rochow)

Effective nuclear charge: 4.05 (Slater); 7.84 (Clementi); 11.11 (Froese–Fischer)

Standard reduction potentials $E^{\ominus}$/V

	VI		IV		II		0
Acid solution			PdO_2	$\xrightarrow{1.263}$	Pd^{2+}	$\xrightarrow{0.915}$	Pd
Alkaline solution	'PdO$_3$'	$\xrightarrow{2.03}$	PdO_2	$\xrightarrow{1.283}$	$Pd(OH)_2$	$\xrightarrow{-0.19}$	Pd

Oxidation states

Pd^0 (d^{10})	$[Pd(PPh_3)_3]$, $[Pd(PF_3)_4]$
Pd^{II} (d^8)	PdO, $[Pd(H_2O)_4]^{2+}$ (aq); PdF_3, $PdCl_2$ etc., $PdCl_4^{2-}$, salts, complexes
Pd^{IV} (d^6)	PdO_2, PdF_4, $PdCl_6^{2-}$

Physical properties

Melting point/K: 1825
Boiling point/K: 3413
ΔH_{fusion}/kJ mol^{-1}: 17.2
ΔH_{vap}/kJ mol^{-1}: 361.5

Thermodynamic properties (298.15 K, 0.1 MPa)

State	$\Delta_f H^{\ominus}$/kJ mol^{-1}	$\Delta_f G^{\ominus}$/kJ mol^{-1}	$S^{\ominus}$/J K^{-1} mol^{-1}	C_p/J K^{-1} mol^{-1}
Solid	0	0	37.57	25.98
Gas	378.2	339.7	167.05	20.786

Density/kg m^{-3}: 12 020 [293 K]; 10 379 [liquid at m.p.]
Thermal conductivity/W m^{-1} K^{-1}: 71.8 [300 K]
Electrical resistivity/Ω m: 10.8×10^{-8} [293 K]
Mass magnetic susceptibility/kg^{-1} m^3: $+6.702 \times 10^{-8}$ (s)
Molar volume/cm^3: 8.85
Coefficient of linear thermal expansion/K^{-1}: 11.2×10^{-6}

Lattice structure (cell dimensions/pm), space group

f.c.c. ($a = 389.08$), Fm3m

X-ray diffraction: mass absorption coefficients (μ/ρ)/cm^2 g^{-1}:
CuK$_\alpha$ 206   MoK$_\alpha$ 24.1

Palladium

Atomic number: 46
Thermal neutron capture cross-section/barns: 6.0 ± 1.0
Number of isotopes (including nuclear isomers): 21
Isotope mass range: $95 \rightarrow 115$

Nuclear properties

Key isotopes

Nuclide	Atomic mass	Natural abundance (%)	Half life $T_{1/2}$	Decay mode and energy (MeV)	Nuclear spin I	Nuclear magnetic moment μ	Uses
^{102}Pd	101.9049	1.02	stable				
^{103}Pd		0	17d	EC(0.56); γ			tracer
^{104}Pd	103.9036	11.14	stable				
^{105}Pd	104.9046	22.33	stable		5/2	-0.642	NMR
^{106}Pd	105.9032	27.33	stable				
^{108}Pd	107.903 89	26.46	stable				
^{109}Pd		0	13.47h	β^-(1.115); γ			tracer
^{110}Pd		11.72	stable				

NMR

^{105}Pd   (Only K_2PdCl_6 recorded)

Relative sensitivity ($^1H = 1.00$) 1.12×10^{-3}
Absolute sensitivity ($^1H = 1.00$) 2.49×10^{-4}
Receptivity ($^{13}C = 1.00$)   1.41
Magnetogyric ratio/rad T^{-1} s^{-1} -0.756×10^7
Quadrupole moment/m^2 $+0.8 \times 10^{-28}$
Frequency ($^1H = 100$ MHz; 2.3488 T)/MHz 4.576

Ground state electron configuration: $[Kr]4d^{10}$
Term symbol: $^1S_0$
Electron affinity ($M \rightarrow M^-$)/kJ mol^{-1}: 98.4

Electron shell properties

Ionization energies/kJ mol^{-1}

1. $M \rightarrow M^+$ 805	6. $M^{5+} \rightarrow M^{6+}$ (8 700)	
2. $M^+ \rightarrow M^{2+}$ 1875	7. $M^{6+} \rightarrow M^{7+}$ (10 700)	
3. $M^{2+} \rightarrow M^{3+}$ 3177	8. $M^{7+} \rightarrow M^{8+}$ (12 700)	
4. $M^{3+} \rightarrow M^{4+}$ (4700)	9. $M^{8+} \rightarrow M^{9+}$ (15 000)	
5. $M^{4+} \rightarrow M^{5+}$ (6300)	10. $M^{9+} \rightarrow M^{10+}$ (17 200)	

Principal lines in atomic spectrum

Wavelength/nm	Species	Sensitivity	Application
244.791	I		AA
247.642	I		AA
276.308*	I		AA
340.458	I	U1	AA, AE
342.124	I	U2	AE
363.470	I	U3	AE

* Lowest energy transition from ground state to nearest empty orbital

Abundance: Earth's crust 0.015 p.p.m.; seawater n.a. ($< 1 \times 10^{10}$ p.p.m.?)
Biological role: None; non-toxic

Atomic number: 15

Relative atomic mass ($^{12}C = 12.0000$): 30.973 762

Chemical properties

White phosphorus (P_4) is soft and flammable, red phosphorus is powdery and usually non-flammable. Neither form reacts with water or dilute acid but alkalis react to form phosphine gas. Occurs as vast deposits of apatite, $Ca_5(PO_4)_3F$. Used in fertilizers, insecticides, etc.

Radii/pm: P^{3+} 44; atomic 93 (white) 115 (red); covalent 110; van der Waals 190; P^{3-} 212

Electronegativity: 2.19 (Pauling); 2.06 (Allred-Rochow)

Effective nuclear charge: 4.80 (Slater); 4.89 (Clementi); 5.28 (Froese–Fischer)

Standard reduction potentials $E^{\ominus}/V$

	V	'IV'	III	'I'	0	$-$II	$-$III

Acid solution

$H_3PO_4 \xrightarrow{-0.933} H_4P_2O_6 \xrightarrow{0.380} H_3PO_3 \xrightarrow{-0.499} H_3PO_2 \xrightarrow{-0.365} P \xrightarrow{-0.100} P_2H_4 \xrightarrow{-0.006} PH_3$

with brackets: -0.276 (over H_3PO_4–H_3PO_3), -0.502 (over H_3PO_3–P), -0.063 (over P–PH_3)

Alkaline solution

$PO_4^{3-} \xrightarrow{-1.12} HPO_3^{2-} \xrightarrow{-1.57} H_2PO_2^- \xrightarrow{-2.05} P \xrightarrow{-0.89} PH_3$

with brackets: -1.73 (over HPO_3^{2-}–P), -1.18 (over $H_2PO_2^-$–PH_3)

Covalent bonds r/pm E/kJ mol^{-1}

	r/pm	E/kJ mol^{-1}
P—H	144	328
P—C	185	264
P—O	164	407
P=O	145	560
P—F	157	490
P—Cl	204	319
P—P	222	209

Oxidation states

P^{-III} PH_3, Ca_3P_2
P^{-II} P_2H_4
P^0 P_4
P^{II} P_2I_4
P^{III} P_4O_6, H_3PO_3 (aq) etc., H_3PO_2 (aq), $H_2PO_3^-$ PCl_3, etc.

P^V P_4O_{10}, H_3PO_4 (aq), $H_2PO_4^-$ (aq) etc., PF_5, PCl_5, $POCl_3$, phosphates

Physical properties

Melting point/K: 317.3 (P_4); 683 (red) under pressure
Boiling point/K: 553 (P_4)
ΔH_{fusion}/kJ mol^{-1}: 2.51 (P_4)
ΔH_{vap}/kJ mol^{-1}: 51.9 (P_4)

Thermodynamic properties (298.15 K, 0.1 MPa)

State	$\Delta_f H^{\ominus}$/kJ mol^{-1}	$\Delta_f G^{\ominus}$/kJ mol^{-1}	$S^{\ominus}$/J K^{-1} mol^{-1}	C_p/J K^{-1} mol^{-1}
Solid (P_4)	0	0	41.09	23.840
Solid (red)	-17.6	-12.1	22.80	21.21
Gas	314.64	278.25	163.193	20.786

Density/kg m^{-3}: 1820 (P_4); 2200 (red); 2690 (black) [293 K];
Thermal conductivity/W m^{-1} K^{-1}: 0.235 (P_4); 12.1 (black) [300 K]
Electrical resistivity/Ω m: 1×10^9 (P_4) [293 K]
Mass magnetic susceptibility/kg^{-1} m^3: -1.1×10^{-8} (P_4); -8.4×10^{-9} (red)
Molar volume/cm^3: 17.02 (P_4)
Coefficient of linear thermal expansion/K^{-1}: 124.5×10^{-6} (P_4)

Lattice structure (cell dimensions/pm), space group

α-P_4 white, cubic ($a = 1851$), $I\bar{4}3m$
β-P_4 white, rhombohedral ($a = 337.7$; $c = 880.6$), $R\bar{3}m$ [high pressure form]
γ-P_4 white, cubic ($a = 237.7$), $Pm3m$ [high pressure form]
Red, cubic ($a = 1131$) $Pm3m$ or $P\bar{4}3$
Hittorf's phosphorus (red), monoclinic ($a = 921$; $b = 915$; $c = 2260$; $\beta = 106.1°$), $P2/c$
Black, orthorhombic ($a = 331.36$; $b = 1047.8$; $c = 437.63$), $Cmca$

X-ray diffraction: mass absorption coefficients (μ/ρ)/cm^2 g^{-1}:
CuK$_\alpha$ 74.1   MoK$_\alpha$ 7.89

Phosphorus

Atomic number: 15
Thermal neutron capture cross-section/barns: 0.190
Number of isotopes (including nuclear isomers): 7
Isotope mass range: $28 \rightarrow 34$

Key isotopes

Nuclide	Atomic mass	Natural abundance (%)	Half life $T_{1/2}$	Decay mode and energy (MeV)	Nuclear spin I	Nuclear magnetic moment μ	Uses
^{31}P	30.973 76	100	stable		1/2	$+1.1317$	NMR
^{32}P	31.9739	0	14.3d	β^- (1.710); no γ	1	-0.2523	tracer, medical
^{33}P	32.9717	0	25d	β^- (0.248); no γ			

NMR ^{31}P

Relative sensitivity ($^1H = 1.00$) 6.63×10^{-2}
Absolute sensitivity ($^1H = 1.00$) 6.63×10^{-2}
Receptivity ($^{13}C = 1.00$)   377
Magnetogyric ratio/rad T^{-1} s^{-1} 10.8289×10^7
Frequency ($^1H = 100$ MHz; 2.3488 T)/MHz   40.481
References: 85% H_3PO_4

Ground state electron configuration: $[Ne]3s^2 3p^3$
Term symbol: $^4S_{3/2}$
Electron affinity ($M \rightarrow M^-$)/kJ mol^{-1}: 71.7

Ionization energies/kJ mol^{-1}

1. $M \rightarrow M^+$ 1011.7	6. $M^{5+} \rightarrow M^{6+}$ 21 268	
2. $M^+ \rightarrow M^{2+}$ 1903.2	7. $M^{6+} \rightarrow M^{7+}$ 25 397	
3. $M^{2+} \rightarrow M^{3+}$ 2912	8. $M^{7+} \rightarrow M^{8+}$ 29 854	
4. $M^{3+} \rightarrow M^{4+}$ 4956	9. $M^{8+} \rightarrow M^{9+}$ 35 867	
5. $M^{4+} \rightarrow M^{5+}$ 6273	10. $M^{9+} \rightarrow M^{10+}$ 40 958	

Principal lines in atomic spectrum

Wavelength/nm	Species	Sensitivity	Application
177.499* (vac)	I		AA but outside
178.287 (vac)	I		range of most
178.768 (vac)	I		equipment
213.547	I		AA
213.618	I		AA
253.565	I	U2	AE
255.328	I	U3	AE

* Lowest energy transition from ground state to nearest empty orbital

Abundance: Earth's crust 1120 p.p.m.; seawater 0.07 p.p.m. (low at surface)
Biological role: Basis of life (part of DNA); phosphate cycle in nature; P_4 very toxic

<table>
<tr><td>**Pt**</td><td>**Atomic number: 78**
Relative atomic mass ($^{12}C = 12.0000$): 195.08</td></tr>
</table>

<table>
<tr><td>**Chemical properties**</td><td>Silvery-white metal, lustrous, malleable, ductile. Extracted from Cu–Ni ores. Unaffected by oxygen and water, only dissolves in aqua-regia and fused alkalis. Used in jewellery, drugs, catalysts.</td></tr>
</table>

Radii/pm: Pt^{2+} 85; Pt^{4+} 70; atomic 138; covalent 129

Electronegativity: 2.28 (Pauling); 1.44 (Allred-Rochow)

Effective nuclear charge: 4.05 (Slater); 10.75 (Clementi); 15.65 (Froese–Fischer)

Standard reduction potentials $E^{\ominus}/V$

VI		IV		II		0
PtO_3	—2.0—	PtO_2	—1.045—	PtO	—0.980—	Pt
		PtO_2	—0.837—	Pt^{2+}	—1.188—	Pt
		$PtCl_6^{2-}$	—0.726—	$PtCl_4^{2-}$	—0.758—	Pt

Oxidation states

Pt^0 (d^{10})	$[Pt(PPh_3)_3]$ $[Pt(PF_3)_4]$
Pt^{II} (d^8)	PtO, $PtCl_2$, $PrBr_3$, PtI_2, $PtCl_4^{2-}$, $[Pt(CN)_4]^{2-}$, complexes
Pt^{IV} (d^6)	PtO_2, $[Pt(OH)_6]^{2-}$ (aq), PtF_4, $PtCl_4$ etc., $PtCl_6^{2-}$ complexes
Pt^V (d^5)	$(PtF_5)_4$, PtF_6^-
Pt^{VI} (d^4)	PtO_3, PtF_6

<table>
<tr><td>**Physical properties**</td><td>**Melting point**/K: 2045
Boiling point/K: 4100 ± 100
ΔH_{fusion}/kJ mol^{-1}: 19.7
ΔH_{vap}/kJ mol^{-1}: 469</td></tr>
</table>

Thermodynamic properties (298.15 K, 0.1 MPa)

State	$\Delta_f H^{\ominus}$/kJ mol^{-1}	$\Delta_f G^{\ominus}$/kJ mol^{-1}	$S^{\ominus}$/J K^{-1} mol^{-1}	C_p/J K^{-1} mol^{-1}
Solid	0	0	41.63	25.86
Gas	565.3	520.5	192.406	25.531

Density/kg m^{-3}: 21 450 [293 K]

Thermal conductivity/W m^{-1} K^{-1}: 71.6 [300 K]

Electrical resistivity/Ω m: 10.6×10^{-8} [293 K]

Mass magnetic susceptibility/kg^{-1} m^3: $+1.301 \times 10^{-8}$ (s)

Molar volume/cm^3: 9.10

Coefficient of linear thermal expansion/K^{-1}: 9.0×10^{-6}

Lattice structure (cell dimensions/pm), space group

f.c.c. ($a = 392.40$), Fm3m

X-ray diffraction: mass absorption coefficients (μ/ρ)/cm^2 g^{-1}:
CuK_{α} 200 MoK_{α} 113

Platinum

Atomic number: 78
Thermal neutron capture cross-section/barns: 9 ± 1
Number of isotopes (including nuclear isomers): 32
Isotope mass range: $173 \rightarrow 200$

Key isotopes

Nuclide	Atomic mass	Natural abundance (%)	Half life $T_{1/2}$	Decay mode and energy (MeV)	Nuclear spin I	Nuclear magnetic moment μ	Uses
^{190}Pt	189.960	0.010	6.9×10^{11}y	α (3.18)			
^{192}Pt	191.9614	0.79	$c.\ 10^{15}$y	α			
^{194}Pt	193.9628	32.9	stable		0		
^{195m}Pt		0	4.1d	IT(0.259); γ		± 0.60	tracer
^{195}Pt	194.9648	33.8	stable		1/2	$+0.6022$	NMR
^{196}Pt	195.9650	25.3	stable		0		
^{197}Pt		0	18h	β^- (0.75); γ	1/2	± 0.5	tracer
^{198}Pt	197.9675	7.2	stable				

NMR

^{195}Pt

Relative sensitivity (^{1}H $= 1.00$) 9.94×10^{-3}
Absolute sensitivity (^{1}H $= 1.00$) 3.36×10^{-3}
Receptivity (^{13}C $= 1.00$)   19.1
Magnetogyric ratio/rad T^{-1} s^{-1} 5.7412×10^7
Frequency (^{1}H $= 100$ MHz; 2.3488 T)/MHz   21.499
Reference: $[Pt(CN)_6]^{2-}$

Ground state electron configuration: $[Xe]4f^{14}5d^96s^1$
Term symbol: $^3D_3$
Electron affinity (M$\rightarrow$M$^-$)/kJ mol^{-1}: 247

Ionization energies/kJ mol^{-1}

1. M $\rightarrow$M$^+$	870	6. M$^{5+}\rightarrow$M^{6+}	(7 200)
2. M$^+\rightarrow$M^{2+}	1791	7. M$^{6+}\rightarrow$M^{7+}	(8 900)
3. M$^{2+}\rightarrow$M^{3+}	(2800)	8. M$^{7+}\rightarrow$M^{8+}	(10 500)
4. M$^{3+}\rightarrow$M^{4+}	(3900)	9. M$^{8+}\rightarrow$M^{9+}	(12 300)
5. M$^{4+}\rightarrow$M^{5+}	(5300)	10. M$^{9+}\rightarrow$M^{10+}	(14 100)

Principal lines in atomic spectrum

Wavelength/nm	Species	Sensitivity	Application
204.937	I		
265.945	I	U2	AA, AE
283.030	I		AA, AE
299.767	I		AA, AE
306.471	I	U1	AA, AE
331.505*	I		

* Lowest energy transition from ground state to nearest empty orbital

Abundance: Earth's crust 0.01 p.p.m.; seawater n.a. ($< 1 \times 10^{-10}$ p.p.m.?)
Biological role: None; non-toxic

<table>
<tr><td>## Pu</td><td>**Atomic number: 94**
Relative atomic mass ($^{12}C = 12.0000$): **(244)**</td></tr>
</table>

<table>
<tr><td>**Chemical properties**</td><td>Radioactive silvery metal obtained in tonne quantities from uranium fuel elements. Attacked by oxygen, steam, and acids, but not alkalis. Used as compact energy source, nuclear fuel, and for nuclear weapons.</td></tr>
</table>

Radii/pm: Pu^{3+} 108; Pu^{4+} 93; Pu^{5+} 87; Pu^{6+} 81; atomic 151 (α form)

Electronegativity: 1.28 (Pauling); 1.22 (Allred-Rochow)

Effective nuclear charge: 1.65 (Slater)

Standard reduction potentials $E^{\ominus}$/V

	VII	VI	V	IV	III	0

Acid solution

$$PuO_2^{2+} \xrightarrow{1.02} PuO_2^{+} \xrightarrow{1.04} Pu^{4+} \xrightarrow{1.01} Pu^{3+} \xrightarrow{1.584} Pu$$

with $PuO_2^{2+} \xrightarrow{1.03} Pu^{4+}$ and $Pu^{4+} \xrightarrow{-1.25} Pu^{3+}$ brackets

Alkaline solution

$$PuO_5^{3-} \xrightarrow{0.95} Pu_2(OH)_3^{-} \xrightarrow{0.3} PuO_2(OH) \xrightarrow{0.9} PuO_2 \xrightarrow{-1.4} Pu(OH)_3 \xrightarrow{2.46} Pu$$

Oxidation states

Pu^{II} (f^6)	PuO, PuH_2
Pu^{III} (f^5)	Pu_2O_3, PuF_3 $PuCl_3$ etc., $[Pu(H_2O)_x]^{3+}$ (aq), Pu^{3+} salts, complexes
Pu^{IV} (f^4)	PuO_2, PuF_4, $[PuCl_6]^{2-}$, $[Pu(H_2O)_x]^{4+}$ (aq) unstable, complexes
Pu^{V} (f^3)	PuO_2^{+} (aq) unstable, $CsPuF_6$
Pu^{VI} (f^2)	PuO_2^{2+} (aq), PuF_6
Pu^{VII} (f^1)	Li_5PuO_6, $[PuO_5]^{3-}$ (aq)

<table>
<tr><td>**Physical properties**</td><td>**Melting point**/K: 914 ΔH_{fusion}/kJ mol^{-1}: 2.8
Boiling point/K: 3505 ΔH_{vap}/kJ mol^{-1}: 343.5</td></tr>
</table>

Thermodynamic properties (298.15 K, 0.1 MPa)

State	$\Delta_f H^{\ominus}$/kJ mol^{-1}	$\Delta_f G^{\ominus}$/kJ mol^{-1}	$S^{\ominus}$/J K^{-1} mol^{-1}	C_p/J K^{-1} mol^{-1}
Solid	0	0	n.a.	n.a.
Gas	n.a.	n.a.	n.a.	n.a.

Density/kg m^{-3}: 19 840 (α) [298 K]; 16 623 [liquid at m.p.]

Thermal conductivity/W m^{-1} K^{-1}: 6.74 [300 K]

Electrical resistivity/Ω m: 146×10^{-8} [273 K]

Mass magnetic susceptibility/kg^{-1} m^3: $+3.17 \times 10^{-8}$ (s)

Molar volume/cm^3: 12.3

Coefficient of linear thermal expansion/K^{-1}: 55×10^{-6}

Lattice structure (cell dimensions/pm), space group

α-Pu monoclinic ($a = 618.3$; $b = 482.2$; $c = 1096.3$; $\beta = 101.79°$), P2$_1$/m
β-Pu monoclinic ($a = 928.4$; $b = 1046.3$; $c = 785.9$; $\beta = 92.13°$), I2/m
γ-Pu orthorhombic ($a = 315.87$; $b = 576.82$; $c = 1016.2$), Fddd
δ-Pu f.c.c. ($a = 463.71$), Fm3m
δ'-Pu tetragonal ($a = 333.9$; $c = 444.6$), I4/mmm
ε-Pu b.c.c. ($a = 363.48$), Im3m
$T(\alpha \rightarrow \beta) = 395$ K; $(\beta \rightarrow \gamma) = 473$ K; $(\gamma \rightarrow \delta) = 583$ K; $(\delta \rightarrow \delta') = 725$ K; $(\delta' \rightarrow \varepsilon) = 753$ K

X-ray diffraction: mass absorption coefficients (μ/ρ)/cm^2 g^{-1}: n.a.

Discovered in 1940 by G. T. Seaborg, A. C. Wahl, and J. W. Kennedy at Berkeley, California, USA

[Named after the planet Pluto]

Plutonium

Atomic number: 94

Thermal neutron capture cross-section/barns: 1.8 ± 0.3 (^{244}Pu)

Number of isotopes (including nuclear isomers): 16

Isotope mass range: $232 \rightarrow 246$

Nuclear properties

Key isotopes

Nuclide	Atomic mass	Natural abundance (%)	Half life $T_{1/2}$	Decay mode and energy (MeV)	Nuclear spin I	Nuclear magnetic moment μ	Uses
^{239}Pu	239.0522	0	24400y	α(5.243), SF, γ	1/2	+0.200	NMR
^{242}Pu	242.0587	0	3.79×10^5y	α(4.98)			
^{244}Pu	244.0642	0	8.2×10^7y	α(4.66), SF			

NMR
^{239}Pu

Relative sensitivity (^{1}H = 1.00) —

Absolute sensitivity (^{1}H = 1.00) —

Receptivity (^{13}C = 1.00) —

Magnetogyric ratio/rad T^{-1} s^{-1} 0.972×10^7

Quadrupole moment/m^2 —

Frequency (^{1}H = 100 MHz; 2.3488 T)/MHz   3.63

Ground state electron configuration: [Rn]5f^{6}7s^2

Term symbol: $^7F_0$

Electron affinity (M$\rightarrow$M$^-$)/kJ mol^{-1}: n.a.

Electron shell properties

Ionization energies/kJ mol^{-1}

1. M $\rightarrow$ M$^+$ 585	6. M^{5+} $\rightarrow$ M^{6+}
2. M$^+$ $\rightarrow$ M^{2+}	7. M^{6+} $\rightarrow$ M^{7+}
3. M^{2+} $\rightarrow$ M^{3+}	8. M^{7+} $\rightarrow$ M^{8+}
4. M^{3+} $\rightarrow$ M^{4+}	9. M^{8+} $\rightarrow$ M^{9+}
5. M^{4+} $\rightarrow$ M^{5+}	10. M^{9+} $\rightarrow$ M^{10+}

Principal lines in atomic spectrum

Wavelength/nm	Species	Sensitivity	Application
321.508	I		
324.416	I		
325.208	I		
327.524	I		
329.256	I		
329.361	I		
329.691	I		

Abundance: Earth's crust: traces in uranium ores; seawater: nil

Biological role: None; toxic due to radioactivity

Po

Atomic number: 84
Relative atomic mass ($^{12}C = 12.0000$): (209)

Chemical properties

Radioactive silver-grey metal produced in gramme amounts from neutron bombardment of bismuth. Soluble in dilute acids. Used as heat source in space equipment and as source of α radiation for research.

Radii/pm: Po^{4+} 65; atomic 167; covalent 153; Po^{2-} 230
Electronegativity: 2.0 (Pauling); 1.76 (Allred-Rochow)
Effective nuclear charge: 6.95 (Slater); 14.22 (Clementi); 18.31 (Froese–Fischer)

Standard reduction potentials $E^{\ominus}$/V

	VI	IV	II	0	−II

Acid solution: PoO_3 —1.51— PoO_2 —1.1— Po^{2+} —0.37— Po —c. −1.0— H_2Po
(1.3 from PoO_3 to Po^{2+}; 0.73 from PoO_2 to Po)

Alkaline solution: PoO_3 —1.48— PoO_3^{2-} —−0.5— Po —c. −1.4— Po^{2-}
(0.16 from PoO_3 to PoO_3^{2-})

Covalent bonds r/pm E/kJ mol^{-1}

Bond	r/pm	E/kJ mol^{-1}
Po–Cl	238	n.a.
Po–Po	335	n.a.

Oxidation states

Po^{-II}	H_2Po, Na_2Po
Po^{II}	PoO, $PoCl_2$, $PoBr_2$
Po^{IV}	PoO_2, PoO_3^{2-}(aq), $PoCl_4$, $PoBr_4$, PoI_4, PoI_6^{2-}
Po^{VI}	PoO_3 ? PoF_6

Physical properties

Melting point/K: 527
Boiling point/K: 1235
ΔH_{fusion}/kJ mol^{-1}: 10
ΔH_{vap}/kJ mol^{-1}: 100.8

Thermodynamic properties (298.15 K, 0.1 MPa)

State	$\Delta_f H^{\ominus}$/kJ mol^{-1}	$\Delta_f G^{\ominus}$/kJ mol^{-1}	$S^{\ominus}$/J K^{-1} mol^{-1}	C_p/J K^{-1} mol^{-1}
Solid	0	0	n.a.	26.1
Gas	146?	n.a.	n.a.	n.a.

Density/kg m^{-3}: 9320 (α) [293 K]
Thermal conductivity/W m^{-1} K^{-1}: 20 [300 K]
Electrical resistivity/Ω m: 140×10^{-8} [293 K]
Mass magnetic susceptibility/kg^{-1} m^3: n.a.
Molar volume/cm^3: 22.4
Coefficient of linear thermal expansion/K^{-1}: 23.0×10^{-6}

Lattice structure (cell dimensions/pm), space group

α-Po cubic ($a = 335.2$), Pm3m
β-Po rhombohedral ($a = 336.6$, $\alpha = 98°13'$), R$\bar{3}$m
$T (\alpha \to \beta) = 309$ K

X-ray diffraction: mass absorption coefficients (μ/ρ)/cm^2 g^{-1}: n.a.

Polonium

Atomic number: 84
Thermal neutron capture cross-section/barns: <0.5 (^{210}Po)
Number of isotopes (including nuclear isomers): 34
Isotope mass range: $192 \rightarrow 218$

Nuclear properties

Key isotopes

Nuclide	Atomic mass	Natural abundance (%)	Half life $T_{1/2}$	Decay mode and energy (MeV)	Nuclear spin I	Nuclear magnetic moment μ	Uses
^{209}Po	208.9825	0	103y	$\alpha(4.98)$ 99%; EC; γ	1/2	+0.77	
^{210}Po	209.9829	trace	138.4d	$\alpha(5.408)$; γ	0		tracer, fuel
^{211}Po	210.9866	trace	0.52s	$\alpha(7.592)$; γ			
^{216}Po	216.0019	trace	0.15s	$\alpha(6.906)$; no γ			
^{218}Po	218.0089	trace	3.05m	$\alpha(6.111)$; no γ			

Ground state electron configuration: $[Xe]4f^{14}5d^{10}6s^26p^4$
Term symbol: $^3P_2$
Electron affinity $(M \rightarrow M^-)$/kJ mol^{-1}: 186

Electron shell properties

Ionization energies/kJ mol^{-1}

1. $M \rightarrow M^+$ 812	6. $M^{5+} \rightarrow M^{6+}$ (7 000)	
2. $M^+ \rightarrow M^{2+}$ (1800)	7. $M^{6+} \rightarrow M^{7+}$ (10 800)	
3. $M^{2+} \rightarrow M^{3+}$ (2700)	8. $M^{7+} \rightarrow M^{8+}$ (12 700)	
4. $M^{3+} \rightarrow M^{4+}$ (3700)	9. $M^{8+} \rightarrow M^{9+}$ (14 900)	
5. $M^{4+} \rightarrow M^{5+}$ (5900)	10. $M^{9+} \rightarrow M^{10+}$ (17 000)	

Principal lines in atomic spectrum

Wavelength/nm	Species	Sensitivity	Application
245.008	I		
255.801*	I		
300.321	I		
417.052	I		

* Lowest energy transition from ground state to nearest empty orbital

Abundance: Earth's crust 3×10^{-10}; seawater nil
Biological role: None; highly toxic due to radioactivity

K	Atomic number: **19**
	Relative atomic mass ($^{12}C = 12.0000$): **39.0983**

Chemical properties

Soft white metal, silvery when cut but reacts rapidly with oxygen and vigorously with water. Large deposits of sylvite (KCl) and sylvinite (NaCl, KCl). Metal obtained from Na + KCl at 1100 K. Used in fertilizers, chemicals, and glass.

Radii/pm: K$^+$ 133; atomic 227; covalent 203; van der Waals 231

Electronegativity: 0.82 (Pauling); 0.91 (Allred-Rochow)

Effective nuclear charge: 2.20 (Slater); 3.50 (Clementi); 4.58 (Froese–Fischer)

Standard reduction potentials $E^{\ominus}/V$

I	0
K$^+$ $\xrightarrow{-2.924}$	K

Oxidation states

K^{-1} (s^2)	solution in liquid ammonia
K^{I} ([Ar])	K$_2$O, K$_2$O$_2$ (peroxide)
	KO$_2$ (superoxide), KO$_3$ (ozonide),
	KOH, [K(H$_2$O)$_4$]$^+$ (aq),
	KH, KF, KCl etc.,
	K$^+$ salts, K$_2$CO$_3$, complexes

Physical properties

Melting point/K: 336.80
Boiling point/K: 1047
ΔH_{fusion}/kJ mol^{-1}: 2.40
ΔH_{vap}/kJ mol^{-1}: 79.1

Thermodynamic properties (298.15 K, 0.1 MPa)

State	$\Delta_f H^{\ominus}$/kJ mol^{-1}	$\Delta_f G^{\ominus}$/kJ mol^{-1}	$S^{\ominus}$/J K^{-1} mol^{-1}	C_p/J K^{-1} mol^{-1}
Solid	0	0	64.18	29.58
Gas	89.24	60.59	160.336	20.786

Density/kg m^{-3}: 862 [293 K]; 828 [liquid at m.p.]
Thermal conductivity/W m^{-1} K^{-1}: 102.4 [300 K]
Electrical resistivity/Ω m: 6.15×10^{-8} [273 K]
Mass magnetic susceptibility/kg^{-1} m^3: $+6.7 \times 10^{-9}$ (s)
Molar volume/cm^3: 45.36
Coefficient of linear thermal expansion/K^{-1}: 83×10^{-6}

Lattice structure (cell dimensions/pm), space group

b.c.c. ($a = 533.4$), Im3m

X-ray diffraction: mass absorption coefficients (μ/ρ)/cm^2 g^{-1}:
CuK$_\alpha$ 143   MoK$_\alpha$ 15.8

Atomic number: 19
Thermal neutron capture cross-section/barns: 2.2 ± 0.2
Number of isotopes (including nuclear isomers): 10
Isotope mass range: $37 \rightarrow 45$

Nuclear properties

Key isotopes

Nuclide	Atomic mass	Natural abundance (%)	Half life $T_{1/2}$	Decay mode and energy (MeV)	Nuclear spin I	Nuclear magnetic moment μ	Uses
^{39}K	38.963 71	93.2581	stable		3/2	+0.3914	NMR
^{40}K	39.974	0.0117	1.28×10^9y	β^- (1.35); EC; γ 4		−1.2981	
^{41}K	40.962	6.7302	stable		3/2	+0.2149	NMR
^{42}K	41.963	0	12.4h	β^- (0.60); γ	2	−1.1424	tracer, medical
^{43}K	42.964	0	22.4h	β^- (1.82); γ	3/2	+0.163	

NMR

	^{39}K	$[^{41}K]$
Relative sensitivity ($^1H = 1.00$)	5.08×10^{-4}	8.40×10^{-5}
Absolute sensitivity ($^1H = 1.00$)	4.73×10^{-4}	5.78×10^{-6}
Receptivity ($^{13}C = 1.00$)	2.69	3.28×10^{-2}
Magnetogyric ratio/rad T^{-1} s^{-1}	1.2483×10^7	0.6851×10^7
Quadrupole moment /m^2	5.5×10^{-30}	6.7×10^{-30}
Frequency ($^1H = 100$ MHz; 2.3488 T)/MHz	4.667	2.561

Reference: K^+ (aq)

Ground state electron configuration: $[Ar]4s^1$
Term symbol: $^2S_{1/2}$
Electron affinity $(M \rightarrow M^-)$/kJ mol^{-1}: 48.3

Electron shell properties

Ionization energies/kJ mol^{-1}

1. $M \rightarrow M^+$	418.8	6. $M^{5+} \rightarrow M^{6+}$	9 649
2. $M^+ \rightarrow M^{2+}$	3051.4	7. $M^{6+} \rightarrow M^{7+}$	11 343
3. $M^{2+} \rightarrow M^{3+}$	4411	8. $M^{7+} \rightarrow M^{8+}$	14 942
4. $M^{3+} \rightarrow M^{4+}$	5877	9. $M^{8+} \rightarrow M^{9+}$	16 964
5. $M^{4+} \rightarrow M^{5+}$	7975	10. $M^{9+} \rightarrow M^{10+}$	48 575

Principal lines in atomic spectrum

Wavelength/nm	Species	Sensitivity	Application
404.414	I	U3	AA, AE
404.720	I	U4	AA, AE
766.491	I	U1	AA, AE
769.898*	I	U2	AA, AE

* Lowest energy transition from ground state to nearest empty orbital

Abundance: Earth's crust 18 400 p.p.m.; seawater 380 p.p.m.
Biological role: Essential; non-toxic

Atomic number: **59**
Relative atomic mass ($^{12}C=12.0000$): **140.90765**

Chemical properties

Soft, malleable, silvery metal of the lanthanide (rare earth) group. Obtained from monazite and bastnaesite. Reacts slowly with oxygen, rapidly with water. Used in alloys, flints, yellow glass for eye protection for welders, etc.

Radii/pm: Pr^{3+} 106; Pr^{4+} 92; atomic 182.8; covalent 165
Electronegativity: 1.13 (Pauling); 1.07 (Allred–Rochow)
Effective nuclear charge: 2.85 (Slater); 7.75 (Clementi); 10.70 (Froese–Fischer)

Standard reduction potentials $E^{\ominus}/V$

	IV		III		0
Acid solution	Pr^{4+}	—3.2—	Pr^{3+}	—-2.35—	Pr
Alkaline solution	PrO_2	—0.8—	$Pr(OH)_3$	—-2.79—	Pr

Oxidation states

Pr^{III} (f^2)	P_2O_3, $Pr(OH)_3$, $[Pr(H_2O)_x]^{3+}$ (aq), Pr^{3+} salts, PrF_3, $PrCl_3$ etc., complexes
Pr^{IV} (f^1)	PrO_2, PrF_4, Na_2PrF_6

Physical properties

Melting point/K: 1204
Boiling point/K: 3785
ΔH_{fusion}/kJ mol^{-1}: 11.3
ΔH_{vap}/kJ mol^{-1}: 357

Thermodynamic properties (298.15 K, 0.1 MPa)

State	$\Delta_f H^{\ominus}$/kJ mol^{-1}	$\Delta_f G^{\ominus}$/kJ mol^{-1}	$S^{\ominus}$/J K^{-1} mol^{-1}	C_p/J K^{-1} mol^{-1}
Solid	0	0	73.2	27.20
Gas	355.6	320.9	189.808	21.359

Density/kg m^{-3}: 6773 [293 K]
Thermal conductivity/W m^{-1} K^{-1}: 12.5 [300 K]
Electrical resistivity/Ω m: 68×10^{-8} [298 K]
Mass magnetic susceptibility/kg^{-1} m^3: $+4.47 \times 10^{-7}$ (s)
Molar volume/cm^3: 20.80
Coefficient of linear thermal expansion/K^{-1}: 6.79×10^{-6}

Lattice structure (cell dimensions/pm), space group

α-Pr h.c.p. ($a=367.25$, $c=1183.5$), P6$_3$/mmc
β-Pr b.c.c. ($a=413$), Im3m

$T(\alpha \rightarrow \beta) = 1065$ K

X-ray diffraction: mass absorption coefficients (μ/ρ)/cm^2 g^{-1}:
CuK$_\alpha$ 363   MoK$_\alpha$ 50.7

Praseodymium

Atomic number: 59
Thermal neutron capture cross-section/barns: 11.5 ± 1.0
Number of isotopes (including nuclear isomers): 15
Isotope mass range: $134 \rightarrow 148$

Nuclear properties

Key isotopes

Nuclide	Atomic mass	Natural abundance (%)	Half life $T_{1/2}$	Decay mode and energy (MeV)	Nuclear spin I	Nuclear magnetic moment μ	Uses
^{141}Pr	140.9074	100	stable		5/2	+4.16	NMR
^{142}Pr		0	19.2h	β^- (2.16); γ	2	−0.26	tracer
^{143}Pr		0	13.59d	β^- (0.933); no γ	7/2		tracer

NMR

^{141}Pr

Relative sensitivity (^{1}H = 1.00)	0.29
Absolute sensitivity (^{1}H = 1.00)	0.29
Receptivity (^{13}C = 1.00)	1.62×10^3
Magnetogyric ratio/rad T^{-1} s^{-1}	7.765×10^7
Quadrupole moment /m^2	-5.9×10^{-30}
Frequency (^{1}H = 100 MHz; 2.3488 T)/MHz	29.291

Ground state electron configuration: $[\text{Xe}]4f^3 6s^2$
Term symbol: $^4I_{9/2}$
Electron affinity (M→M$^-$)/kJ mol^{-1}: $\leqslant 50$

Electron shell properties

Ionization energies/kJ mol^{-1}

1. M →M$^+$	523.1	6. M^{5+}→M^{6+}	
2. M$^+$→M^{2+}	1018	7. M^{6+}→M^{7+}	
3. M^{2+}→M^{3+}	2086	8. M^{7+}→M^{8+}	
4. M^{3+}→M^{4+}	3761	9. M^{8+}→M^{9+}	
5. M^{4+}→M^{5+}	5543	10. M^{9+}→M^{10+}	

Principal lines in atomic spectrum

Wavelength/nm	Species	Sensitivity	Application
406.282	II		AE
417.938	II		AE
422.535	II		AE
495.137	I		AA
491.402	I		AA

Abundance: Earth's crust 9.1 p.p.m.; seawater 2×10^{-7} p.p.m.
Biological role: None; toxicity low

Pm

Atomic number: **61**

Relative atomic mass (^{12}C$=12.0000$): **(145)**

Chemical properties

Radioactive metal of the lanthanide (rare earth) group, obtained in mg quantities from the fission products of nuclear reactors. Used in specialized miniature batteries.

Radii/pm: Pm^{3+} 106; atomic 181.0

Electronegativity: n.a. (Pauling); 1.07 (Allred-Rochow)

Effective nuclear charge: 2.85 (Slater); 9.40 (Clementi); 10.94 (Froese–Fischer)

Standard reduction potentials $E^{\ominus}$/V

	III		0
Acid solution	Pm^{3+}	$\underline{-2.29}$	Pm
Alkaline solution	Pm(OH)$_3$	$\underline{-2.76}$	Pm

Oxidation states

Pm^{III} (f^4)	Pm$_2$O$_3$, Pm(OH)$_3$, [Pm(H$_2$O)$_x$]$^{3+}$ (aq), PmF$_3$, some complexes

Physical properties

Melting point/K: 1441
Boiling point/K: *c.* 3000
ΔH_{fusion}/kJ mol^{-1}: 12.6
ΔH_{vap}/kJ mol^{-1}: n.a.

Thermodynamic properties (298.15 K, 0.1 MPa)

State	$\Delta_f H^{\ominus}$/kJ mol^{-1}	$\Delta_f G^{\ominus}$/kJ mol^{-1}	$S^{\ominus}$/J K^{-1} mol^{-1}	C_p/J K^{-1} mol^{-1}
Solid	0	0	n.a.	26.8
Gas	n.a.	n.a.	187.101	24.255

Density/kg m^{-3}: 7220 [298 K]
Thermal conductivity/W m^{-1} K^{-1}: 17.9 (est.) [300 K]
Electrical resistivity/Ω m: 50×10^{-8} (est.) [273 K]
Mass magnetic susceptibility/kg^{-1} m^3: n.a.
Molar volume/cm^3: 20.1
Coefficient of linear thermal expansion/K^{-1}: n.a.

Lattice structure (cell dimensions/pm), space group

Hexagonal

X-ray diffraction: mass absorption coefficients (μ/ρ)/cm^2 g^{-1}:
CuK$_\alpha$ 386   MoK$_\alpha$ 55.9

Produced in 1945 by J. A. Marinsky, L. E. Glendenin, and C. D. Coryell at Oak Ridge, USA

[Greek, *Prometheus*, who stole fire from the gods]

Promethium

Atomic number: 61

Thermal neutron capture cross-section/barns: 8400 ± 1680 (^{146}Pm)

Number of isotopes (including nuclear isomers): 14

Isotope mass range: $141 \rightarrow 154$

Key isotopes

Nuclide	Atomic mass	Natural abundance (%)	Half life $T_{1/2}$	Decay mode and energy (MeV)	Nuclear spin I	Nuclear magnetic moment μ	Uses
^{145}Pm	144.9128	0	17.7y	EC(0.14); γ			
^{146}Pm		0	4.4y	EC(1.72); β^-(1.53); γ			
^{147}Pm	146.9152	0	2.62y	β^-(0.225); no γ	7/2	$+2.62$	tracer
^{149}Pm		0	53.1h	β^-(1.06); γ	7/2	± 3.3	tracer
^{151}Pm		0	28h	β^-(1.19); γ	5/2	± 1.6	combination

NMR

^{147}Pm

Relative sensitivity (^{1}H $= 1.00$)   —

Absolute sensitivity (^{1}H $= 1.00$)   —

Receptivity (^{13}C $= 1.00$)   —

Magnetogyric ratio/rad T^{-1} s^{-1} 3.613×10^7

Quadrupole moment /m^2 0.67×10^{-28}

Frequency (^{1}H $= 100$ MHz; 2.3488 T)/MHz   13.51

Ground state electron configuration: [Xe]$4f^5 6s^2$

Term symbol: $^6H_{5/2}$

Electron affinity (M$\rightarrow$M$^-$)/kJ mol^{-1}: $\leqslant 50$

Ionization energies/kJ mol^{-1}

1. M $\rightarrow$ M$^+$	535.9	6. M$^{5+} \rightarrow$ M^{6+}	
2. M$^+ \rightarrow$ M^{2+}	1052	7. M$^{6+} \rightarrow$ M^{7+}	
3. M$^{2+} \rightarrow$ M^{3+}	2150	8. M$^{7+} \rightarrow$ M^{8+}	
4. M$^{3+} \rightarrow$ M^{4+}	3970	9. M$^{8+} \rightarrow$ M^{9+}	
5. M$^{4+} \rightarrow$ M^{5+}		10. M$^{9+} \rightarrow$ M^{10+}	

Principal lines in atomic spectrum

Wavelength/nm	Species	Sensitivity	Application
371.172	II		
391.026	II		
395.774	II		
398.074	II		
441.796	II		
652.045	II		

Abundance: Earth's crust, traces in uranium ores; seawater nil

Biological role: None; toxic due to radioactivity

Pa

Atomic number: **91**

Relative atomic mass (^{12}C = 12.0000): **231.03588**

Chemical properties

Radioactive, silvery metal found naturally in uranium ores and produced in gramme quantities from uranium fuel elements. Attacked by oxygen, steam, and acids, but not by alkalis. Little used.

Radii/pm: Pa^{3+} 113; Pa^{4+} 98; Pa^{5+} 89; atomic 160.6

Electronegativity: 1.5 (Pauling); 1.14 (Allred-Rochow)

Effective nuclear charge: 1.80 (Slater)

Standard reduction potentials $E^{\ominus}$/V

	V		IV		0
Acid solution	$PaO(OH)^{2+}$	$\xrightarrow{-0.1}$	Pa^{4+}	$\xrightarrow{-1.46}$	Pa

(V → 0: −1.19)

Oxidation states

Pa^{III} (f^2)	PaI_3
Pa^{IV} (f^1)	PaO_2, $[Pa(H_2O)_x]^{4+}$ (aq), PaF_4, $PaCl_4$, etc.
$\mathbf{Pa^V}$ (f^0, [Rn])	Pa_2O_5, PaO_2^+ (aq), PaF_5, $PaCl_5$, etc., $[PaF_6]^-$, $[PaF_7]^{2-}$, $[PaF_8]^{3-}$

Physical properties

Melting point/K: 2113

Boiling point/K: *c.* 4300

ΔH_{fusion}/kJ mol^{-1}: 16.7

ΔH_{vap}/kJ mol^{-1}: 481

Thermodynamic properties (298.15 K, 0.1 MPa)

State	$\Delta_f H^{\ominus}$/kJ mol^{-1}	$\Delta_f G^{\ominus}$/kJ mol^{-1}	$S^{\ominus}$/J K^{-1} mol^{-1}	C_p/J K^{-1} mol^{-1}
Solid	0	0	51.9	28
Gas	607	563	198.05	22.93

Density/kg m^{-3}: 15 370 (est.)

Thermal conductivity/W m^{-1} K^{-1}: 47 (est.) [300 K]

Electrical resistivity/Ω m: 17.7×10^{-8} [273 K]

Mass magnetic susceptibility/kg^{-1} m^3: n.a.

Molar volume/cm^3: 15.0

Coefficient of linear thermal expansion/K^{-1}: 7.3×10^{-6}

Lattice structure (cell dimensions/pm), space group

Tetragonal (a = 393.2, c = 323.8). I4/mmm

X-ray diffraction: mass absorption coefficients (μ/ρ)/cm^2 g^{-1}:
n.a.

Discovered in 1917 by Otto Hahn and Lise Meitner at Berlin, by
K. Fajans at Karlsruhe, Germany, and by F. Soddy,
J. A. Cranston, and A. Fleck at Glasgow, Scotland
[Greek, *protos* = first]

Protactinium

Atomic number: 91

Thermal neutron capture cross-section/barns: 200 ± 10 (^{231}Pa)

Number of isotopes (including nuclear isomers): 14

Isotope mass range: $225 \rightarrow 237$

Nuclear properties

Key isotopes

Nuclide	Atomic mass	Natural abundance (%)	Half life $T_{1/2}$	Decay mode and energy (MeV)	Nuclear spin I	Nuclear magnetic moment μ	Uses
^{231}Pa	231.0359	trace	3.26×10^4y	α(5.148)	3/2	± 1.98	
^{233}Pa	233.040	0	27.0d	β^-(0.571)	3/2	$+3.4$	tracer
^{234}Pa	234.043	trace	6.75h	β^-(2.23); γ			

Ground state electron configuration: $[Rn]5f^2 6d^1 7s^2$

Term symbol: $^4K_{11/2}$

Electron affinity $(M \rightarrow M^-)$/kJ mol^{-1}: n.a.

Electron shell properties

Ionization energies/kJ mol^{-1}

1. $M \rightarrow M^+$ 568	6. $M^{5+} \rightarrow M^{6+}$
2. $M^+ \rightarrow M^{2+}$	7. $M^{6+} \rightarrow M^{7+}$
3. $M^{2+} \rightarrow M^{3+}$	8. $M^{7+} \rightarrow M^{8+}$
4. $M^{3+} \rightarrow M^{4+}$	9. $M^{8+} \rightarrow M^{9+}$
5. $M^{4+} \rightarrow M^{5+}$	10. $M^{9+} \rightarrow M^{10+}$

Principal lines in atomic spectrum

Wavelength/nm	Species	Sensitivity	Application
395.785	II		
398.223	I		
694.572	I		
711.489	I		
736.825	I		
762.679	I		
763.518	I		
766.934	I		

Abundance: Earth's crust, trace in uranium ores; seawater 2×10^{-12} p.p.m.

Biological role: None; toxic due to radioactivity

Ra

Atomic number: 88
Relative atomic mass ($^{12}C = 12.0000$): 226.0254

Chemical properties

Radioactive element found naturally in uranium ores. Silvery, lustrous, soft. Annual production $c.$ 100 g. Formerly used in cancer therapy and for luminous paint; both uses now rare. Reacts with oxygen and water.

Radii/pm: Ra^{2+} 152; atomic 223
Electronegativity: 0.89 (Pauling); 0.97 (Allred-Rochow)
Effective nuclear charge: 1.65 (Slater)

Standard reduction potentials $E^{\ominus}$/V

II		0
Ra^{2+}	$\underline{-2.916}$	Ra
RaO	$\underline{-1.319}$	Ra

Oxidation states

Ra^{II} ([Rn])	RaO, $Ra(OH)_2$, $[Ra(H_2O)_x]^{2+}$ (aq), Ra^{2+} salts

Physical properties

Melting point/K: 973
Boiling point/K: 1413
ΔH_{fusion}/kJ mol^{-1}: 7.15
ΔH_{vap}/kJ mol^{-1}: 136.7

Thermodynamic properties (298.15 K, 0.1 MPa)

State	$\Delta_f H^{\ominus}$/kJ mol^{-1}	$\Delta_f G^{\ominus}$/kJ mol^{-1}	$S^{\ominus}$/J K^{-1} mol^{-1}	C_p/J K^{-1} mol^{-1}
Solid	0	0	71	27.1
Gas	159	130	176.47	20.79

Density/kg m^{-3}: $c.$ 5000 [293 K]
Thermal conductivity/W m^{-1} K^{-1}: 18.6 (est.) [300 K]
Electrical resistivity/Ω m: 100×10^{-8} [273 K]
Mass magnetic susceptibility/kg^{-1} m^3: n.a.
Molar volume/cm^3: 45.2
Coefficient of linear thermal expansion/K^{-1}: 20.2×10^{-6}

Lattice structure (cell dimensions/pm), space group

b.c.c. ($a = 515$)

X-ray diffraction: mass absorption coefficients (μ/ρ)/cm^2 g^{-1}:
CuK$_\alpha$ 304   MoK$_\alpha$ 172

Radium

Atomic number: 88
Thermal neutron capture cross-section/barns: 20 ± 3 (^{226}Ra)
Number of isotopes (including nuclear isomers): 16
Isotope mass range: $213 \rightarrow 230$ (^{218}Ra and ^{219}Ra unknown)

Nuclear properties

Key isotopes

Nuclide	Atomic mass	Natural abundance (%)	Half life $T_{1/2}$	Decay mode and energy (MeV)	Nuclear spin I	Nuclear magnetic moment μ	Uses
^{223}Ra	226.0186	some	11.43d	α(5.977); γ			
^{224}Ra	224.0202	some	3.64d	α(5.787); γ			
^{226}Ra	226.0254	some	1602y	α(4.781); γ			tracer, medical
^{228}Ra	228.031	some	5.77y	β^-(0.055); no γ			

Ground state electron configuration: [Rn]$7s^2$
Term symbol: $^1S_0$
Electron affinity (M$\rightarrow$M$^-$)/kJ mol^{-1}: n.a.

Electron shell properties

Ionization energies/kJ mol^{-1}

1. M $\rightarrow$M$^+$	509.3	6. M^{5+}$\rightarrow$M^{6+} (7 300)
2. M$^+$$\rightarrow$M^{2+}	979.0	7. M^{6+}$\rightarrow$M^{7+} (8 600)
3. M^{2+}$\rightarrow$M^{3+}	(3300)	8. M^{7+}$\rightarrow$M^{8+} (9 900)
4. M^{3+}$\rightarrow$M^{4+}	(4400)	9. M^{8+}$\rightarrow$M^{9+} (13 500)
5. M^{4+}$\rightarrow$M^{5+}	(5700)	10. M^{9+}$\rightarrow$M^{10+} (15 100)

Principal lines in atomic spectrum

Wavelength/nm	Species	Sensitivity	Application
381.442	II	V1	AE
468.228	II	V2	AE
482.591	I	U1	AE
714.121*	I		

* Lowest energy transition from ground state to nearest empty orbital

Abundance: Earth's crust 10^{-6} p.p.m.; seawater 3×10^{-11} p.p.m.
Biological role: None; toxic due to radioactivity

Rn	Atomic number: 86
	Relative atomic mass ($^{12}C = 12.0000$): **(222)**

Chemical properties

Colourless, odourless gas produced by ^{226}Ra. Little studied because of hazardous radiation which destroys any compounds that are formed. Chemically should be like xenon. May be health hazard in certain localities.

Radii/pm: n.a.

Electronegativity: n.a. (Pauling); n.a. (Allred–Rochow)

Effective nuclear charge: 8.25 (Slater); 16.08 (Clementi); 20.84 (Froese–Fischer)

Solubility in water at 293 K, 230 cm^3 per dm^3

Standard reduction potentials $E^{\ominus}$/V

n.a.

Covalent bonds	r/pm	E/kJ mol^{-1}	Oxidation states
Rn–F		n.a.	Rn^{II} RnF$_2$

Physical properties

Melting point/K: 202
Boiling point/K: 211.4
ΔH_{fusion}/kJ mol^{-1}: 2.7 (est.)
ΔH_{vap}/kJ mol^{-1}: 18.1

Thermodynamic properties (298.15 K, 0.1 MPa)

State	$\Delta_f H^{\ominus}$/kJ mol^{-1}	$\Delta_f G^{\ominus}$/kJ mol^{-1}	$S^{\ominus}$/J K^{-1} mol^{-1}	C_p/J K^{-1} mol^{-1}
Gas	0	0	176.21	20.786

Density/kg m^{-3}: n.a. [solid]; 4400 [liquid, b.p.]; 9.73 [gas, 273 K]
Thermal conductivity/W m^{-1} K^{-1}: 0.003 64 (est.) [300 K]$_g$
Mass magnetic susceptibility/kg^{-1} m^3: n.a.
Molar volume/cm^3: 50.5 [211 K]

Lattice structure (cell dimensions/pm)

f.c.c.

X-ray diffraction: mass absorption coefficients (μ/ρ)/cm^2 g^{-1}:
n.a.

Radon

Atomic number: 86
Thermal neutron capture cross-section/barns: 0.72 ± 0.07 (^{222}Rn)
Number of isotopes (including nuclear isomers): 20
Isotope mass range: $204 \rightarrow 224$

Nuclear properties

Key isotopes

Nuclide	Atomic mass	Natural abundance (%)	Half life $T_{1/2}$	Decay mode and energy (MeV)	Nuclear spin I	Nuclear magnetic moment μ	Uses
^{219}Rn	219.0095	trace	4.0s	α(6.944); γ			
^{220}Rn	220.0114	trace	55s	α(6.405); γ			
^{222}Rn	222.0175	trace	3.82d	α(5.587); γ			tracer, medical

Ground state electron configuration: $[Xe]4f^{14}5d^{10}6s^26p^6$
Term symbol: $^1S_0$
Electron affinity ($M \rightarrow M^-$)/kJ mol^{-1}: -41 (calc.)

Electron shell properties

Ionization energies/kJ mol^{-1}

1.	$M \rightarrow M^+$ 1037	6.	$M^{5+} \rightarrow M^{6+}$
2.	$M^+ \rightarrow M^{2+}$	7.	$M^{6+} \rightarrow M^{7+}$
3.	$M^{2+} \rightarrow M^{3+}$	8.	$M^{7+} \rightarrow M^{8+}$
4.	$M^{3+} \rightarrow M^{4+}$	9.	$M^{8+} \rightarrow M^{9+}$
5.	$M^{4+} \rightarrow M^{5+}$	10.	$M^{9+} \rightarrow M^{10+}$

Principal lines in atomic spectrum

Wavelength/nm	Species	Sensitivity	Application
183.048* (vac)	I		
434.960	I		
705.542	I	U3	AE
745.000	I	U2	AE

* Lowest energy transition from ground state to nearest empty orbital

Abundance: Atmosphere trace; Earth's crust 1.7×10^{-10} p.p.m.; seawater 9×10^{-15} p.p.m.
Biological role: None; toxic due to radioactivity

Re	**Atomic number:** 75
	Relative atomic mass ($^{12}C = 12.0000$): **186.207**

Chemical properties

Silvery metal, usually obtained as grey powder; extracted from flue dusts of Mo smelters. Resists corrosion and oxidation but slowly tarnishes in moist air. Dissolves in nitric and sulfuric acids. Used in filaments, thermistors, and catalysts.

Radii/pm: Re^{4+} 72; Re^{6+} 61; Re^{7+} 60; atomic 137.0; covalent 128

Electronegativity: 1.9 (Pauling); 1.46 (Allred-Rochow)

Effective nuclear charge: 3.60 (Slater); 10.12 (Clementi); 14.62 (Froese–Fischer)

Standard reduction potentials $E^{\ominus}/V$

Oxidation states

Re^{-III} (d^{10}) $[Re(CO)_4]^{3-}$
Re^{-I} (d^8) $[Re(CO)_5]^-$
Re^0 (d^7) $Re_2(CO)_{10}$
Re^I (d^6) $Re(CO)Cl$, $K_5Re(CN)_6$
Re^{II} (d^5) ReF_2, $ReCl_2$ etc.
Re^{III} (d^4) $Re_2O_3 \cdot xH_2O$, Re_3Cl_9,
  Re_3Br_9, Re_3I_9, $Re_2Cl_8^{2-}$,
  $[Re(CN)_7]^{4-}$, complexes

Re^{IV} (d^3) ReO_2, ReF_4, $ReCl_4$, etc., complexes
Re^V (d^2) Re_2O_5, ReF_5, $ReCl_5$, $ReBr_5$, $ReOF_3$, complexes
Re^{VI} (d^1) ReO_3, ReF_6, $ReCl_6$, ReF_8^{2-}, $ReOCl_4$, complexes
Re^{VII} (d^0, f^{14}) Re_2O_7, ReF_7, ReO_3F, $ReOF_5$, ReO_4^- (aq), $[ReH_9]^{2-}$, complexes

---

Physical properties

Melting point/K: 3453

Boiling point/K: 5900

ΔH_{fusion}/kJ mol^{-1}: 33.1

ΔH_{vap}/kJ mol^{-1}: 704.25

Thermodynamic properties (298.15 K, 0.1 MPa)

State	$\Delta_f H^{\ominus}$/kJ mol^{-1}	$\Delta_f G^{\ominus}$/kJ mol^{-1}	$S^{\ominus}$/J K^{-1} mol^{-1}	C_p/J K^{-1} mol^{-1}
Solid	0	0	36.86	25.48
Gas	769.9	724.6	188.938	20.786

Density/kg m^{-3}: 21 020 [293 K]; 18 900 [liquid at m.p.]

Thermal conductivity/W m^{-1} K^{-1}: 47.9 [300 K]

Electrical resistivity/Ω m: 19.3×10^{-8} [293 K]

Mass magnetic susceptibility/kg^{-1} m^3: $+4.56 \times 10^{-9}$ (s)

Molar volume/cm^3: 8.86

Coefficient of linear thermal expansion/K^{-1}: 6.63×10^{-6}

Lattice structure (cell dimensions/pm), space group

h.c.p. ($a = 276.09$; $c = 445.76$), P6$_3$/mmc

X-ray diffraction: mass absorption coefficients (μ/ρ)/cm^2 g^{-1}:
CuK$_\alpha$ 179   MoK$_\alpha$ 103

Atomic number: 75
Thermal neutron capture cross-section/barns: 85 ± 5
Number of isotopes (including nuclear isomers): 20
Isotope mass range: $177 \rightarrow 192$

Key isotopes

Nuclide	Atomic mass	Natural abundance (%)	Half life $T_{1/2}$	Decay mode and energy (MeV)	Nuclear spin I	Nuclear magnetic moment μ	Uses
^{185}Re	184.9530	37.40	stable		5/2	$+3.172$	NMR
^{186}Re	0		88.9h	β^-(1.071); EC;			
				γ	1	$+1.73$	tracer
^{187}Re	186.9560	62.60	4.3×10^{10}y	β^-(<0.01)	5/2	$+3.204$	NMR
^{188}Re	0		16.7h	β^-(2.116); γ	1	$+1.78$	tracer

NMR

	$[^{185}\text{Re}]$	^{187}Re
Relative sensitivity (^{1}H = 1.00)	0.13	0.13
Absolute sensitivity (^{1}H = 1.00)	4.93×10^{-2}	8.62×10^{-2}
Receptivity (^{13}C = 1.00)	280	490
Magnetogyric ratio/rad T^{-1} s^{-1}	6.0255×10^7	6.0862×10^7
Quadrupole moment/m^2	2.8×10^{-28}	2.6×10^{-28}
Frequency (^{1}H = 100 MHz; 2.3488 T)/MHz	22.513	22.744

Reference: NaReO$_4$ (aq)

Ground state electron configuration: [Xe]$4f^{14}5d^26s^2$
Term symbol: $^6S_{5/2}$
Electron affinity (M$\rightarrow$M$^-$)/kJ mol^{-1}: 37

Ionization energies/kJ mol^{-1}

1. M $\rightarrow$ M$^+$	760	6. M$^{5+}\rightarrow$M^{6+}	(6300)
2. M$^+\rightarrow$M^{2+}	1260	7. M$^{6+}\rightarrow$M^{7+}	(7600)
3. M$^{2+}\rightarrow$M^{3+}	2510	8. M$^{7+}\rightarrow$M^{8+}	
4. M$^{3+}\rightarrow$M^{4+}	3640	9. M$^{8+}\rightarrow$M^{9+}	
5. M$^{4+}\rightarrow$M^{5+}	(4900)	10. M$^{9+}\rightarrow$M^{10+}	

Principal lines in atomic spectrum

Wavelength/nm	Species	Sensitivity	Application
345.188	I		AA
346.047	I	U1	AA, AE
346.473	I		AA
488.917	I	U2	AE
527.556*	I		

* Lowest energy transition from ground state to nearest empty orbital

Abundance: Earth's crust 0.0007 p.p.m.; seawater $<1 \times 10^{-10}$ p.p.m.
Biological role: None; toxicity data n.a.

Rh

Atomic number: 45

Relative atomic mass ($^{12}C = 12.0000$): **102.90550**

Chemical properties

Rare; lustrous, silvery, hard metal of the platinum group. Certain Ni–Cu ores contain 0.1% rhodium. Stable in air up to 875 K; inert to all acids; attacked by fused alkalis. Used as a catalyst.

Radii/pm: Rh^{2+} 86; Rh^{3+} 75; Rh^{4+} 67; atomic 134.5; covalent 125

Electronegativity: 2.28 (Pauling); 1.45 (Allred-Rochow)

Effective nuclear charge: 3.90 (Slater); 7.64 (Clementi); 10.85 (Froese–Fischer)

Standard reduction potentials $E^{\ominus}/V$

III		0
Rh^{3+}	_0.76_	Rh

Oxidation states

Rh^{-I} (d^{10})	$[Rh(CO)_4]^-$
Rh^0 (d^9)	$[Rh_4(CO)_{12}]$
Rh^I (d^8)	$[Rh(PPh_3)_3]^+$
Rh^{II} (d^7)	RhO
Rh^{III} (d^6)	Rh_2O_3 RhF_3, $RhCl_3$ etc., $[RhCl_6]^{3-}$
	$[Rh(H_2O)_x]^{3+}$ (aq)
Rh^{IV} (d^5)	RhO_2, RhF_4, $[RhCl_6]^{2-}$
Rh^V (d^4)	$[RhF_5]_4$, $[RhF_6]^-$
Rh^{VI} (d^3)	RhF_6

Physical properties

Melting point/K: 2239

Boiling point/K: 4000

ΔH_{fusion}/kJ mol^{-1}: 21.55

ΔH_{vap}/kJ mol^{-1}: 494.34

Thermodynamic properties (298.15 K, 0.1 MPa)

State	$\Delta_f H^{\ominus}$/kJ mol^{-1}	$\Delta_f G^{\ominus}$/kJ mol^{-1}	$S^{\ominus}$/J K^{-1} mol^{-1}	C_p/J K^{-1} mol^{-1}
Solid	0	0	31.51	24.98
Gas	556.9	510.8	185.808	21.012

Density/kg m^{-3}: 12 410 [293 K]; 10 650 [liquid at m.p.]

Thermal conductivity/W m^{-1} K^{-1}: 150 [300 K]

Electrical resistivity/Ω m: 4.51×10^{-8} [293 K]

Mass magnetic susceptibility/kg^{-1} m^3: $+1.36 \times 10^{-8}$ (s)

Molar volume/cm^3: 8.29

Coefficient of linear thermal expansion/K^{-1}: 8.40×10^{-6}

Lattice structure (cell dimensions/pm), space group

f.c.c. ($a = 380.36$), Fm3m

X-ray diffraction: mass absorption coefficients (μ/ρ)/cm^2 g^{-1}: CuK$_\alpha$ 194   MoK$_\alpha$ 22.6

Rhodium

Atomic number: 45

Thermal neutron capture cross-section/barns: 150 ± 5

Number of isotopes (including nuclear isomers): 20

Isotope mass range: $97 \rightarrow 110$

Nuclear properties

Key isotopes

Nuclide	Atomic mass	Natural abundance (%)	Half life $T_{1/2}$	Decay mode and energy (MeV)	Nuclear spin I	Nuclear magnetic moment μ	Uses
^{103}Rh	102.9048	100	stable		1/2	-0.0883	NMR
^{105}Rh		0	35.88h	β^-(0.565); γ			tracer

NMR

^{103}Rh

Relative sensitivity (^{1}H = 1.00)	3.11×10^{-5}
Absolute sensitivity (^{1}H = 1.00)	3.11×10^{-5}
Receptivity (^{13}C = 1.00)	0.177
Magnetogyric ratio/rad T^{-1} s^{-1}	-0.8520×10^7
Frequency (^{1}H = 100 MHz; 2.3488 T)/MHz	3.172

Reference: mer-[RhCl$_3$(SMe$_2$)$_3$]

Ground state electron configuration: [Kr]4d^{8}5s^1

Term symbol: $^4F_{9/2}$

Electron affinity (M$\rightarrow$M$^-$)/kJ mol^{-1}: 162

Electron shell properties

Ionization energies/kJ mol^{-1}

1. M $\rightarrow$M$^+$	720	6. M$^{5+}\rightarrow$M^{6+}	(8 200)
2. M$^+\rightarrow$M^{2+}	1744	7. M$^{6+}\rightarrow$M^{7+}	(10 100)
3. M$^{2+}\rightarrow$M^{3+}	2997	8. M$^{7+}\rightarrow$M^{8+}	(12 200)
4. M$^{3+}\rightarrow$M^{4+}	(4400)	9. M$^{8+}\rightarrow$M^{9+}	(14 200)
5. M$^{4+}\rightarrow$M^{5+}	(6500)	10. M$^{9+}\rightarrow$M^{10+}	(22 000)

Principal lines in atomic spectrum

Wavelength/nm	Species	Sensitivity	Application
332.309	I		AE
339.685	I		AA, AE
343.489	I	U1	AA, AE
350.252	I		AA
365.799	I		AA, AE
369.236*	I		AA, AE

* Lowest energy transition from ground state to nearest empty orbital

Abundance: Earth's crust 0.0001 p.p.m.; seawater $< 1 \times 10^{-10}$ p.p.m.

Biological role: None; toxicity n.a.; suspected carcinogenic

<table>
<tr><td>**Rb**</td><td>**Atomic number: 37**
Relative atomic mass ($^{12}C = 12.0000$): **85.4678**</td></tr>
</table>

Chemical properties

Very soft metal with silvery white lustre when cut. Ignites in air and reacts violently with water. Present in minerals such as lepidolite, pollucite and carnalite. Finds little use outside research.

Radii/pm: Rb^+ 1.49; atomic 247.5; van der Waals 244
Electronegativity: 0.82 (Pauling); 0.89 (Allred-Rochow)
Effective nuclear charge: 2.20 (Slater); 4.98 (Clementi); 6.66 (Froese–Fischer)

Standard reduction potentials $E^{\ominus}/V$

$$\begin{array}{ccc} I & & 0 \\ Rb^+ & \underline{\quad -2.924 \quad} & Rb \end{array}$$

Oxidation states

Rb^{-1} (s^2)	solution of metal in liquid ammonia
Rb^I ([Kr])	Rb_2O, Rb_2O_2 (peroxide),
	RbO_2 (superoxide),
	RbOH, RbH, RbF, RbCl etc.,
	$[Rb(H_2O)_x]^+$ (aq),
	Rb_2CO_3, many salts,
	some complexes

Physical properties

Melting point/K: 312.2
Boiling point/K: 961
ΔH_{fusion}/kJ mol^{-1}: 2.20
ΔH_{vap}/kJ mol^{-1}: 75.7

Thermodynamic properties (298.15 K, 0.1 MPa)

State	$\Delta_f H^{\ominus}$/kJ mol^{-1}	$\Delta_f G^{\ominus}$/kJ mol^{-1}	$S^{\ominus}$/J K^{-1} mol^{-1}	C_p/J K^{-1} mol^{-1}
Solid	0	0	76.78	31.062
Gas	80.88	53.06	170.089	20.786

Density/kg m^{-3}: 1532 [293 K]; 1475 [liquid at m.p.]
Thermal conductivity/W m^{-1} K^{-1}: 58.2 [300 K]
Electrical resistivity/Ω m: 12.5×10^{-8} [293 K]
Mass magnetic susceptibility/kg^{-1} m^3: $+2.49 \times 10^{-9}$ (s)
Molar volume/cm^3: 55.79
Coefficient of linear thermal expansion/K^{-1}: 90×10^{-6}

Lattice structure (cell dimensions/pm), space group

b.c.c. ($a = 562$), Im3m

X-ray diffraction: mass absorption coefficients (μ/ρ)/cm^2 g^{-1}:
CuK_{α} 117 MoK_{α} 90.0

Rubidium

Atomic number: 37
Thermal neutron capture cross-section/barns: 0.5 ± 0.1
Number of isotopes (including nuclear isomers): 20
Isotope mass range: $79 \rightarrow 95$

Nuclear
properties

Key isotopes

Nuclide	Atomic mass	Natural abundance (%)	Half life $T_{1/2}$	Decay mode and energy (MeV)	Nuclear spin I	Nuclear magnetic moment μ	Uses
^{83}Rb		0	83d	EC(0.83); γ	5/2	+1.4	tracer
^{85}Rb	84.9117	72.17	stable		5/2	+1.3524	NMR
^{86}Rb	85.911	0	18.66d	β^-(1.78); γ	2	−1.691	tracer
^{87}Rb	86.909	27.83	5×10^{11}y	β^-(0.274); no γ	3/2	+2.750	NMR

NMR

	$[^{85}\text{Rb}]$	^{87}Rb
Relative sensitivity (^{1}H = 1.00)	1.05×10^{-2}	0.17
Absolute sensitivity (^{1}H = 1.00)	7.57×10^{-3}	4.87×10^{-2}
Receptivity (^{13}C = 1.00)	43	277
Magnetogyric ratio/rad T^{-1} s^{-1}	2.5828×10^7	8.7532×10^7
Quadrupole moment/m^2	0.25×10^{-28}	0.12×10^{-28}
Frequency (^{1}H = 100 MHz; 2.3488 T)/MHz	9.655	32.721

Reference: RbCl (aq)

Ground state electron configuration: $[\text{Kr}]5s^1$
Term symbol: $^2S_{1/2}$
Electron affinity $(M \rightarrow M^-)$/kJ mol^{-1}: 46.9

Electron shell properties

Ionization energies/kJ mol^{-1}

1.	$M \rightarrow M^+$	403.0	6. $M^{5+} \rightarrow M^{6+}$ 8 140
2.	$M^+ \rightarrow M^{2+}$	2632	7. $M^{6+} \rightarrow M^{7+}$ 9 570
3.	$M^{2+} \rightarrow M^{3+}$	3900	8. $M^{7+} \rightarrow M^{8+}$ 13 100
4.	$M^{3+} \rightarrow M^{4+}$	5080	9. $M^{8+} \rightarrow M^{9+}$ 14 800
5.	$M^{4+} \rightarrow M^{5+}$	6850	10. $M^{9+} \rightarrow M^{10+}$ 26 740

Principal lines in atomic spectrum

Wavelength/nm	Species	Sensitivity	Application
420.180	I	U3	AA, AE
421.553	I	U4	AA, AE
780.027	I	U1	AA, AE
794.760*	I	U2	AA, AE

* Lowest energy transition from ground state to nearest empty orbital

Abundance: Earth's crust 78 p.p.m.; seawater 0.12 p.p.m.
Biological role: None; non-toxic, except in large doses; stimulatory

| **Ru** | Atomic number: **44**
Relative atomic mass (^{12}C = 12.0000): **101.07** |

Chemical properties

Lustrous, silvery metal of the platinum group. Found in the free state but obtained from the wastes of Ni refining. Unaffected by air, water and acids but dissolves in molten alkali. Used to harden Pt and Pd, and as catalyst.

Radii/pm: Ru^{3+} 77; Ru^{4+} 65; Ru^{8+} 54; atomic 134; covalent 124

Electronegativity: 2.2 (Pauling); 1.42 (Allred-Rochow)

Effective nuclear charge: 3.75 (Slater); 7.45 (Clementi); 10.57 (Froese–Fischer)

Standard reduction potentials $E^{\ominus}$/V

| VIII | VII | VI | IV | III | II | 0 |

Oxidation states

Ru^{-II}	(d^{10})	rare $[Ru(CO)_4]^{2-}$
Ru	(d^8)	rare $Ru(CO)_5$
Ru^I	(d^7)	some complexes
Ru^{II}	(d^6)	$[Ru(H_2O)_6]^{2+}$ (aq), $RuCl_2$, $RuBr_2$, RuI_2, $[Ru(CN)_6]^{2-}$, complexes
Ru^{III}	(d^5)	Ru_2O_3, $[Ru(H_2O)_6]^{3+}$ (aq), RuF_3, $RuCl_3$, etc., $RuCl_6^{3-}$
Ru^{IV}	(d^4)	RuO_2, RuF_4, $RuCl_6^{2-}$
Ru^V	(d^3)	RuF_5, RuF_6^-
Ru^{VI}	(d^2)	RuO_3, RuO_4^{2-} (aq), RuF_6
Ru^{VII}	(d^1)	RuO_4^- (aq)
Ru^{VIII}	$(d^0, [Kr])$	RuO_4

Physical properties

Melting point/K: 2583

Boiling point/K: 4173

ΔH_{fusion}/kJ mol^{-1}: 23.7

ΔH_{vap}/kJ mol^{-1}: 567

Thermodynamic properties (298.15 K, 0.1 MPa)

State	$\Delta_f H^{\ominus}$/kJ mol^{-1}	$\Delta_f G^{\ominus}$/kJ mol^{-1}	$S^{\ominus}$/J K^{-1} mol^{-1}	C_p/J K^{-1} mol^{-1}
Solid	0	0	28.53	24.06
Gas	642.7	595.8	186.507	21.522

Density/kg m^{-3}: 12 370 [293 K]; 10 900 [liquid at m.p.]

Thermal conductivity/W m^{-1} K^{-1}: 117 [300 K]

Electrical resistivity/Ω m: 7.6×10^{-8} [273 K]

Mass magnetic susceptibility/kg^{-1} m^3: $+5.37 \times 10^{-9}$ (s)

Molar volume/cm^3: 8.14

Coefficient of linear thermal expansion/K^{-1}: 9.1×10^{-6}

Lattice structure (cell dimensions/pm), space group

h.c.p. (a = 270.58; c = 428.11), P6$_3$/mmc

X-ray diffraction: mass absorption coefficients (μ/ρ)/cm^2 g^{-1}:
CuK$_\alpha$ 183   MoK$_\alpha$ 21.1

Ruthenium

Atomic number: 44

Thermal neutron capture cross-section/barns: 3.0 ± 0.8

Number of isotopes (including nuclear isomers): 16

Isotope mass range: $93 \rightarrow 108$

Nuclear properties

Key isotopes

Nuclide	Atomic mass	Natural abundance (%)	Half life $T_{1/2}$	Decay mode and energy (MeV)	Nuclear spin I	Nuclear magnetic moment μ	Uses
^{96}Ru	95.9076	5.52	stable				
^{97}Ru		0	2.88d	EC(c. 1.2); γ			tracer
^{98}Ru	97.9055	1.88	stable				
^{99}Ru	98.9061	12.7	stable		5/2	−0.62	NMR
^{100}Ru	99.9030	12.6	stable				
^{101}Ru		17.0	stable		5/2	−0.68	NMR
^{102}Ru	101.9037	31.6	stable				
^{103}Ru		0	39.6d	β^- (0.74); γ			tracer
^{104}Ru	103.9055	18.7	stable				
^{106}Ru		0	367d	β^- (0.0394); no γ			tracer, medical

NMR

	^{99}Ru	^{101}Ru
Relative sensitivity (^{1}H = 1.00)	1.95×10^{-4}	1.41×10^{-3}
Absolute sensitivity (^{1}H = 1.00)	2.48×10^{-5}	2.40×10^{-4}
Receptivity (^{13}C = 1.00)	0.83	1.56
Magnetogyric ratio/rad T^{-1} s^{-1}	-1.2343×10^7	-1.3834×10^7
Quadrupole moment/m^2	0.076×10^{-28}	0.44×10^{-28}
Frequency (^{1}H = 100 MHz; 2.3488 T)/MHz	3.389	4.941

Reference: RuO$_4$

Ground state electron configuration: [Kr]4d^{7}5s^1

Term symbol: ^{5}F$_5$

Electron affinity (M→M$^-$)/kJ mol^{-1}: 146

Electron shell properties

Ionization energies/kJ mol^{-1}

1. M →M$^+$ 711	6. M^{5+}→M^{6+} (7 800)
2. M$^+$→M^{2+} 1617	7. M^{6+}→M^{7+} (9 600)
3. M^{2+}→M^{3+} 2747	8. M^{7+}→M^{8+} (11 500)
4. M^{3+}→M^{4+} (4500)	9. M^{8+}→M^{9+} (18 700)
5. M^{4+}→M^{5+} (6100)	10. M^{9+}→M^{10+} (20 900)

Principal lines in atomic spectrum

Wavelength/nm	Species	Sensitivity	Application
343.674	I	U2	AE
349.894	I	U1	AA, AE
359.618	I	U3	AE
372.803	I		AA
396.490*	I		

* Lowest energy transition from ground state to nearest empty orbital

Abundance: Earth's crust 0.0001 p.p.m.; seawater n.a., $<1 \times 10^{-10}$ p.p.m.

Biological role: None; toxicity n.a.

Sm

Atomic number: 62
Relative atomic mass ($^{12}C = 12.0000$): 150.36

Chemical properties

Silvery white metal of the lanthanide (rare earth) group. Obtained from monazite. Relatively stable in dry air but in moist air an oxide coating forms. Used in special glass, catalysts, ceramics, and electronics.

Radii/pm: Sm^{2+} 111; Sm^{3+} 100; atomic 180.2; covalent 166

Electronegativity: 1.17 (Pauling); 1.07 (Allred–Rochow)

Effective nuclear charge: 2.85 (Slater); 8.01 (Clementi); 11.06 (Froese–Fischer)

Standard reduction potentials $E^{\ominus}/V$

	III		II		0
			$\xrightarrow{-2.30}$		
Acid solution	Sm^{3+}	$\xrightarrow{-1.55}$	Sm^{2+}	$\xrightarrow{-2.67}$	Sm
Alkaline solution	$Sm(OH)_3$		$\xrightarrow{-2.80}$		Sm

Oxidation states

Sm^{II} (f^6)	SmO, SmS, SmF_2, $SmCl_2$, etc.
Sm^{III} (f^5)	Sm_2O_3, $Sm(OH)_3$, SmF_3, $SmCl_3$ etc., $[Sm(H_2O)_x]^{3+}$ (aq), Sm^{3+} salts, complexes

Physical properties

Melting point/K: 1350
Boiling point/K: 2064
ΔH_{fusion}/kJ mol^{-1}: 10.9
ΔH_{vap}/kJ mol^{-1}: 164.8

Thermodynamic properties (298.15 K, 0.1 MPa)

State	$\Delta_f H^{\ominus}$/kJ mol^{-1}	$\Delta_f G^{\ominus}$/kJ mol^{-1}	$S^{\ominus}$/J K^{-1} mol^{-1}	C_p/J K^{-1} mol^{-1}
Solid	0	0	69.58	29.54
Gas	206.7	172.8	183.042	30.355

Density/kg m^{-3}: 7520 [293 K]
Thermal conductivity/W m^{-1} K^{-1}: 13.3 [300 K]
Electrical resistivity/Ω m: 88.0×10^{-8} [298 K]
Mass magnetic susceptibility/kg^{-1} m^3: $+1.52 \times 10^{-7}$ (s)
Molar volume/cm^3: 20.00
Coefficient of linear thermal expansion/K^{-1}: 10.4×10^{-6}

Lattice structure (cell dimensions/pm), space group

α-Sm rhombohedral ($a = 899.6$, $\alpha = 23°13'$), $R\bar{3}m$
β-Sm cubic ($a = 407$), Im3m

$T(\alpha \rightarrow \beta) = 1190$ K

High pressure form: h.c.p. ($a = 361.8$, $c = 1166$), $P6_3/mmc$

X-ray diffraction: mass absorption coefficients (μ/ρ)/cm^2 g^{-1}:
CuK$_\alpha$ 397   MoK$_\alpha$ 58.6

Discovered by P.-E. Lecoq de Boisbaudran in 1879 at Paris, France

[Named after the mineral Samarskite]

Samarium

Nuclear properties

Atomic number: 62
Thermal neutron capture cross-section/barns: 5820 ± 100
Number of isotopes (including nuclear isomers): 17
Isotope mass range: $142 \rightarrow 157$

Key isotopes

Nuclide	Atomic mass	Natural abundance (%)	Half life $T_{1/2}$	Decay mode and energy (MeV)	Nuclear spin I	Nuclear magnetic moment μ	Uses
^{144}Sm	143.9115	3.1	stable				
^{146}Sm	145.9129	0	7×10^7y	$\alpha(2.46)$			
^{147}Sm	146.9147	15.1	1.05×10^{11}y	$\alpha(2.314)$	7/2	-0.813	NMR
^{148}Sm	147.9146	11.3	$>12 \times 10^{14}$y	$\alpha(2.001)$			
^{149}Sm	148.9169	13.9	$>1 \times 10^{15}$y	α	7/2	-0.670	NMR
^{150}Sm	149.9170	7.4	stable				
^{152}Sm	151.9195	26.6	stable				
^{153}Sm		0	46.8h	$\beta^-(0.801); \gamma$	3/2	-0.0217	tracer
^{154}Sm	153.9220	22.6	stable				

NMR

	^{147}Sm	^{149}Sm
Relative sensitivity (^{1}H = 1.00)	1.48×10^{-3}	7.47×10^{-4}
Absolute sensitivity (^{1}H = 1.00)	2.21×10^{-4}	1.03×10^{-4}
Receptivity (^{13}C = 1.00)	1.28	0.665
Magnetogyric ratio/rad T^{-1} s^{-1}	-1.1124×10^7	-0.9175×10^7
Quadrupole moment/m^2	-0.208×10^{-28}	6×10^{-30}
Frequency (^{1}H = 100 MHz; 2.3488 T)/MHz	4.128	3.289

Electron shell properties

Ground state electron configuration: [Xc]4f^{6}6s^2
Term symbol: $^7F_0$
Electron affinity (M$\rightarrow$M$^-$)/kJ mol^{-1}: $\leqslant 50$

Ionization energies/kJ mol^{-1}

1. M $\rightarrow$M$^+$	543.3	6. M$^{5+}\rightarrow$M^{6+}	
2. M $^+\rightarrow$M^{2+}	1068	7. M$^{6+}\rightarrow$M^{7+}	
3. M$^{2+}\rightarrow$M^{3+}	2260	8. M$^{7+}\rightarrow$M^{8+}	
4. M$^{3+}\rightarrow$M^{4+}	3990	9. M$^{8+}\rightarrow$M^{9+}	
5. M$^{4+}\rightarrow$M^{5+}		10. M$^{9+}\rightarrow$M^{10+}	

Principal lines in atomic spectrum

Wavelength/nm	Species	Sensitivity	Application
356.827	II		
359.260	II		
429.674	I		AA
442.434	II	V1	AE
443.432	II	V2	AE
476.027	I		AA

Abundance: Earth's crust 7.0 p.p.m.; seawater 2×10^{-7} p.p.m.
Biological role: None; toxicity n.a., but probably low

<table>
<tr><td>**Sc**</td><td>**Atomic number: 21**
Relative atomic mass ($^{12}C = 12.0000$): 44.955910</td></tr>
</table>

Chemical properties

Soft, silvery-white metal. Tarnishes in air and burns easily. Reacts with water to form hydrogen gas. Forms salts with acids. Little used.

Radii/pm: Sc^{3+} 83; atomic 160.6; covalent 144

Electronegativity: 1.36 (Pauling); 1.20 (Allred-Rochow)

Effective nuclear charge: 3.00 (Slater); 4.63 (Clementi); 6.06 (Froese–Fischer)

Standard reduction potentials $E^{\ominus}/V$

III		0
Sc^{3+}	$\xrightarrow{-2.03}$	Sc

Oxidation states

Sc^{II} (d^1)	$CsScCl_3$
Sc^{III} ([Ar])	Sc^2O_3, $ScO.OH$,
	'$Sc(OH)_3$', $Sc(OH)_6^{3-}$,
	ScF_3, $ScCl_3$, etc.,
	ScF_6^{3-}, complexes

ScH_2 is probably $Sc^{III}H^-$ with complex bonding

Physical properties

Melting point/K: 1814
Boiling point/K: 3104
ΔH_{fusion}/kJ mol^{-1}: 15.9
ΔH_{vap}/kJ mol^{-1}: 376.1

Thermodynamic properties (298.15 K, 0.1 MPa)

State	$\Delta_f H^{\ominus}$/kJ mol^{-1}	$\Delta_f G^{\ominus}$/kJ mol^{-1}	$S^{\ominus}$/J K^{-1} mol^{-1}	C_p/J K^{-1} mol^{-1}
Solid	0	0	34.64	25.52
Gas	377.8	336.03	174.79	22.09

Density/kg m^{-3}: 2989 [273 K]
Thermal conductivity/W m^{-1} K^{-1}: 15.8 [300 K]
Electrical resistivity/Ω m: 61.0×10^{-8} [295 K]
Mass magnetic susceptibility/kg^{-1} m^3: $+8.8 \times 10^{-8}$ (s)
Molar volume/cm^3: 15.04
Coefficient of linear thermal expansion/K^{-1}: 10.0×10^{-6}

Lattice structure (cell dimensions/pm), space group

α-Sc h.c.p. ($a = 330.90$, $c = 527.3$), $P6_3/mmc$
β-Sc cubic, Im3m

$T(\alpha \rightarrow \beta) = 1223$ K

X-ray diffraction: mass absorption coefficients (μ/ρ)/cm^2 g^{-1}:
CuK$_\alpha$ 184   MoK$_\alpha$ 21.1

Scandium

Atomic number: 21
Thermal neutron capture cross-section/barns: 25 ± 2
Number of isotopes (including nuclear isomers): 15
Isotope mass range: $40 \rightarrow 50$

Nuclear properties

Key isotopes

Nuclide	Atomic mass	Natural abundance (%)	Half life $T_{1/2}$	Decay mode and energy (MeV)	Nuclear spin I	Nuclear magnetic moment μ	Uses
^{44}Sc		0	3.92h	$\beta^-(3.647); \gamma$	2	$+2.56$	tracer
^{45}Sc	44.95592	100	stable		7/2	$+4.7559$	NMR
^{46}Sc	45.955	0	83.80d	$\beta^-(2.367); \gamma$	4	$+3.03$	tracer
^{47}Sc		0	3.43d	$\beta^-(0.60); \gamma$	7/2	$+5.34$	tracer

NMR

^{45}Sc

Relative sensitivity ($^1H = 1.00$)	0.30
Absolute sensitivity ($^1H = 1.00$)	0.30
Receptivity ($^{13}C = 1.00$)	1710
Magnetogyric ratio/rad T^{-1} s^{-1}	6.4982×10^7
Quadrupole moment m^2	-0.22×10^{-28}
Frequency ($^1H = 100$ MHz; 2.3488 T)/MHz	24.290

Reference: $Sc(ClO_4)_3$ (aq)

Ground state electron configuration: $[Ar]3d^14s^2$
Term symbol: $^2D_{3/2}$
Electron affinity $(M \rightarrow M^-)$/kJ mol^{-1}: -70

Electron shell properties

Ionization energies/kJ mol^{-1}

1. $M \rightarrow M^+$	631	6. $M^{5+} \rightarrow M^{6+}$	10720
2. $M^+ \rightarrow M^{2+}$	1235	7. $M^{6+} \rightarrow M^{7+}$	13320
3. $M^{2+} \rightarrow M^{3+}$	2389	8. $M^{7+} \rightarrow M^{8+}$	15310
4. $M^{3+} \rightarrow M^{4+}$	7089	9. $M^{8+} \rightarrow M^{9+}$	17369
5. $M^{4+} \rightarrow M^{5+}$	8844	10. $M^{9+} \rightarrow M^{10+}$	21740

Principal lines in atomic spectrum

Wavelength/nm	Species	Sensitivity	Application
361.384	II	V2	AE
390.749	I	U2	AA, AE
391.181	I	U1	AA, AE
402.040	I	U4	AA, AE
402.369	I	U3	AA, AE
637.882*	I		

* Lowest energy transition from ground state to nearest empty orbital

Abundance: Earth's crust 25 p.p.m.; seawater 4×10^{-5} p.p.m.
Biological role: None; toxicity n.a.; suspected carcinogen

Se	**Atomic number: 34**
	Relative atomic mass ($^{12}C = 12.0000$): **78.96**

Chemical properties

Obtained as silvery metallic allotrope or red amorphous powder, which is less stable. Found in sulfide ores. Burns in air, unaffected by water, dissolves in concentrated nitric acid and alkalis. Used in photoelectric cells, photocopiers, solar cells, and semiconductors.

Radii/pm: Se^{4+} 69; atomic 215.2 (grey); covalent 117; van der Waals 200; Se^{2-} 191

Electronegativity: 2.55 (Pauling); 2.48 (Allred-Rochow)

Effective nuclear charge: 6.95 (Slater); 8.29 (Clementi); 9.96 (Froese–Fischer)

Standard reduction potentials $E^{\ominus}/V$

	VI		IV		0		$-$II
Acid solution	SeO_4^{2-}	1.1	H_2SeO_3	0.74	Se	-0.11	H_2Se
Alkaline solution	SeO_4^{2-}	0.03	SeO_3^{2-}	-0.36	Se	-0.67	Se^{2-}

Covalent bonds r/pm E/kJ mol^{-1}

Bond	r/pm	E/kJ mol^{-1}
Se–H	146	305
Se–C	198	245
Se–O	161	343
Se–F	170	285
Se–Cl	220	245
Se–Se (Se$_8$)	232	330

Oxidation states

Se^{-II}	H_2Se
Se^{0-I}	Se cluster cations, e.g. Se_4^{2+}, Se_8^{2+}
Se^{I}	Se_2Cl_2, Se_2Br_2
Se^{II}	?
Se^{IV}	SeO_2, H_2SeO_3, SeO_3^{2-} (aq), $SeOF_2$, $SeOCl_4$, $SeOBr_2$, SeF_4, $SeCl_4$, $SeBr_4$, $SeBr_6^{2-}$
Se^{VI}	SeO_3 ? H_2SeO_4, SeO_4^{2-} (aq), SeO_2F_2, SeO_2Cl_2, SeF_6

Physical properties

Melting point/K: 490
Boiling point/K: 958.1
ΔH_{fusion}/kJ mol^{-1}: 5.1
ΔH_{vap}/kJ mol^{-1}: 90

Thermodynamic properties (298.15 K, 0.1 MPa)

State	$\Delta_f H^{\ominus}$/kJ mol^{-1}	$\Delta_f G^{\ominus}$/kJ mol^{-1}	$S^{\ominus}$/J K^{-1} mol^{-1}	C_p/J K^{-1} mol^{-1}
Solid (α)	0	0	42.442	25.363
Gas	227.07	187.03	176.72	20.820

Density/kg m^{-3}: 4790 (grey) [293 K]; 3987 [liquid at m.p.]
Thermal conductivity/W m^{-1} K^{-1}: 2.04 [300 K]
Electrical resistivity/Ω m: 0.01 [293 K]
Mass magnetic susceptibility/kg^{-1} m^3: -4.0×10^{-9} (s)
Molar volume/cm^3: 16.48
Coefficient of linear thermal expansion/K^{-1}: 36.9×10^{-6}

Lattice structure (cell dimensions/pm), space group

Grey hexagonal ($a = 436.56$, $c = 495.90$), P3$_1$21, metallic form
α-Se monoclinic ($a = 906.4$, $b = 907.2$, $c = 115.6$, $\beta = 90°52'$), P2$_1$/a, Se$_8$
β-Se monoclinic ($a = 1285$, $b = 807$, $c = 931$, $\beta = 93°8'$), P2$_1$/a, Se$_8$
α'-Se cubic ($a = 297.0$), Pm3m
β'-Se cubic ($a = 604$), Fd3m

X-ray diffraction: mass absorption coefficients (μ/ρ)/cm^2 g^{-1}:
CuK$_\alpha$ 91.4   MoK$_\alpha$ 74.7

Selenium

Atomic number: 34
Thermal neutron capture cross-section/barns: 12.2 ± 0.6
Number of isotopes (including nuclear isomers): 20
Isotope mass range: $70 \rightarrow 85$

Nuclear properties

Key isotopes

Nuclide	Atomic mass	Natural abundance (%)	Half life $T_{1/2}$	Decay mode and energy (MeV)	Nuclear spin I	Nuclear magnetic moment μ	Uses
^{74}Se	73.9225	0.9	stable		0		
^{75}Se	74.923	0	120.4d	β^- (0.865); γ	5/2		tracer, medical
^{76}Se	75.9192	9.0	stable		0		
^{77}Se	76.9199	7.6	stable		1/2	+0.534	NMR
^{78}Se	77.9173	23.5	stable		0		
^{80}Se	79.9165	49.6	stable		0		
^{82}Se	81.9167	9.4	stable		0		

NMR

^{77}Se

Relative sensitivity (^{1}H = 1.00)	6.93×10^{-3}
Absolute sensitivity (^{1}H = 1.00)	5.25×10^{-4}
Receptivity (^{13}C = 1.00)	2.98
Magnetogyric ratio/rad T^{-1} s^{-1}	5.1018×10^7
Frequency (^{1}H = 100 MHz; 2.3488 T)/MHz	19.092
Reference: $S(CH_3)_2$	

Ground state electron configuration: $[Ar]3d^{10}4s^24p^4$
Term symbol: $^3P_2$
Electron affinity ($M \rightarrow M^-$)/kJ mol^{-1}: 195.0

Electron shell properties

Ionization energies/kJ mol^{-1}

1. $M \rightarrow M^+$	940.9	6. $M^{5+} \rightarrow M^{6+}$ 7883
2. $M^+ \rightarrow M^{2+}$	2044	7. $M^{6+} \rightarrow M^{7+}$ 14990
3. $M^{2+} \rightarrow M^{3+}$	2974	8. $M^{7+} \rightarrow M^{8+}$ (19500)
4. $M^{3+} \rightarrow M^{4+}$	4144	9. $M^{8+} \rightarrow M^{9+}$ (23300)
5. $M^{4+} \rightarrow M^{5+}$	6590	10. $M^{9+} \rightarrow M^{10+}$ (27200)

Principal lines in atomic spectrum

Wavelength/nm	Species	Sensitivity	Application
196.026	I		AA
203.985	I	U2	AA, AE
206.279	I	U3	AA, AE
207.479*	I		
1032.726			

* Lowest energy transition from ground state to nearest empty orbital

Abundance: Earth's crust 0.05 p.p.m.; seawater 0.004 p.p.m.
Biological role: Essential; toxic in excess of dietary requirement; stimulatory; carcinogenic; teratogenic

Chemical properties

Black amorphous Si obtained by reduction of sand (SiO_2) with carbon; ultrapure semiconductor grade crystals are blue–grey metallic. Bulk Si unreactive towards oxygen, water, acids (except HF), but dissolves in hot alkali. Used in semiconductors, alloys, polymers.

Radii/pm: Si^{4+} 26; atomic 117; covalent 117; van der Waals 200; Si^{4-} 271

Electronegativity: 1.90 (Pauling); 1.74 (Allred-Rochow)

Effective nuclear charge: 4.15 (Slater); 4.29 (Clementi); 4.48 (Froese–Fischer)

Standard reduction potentials $E^{\ominus}/V$

	IV		II		0		$-IV$
Acid solution	SiO_2	-0.967	'SiO'	-0.808	Si	-0.143	SiH_4

[Alkaline solutions contain many different forms]

Covalent bonds

	r/pm	E/kJ mol^{-1}
Si–H	148.0	326
Si–C	187	301
Si–O	151	486
Si–F	155	582
Si–Cl	202	391
Si–Si	232	226

Oxidation states

Si^{II}	SiF_2 (gas)
Si^{IV}	SiO_2, 'H_4SiO_4', silicates, zeolites etc., SiH_4 etc., SiF_4, $SiCl_4$ etc., SiF_6^{2-}, metal silicides, e.g. Ca_2Si, $CaSi$, organosilicon compounds

Physical properties

Melting point/K: 1683
Boiling point/K: 2628
ΔH_{fusion}/kJ mol^{-1}: 39.6
ΔH_{vap}/kJ mol^{-1}: 383.3

Thermodynamic properties (298.15 K, 0.1 MPa)

State	$\Delta_f H^{\ominus}$/kJ mol^{-1}	$\Delta_f G^{\ominus}$/kJ mol^{-1}	$S^{\ominus}$/J K^{-1} mol^{-1}	C_p/J K^{-1} mol^{-1}
Solid	0	0	18.83	20.00
Gas	455.6	411.3	167.97	22.251

Density/kg m^{-3}: 2329 [293 K]; 2525 [liquid at m.p.]
Thermal conductivity/W m^{-1} K^{-1}: 148 [300 K]
Electrical resistivity/Ω m: 0.001 [273 K]
Mass magnetic susceptibility/kg^{-1} m^3: -1.8×10^{-9} (s)
Molar volume/cm^3: 12.06
Coefficient of linear thermal expansion/K^{-1}: 4.2×10^{-6}

Lattice structure (cell dimensions/pm), space group

Cubic ($a = 543.07$), Fd3m, diamond structure
High pressure forms: ($a = 468.6$, $c = 258.5$), I4$_1$/amd
($a = 664$), Ia3
($a = 380$, $c = 628$) P6$_3$mc

X-ray diffraction: mass absorption coefficients (μ/ρ)/cm^2 g^{-1}:
CuK$_\alpha$ 60.6   MoK$_\alpha$ 6.44

Silicon

Atomic number: 14
Thermal neutron capture cross-section/barns: 0.160 ± 0.020
Number of isotopes (including nuclear isomers): 8
Isotope mass range: $25 \rightarrow 31$

Nuclear properties

Key isotopes

Nuclide	Atomic mass	Natural abundance (%)	Half life $T_{1/2}$	Decay mode and energy (MeV)	Nuclear spin I	Nuclear magnetic moment μ	Uses
^{28}Si	27.97693	92.23	stable		0		
^{29}Si	28.97649	4.67	stable		1/2	-0.55526	NMR
^{30}Si	29.97376	3.10	stable		0		
^{32}Si	31.9740	0	*c.* 650y	β^- (0.21); no γ			tracer

NMR

^{29}Si

Relative sensitivity (^{1}H = 1.00)	7.84×10^{-3}
Absolute sensitivity (^{1}H = 1.00)	3.69×10^{-4}
Receptivity (^{13}C = 1.00)	2.09
Magnetogyric ratio/rad T^{-1} s^{-1}	-5.3146×10^7
Frequency (^{1}H = 100 MHz; 2.3488 T)/MHz	19.865

Reference: $Si(CH_3)_4$

Ground state electron configuration: $[Ne]3s^2 3p^2$
Term symbol: $^3P_0$
Electron affinity ($M \rightarrow M^-$)/kJ mol^{-1}: 133.6

Electron shell properties

Ionization energies/kJ mol^{-1}

1. $M \rightarrow M^+$	786.5	6. $M^{5+} \rightarrow M^{6+}$ 19784
2. $M^+ \rightarrow M^{2+}$	1577.1	7. $M^{6+} \rightarrow M^{7+}$ 23786
3. $M^{2+} \rightarrow M^{3+}$	3231.4	8. $M^{7+} \rightarrow M^{8+}$ 29252
4. $M^{3+} \rightarrow M^{4+}$	4355.5	9. $M^{8+} \rightarrow M^{9+}$ 33876
5. $M^{4+} \rightarrow M^{5+}$	16091	10. $M^{9+} \rightarrow M^{10+}$ 38732

Principal lines in atomic spectrum

Wavelength/nm	Species	Sensitivity	Application
250.690	I	U4	AA, AE
251.431*	I		AA
251.611	I	U3	AA, AE
252.851	I	U2	AA, AE
288.151	I	U1	AE
other lines of equal intensity at higher wavelengths			

* Lowest energy transition from ground state to nearest empty orbital

Abundance: Earth's crust 272000 p.p.m.; seawater 3.0 p.p.m.
Biological role: Essential; non-toxic but some silicates are carcinogenic (e.g. asbestos)

<table>
<tr><td>

Ag

</td><td>

Atomic number: 47
Relative atomic mass ($^{12}C = 12.0000$): 107.8682

</td></tr>
</table>

Chemical properties

Soft, malleable metal with characteristic silver sheen. Stable to water and oxygen but attacked by sulfur compounds in air to form black sulfide layer. Dissolves in sulfuric and nitric acid. Used in photography, silverware, jewellery, electrical industry, and glass (mirrors).

Radii/pm: Ag^+ 113; Ag^{2+} 89; atomic 144.4; covalent 134
Electronegativity: 1.93 (Pauling); 1.42 (Allred-Rochow)
Effective nuclear charge: 4.20 (Slater); 8.03 (Clementi); 11.35 (Froese–Fischer)

Standard reduction potentials $E^{\ominus}/V$

	III		II		I		0

Acid solution Ag_2O_3 $\xrightarrow{1.360}$ Ag^{2+} $\xrightarrow{1.980}$ Ag^+ $\xrightarrow{0.7991}$ Ag
$\xrightarrow{1.670}$ (Ag₂O₃ → Ag⁺)
$\xrightarrow{1.569}$ (Ag₂O₃ → AgO)
AgO $\xrightarrow{1.772}$ Ag^+
$\xrightarrow{1.398}$ Ag_2O $\xrightarrow{1.173}$

Alkaline solution Ag_2O_3 $\xrightarrow{0.793}$ AgO $\xrightarrow{0.604}$ Ag_2O $\xrightarrow{0.342}$ Ag
$\xrightarrow{1.757}$
Ag^+
$\xrightarrow{1.711}$ Ag_2O_2

Oxidation states

Ag^0 ($d^{10}s^1$)	rare $Ag(CO)_3$ at 10 K
Ag (d^{10})	Ag_2O, $Ag(OH)_2^-$ (aq), $Ag(H_2O)_4^+$ (aq), AgF, AgCl etc., Ag^+ salts, e.g. $AgNO_3$, Ag_2S, $Ag(CN)_2^-$, and other complexes
Ag^{II} (d^9)	AgF_2, $[Ag(C_5H_5N)_2]^+$, AgO is not Ag^{II} but $Ag^IAg^{III}O_2$
Ag^{III} (d^8)	rare AgF_4^-, AgF_6^{3-}

Physical properties

Melting point/K: 1235.08
Boiling point/K: 2485
ΔH_{fusion}/kJ mol^{-1}: 11.3
ΔH_{vap}/kJ mol^{-1}: 257.7

Thermodynamic properties (298.15 K, 0.1 MPa)

State	$\Delta_f H^{\ominus}$/kJ mol^{-1}	$\Delta_f G^{\ominus}$/kJ mol^{-1}	$S^{\ominus}$/J K^{-1} mol^{-1}	C_p/J K^{-1} mol^{-1}
Solid	0	0	42.55	25.351
Gas	284.55	245.65	172.997	20.786

Density/kg m^{-3}: 10 500 [293 K]; 9345 [liquid at m.p.]
Thermal conductivity/W m^{-1} K^{-1}: 429 [300 K]
Electrical resistivity/Ω m: 1.59×10^{-8} [293 K]
Mass magnetic susceptibility/kg^{-1} m^3: -2.27×10^{-9} (s)
Molar volume/cm^3: 10.27
Coefficient of linear thermal expansion/K^{-1}: 19.2×10^{-6}

Lattice structure (cell dimensions/pm), space group

f.c.c. ($a = 408.626$), Fm3m)

X-ray diffraction: mass absorption coefficients (μ/ρ)/cm^2 g^{-1}:
CuK$_\alpha$ 218   MoK$_\alpha$ 25.8

Silver

Atomic number: 47
Thermal neutron capture cross-section/barns: 63.8 ± 0.6
Number of isotopes (including nuclear isomers): 27
Isotope mass range: $102 \rightarrow 117$

**Nuclear
properties**

Key isotopes

Nuclide	Atomic mass	Natural abundance (%)	Half life $T_{1/2}$	Decay mode and energy (MeV)	Nuclear spin I	Nuclear magnetic moment μ	Uses
^{107}Ag	106.905 09	51.83	stable		1/2	-0.1135	NMR
^{109}Ag	108.9047	48.17	stable		1/2	-0.1305	NMR
^{110m}Ag		0	253d	β^- (c. 2.9); γ	6	$+3.604$	tracer
^{111}Ag		0	7.5d	β^- (1.05); γ	1/2	-0.145	tracer

NMR

	^{107}Ag	^{109}Ag
Relative sensitivity (^{1}H = 1.00)	6.62×10^{-5}	1.01×10^{-4}
Absolute sensitivity (^{1}H = 1.00)	3.43×10^{-5}	4.86×10^{-5}
Receptivity (^{13}C = 1.00)	0.195	0.276
Magnetogyric ratio/rad T^{-1} s^{-1}	-1.0828×10^7	-1.2448×10^7
Frequency (^{1}H = 100 MHz; 2.3488 T)/MHz	4.046	4.652

Reference: Ag$^+$ (aq)

Ground state electron configuration: $[Kr]4d^{10}5s^1$
Term symbol: $^2S_{1/2}$
Electron affinity $(M \rightarrow M^-)$/kJ mol^{-1}: 125.7

**Electron
shell
properties**

Ionization energies/kJ mol^{-1}

1. $M \rightarrow M^+$	731.0		6. $M^{5+} \rightarrow M^{6+}$	(8 600)
2. $M^+ \rightarrow M^{2+}$	2073		7. $M^{6+} \rightarrow M^{7+}$	(11 200)
3. $M^{2+} \rightarrow M^{3+}$	3361		8. $M^{7+} \rightarrow M^{8+}$	(13 400)
4. $M^{3+} \rightarrow M^{4+}$	(5000)		9. $M^{8+} \rightarrow M^{9+}$	(15 600)
5. $M^{4+} \rightarrow M^{5+}$	(6700)		10. $M^{9+} \rightarrow M^{10+}$	(18 000)

Principal lines in atomic spectrum

Wavelength/nm	Species	Sensitivity	Application
328.068	I	U1	AA, AE
338.289*	I	U2	AA, AE
520.907	I	U3	AE
546.549	I	U4	AE

* Lowest energy transition from ground state to nearest empty orbital

Abundance: Earth's crust 0.08 p.p.m.; seawater 0.0003 p.p.m.
Biological role: None; toxic to lower organisms but not animals; suspected carcinogen

<table>
<tr><td>**Na**</td><td>**Atomic number: 11**
Relative atomic mass ($^{12}C = 12.0000$): **22.989 768**</td></tr>
</table>

<table>
<tr><td>**Chemical properties**</td><td>Soft, silvery white metal which oxidizes rapidly when cut. Reacts vigorously with water. Produced in large quantities and used as metal in heat exchangers in atomic reactors. Large deposits of NaCl and trona. NaCl is key industrial chemical, used to make Cl_2, NaOH, etc.</td></tr>
</table>

Radii/pm: Na^+ 98; atomic 153.7; van der Waals 231

Electronegativity: 0.93 (Pauling); 1.01 (Allred-Rochow)

Effective nuclear charge: 2.20 (Slater); 2.51 (Clementi); 3.21 (Froese–Fischer)

Standard reduction potentials $E^{\ominus}/V$

I		0
Na^+	$\xrightarrow{-2.713}$	Na

Oxidation states

Na^{-I} (s^2)	Na metal solutions in liquid NH_3
Na^I ([Ne])	Na_2O, Na_2O_2 (peroxide), NaOH, NaH, NaF, NaCl etc., $[Na(H_2O)_4]^+$ (aq), $NaHCO_3$, Na_2CO_3, Na^+ salts, some complexes

<table>
<tr><td>**Physical properties**</td><td>**Melting point**/K: 370.96
Boiling point/K: 1156.1
ΔH_{fusion}/kJ mol^{-1}: 2.64
ΔH_{vap}/kJ mol^{-1}: 99.2</td></tr>
</table>

Thermodynamic properties (298.15 K, 0.1 MPa)

State	$\Delta_f H^{\ominus}$/kJ mol^{-1}	$\Delta_f G^{\ominus}$/kJ mol^{-1}	$S^{\ominus}$/J K^{-1} mol^{-1}	C_p/J K^{-1} mol^{-1}
Solid	0	0	51.21	28.24
Gas	107.32	76.761	153.712	20.786

Density/kg m^{-3}: 971 [293 K]; 928 [liquid at m.p.]

Thermal conductivity/W m^{-1} K^{-1}: 141 [300 K]

Electrical resistivity/Ω m: 4.2×10^{-8} [273 K]

Mass magnetic susceptibility/kg^{-1} m^3: $+8.8 \times 10^{-9}$ (s)

Molar volume/cm^3: 23.68

Coefficient of linear thermal expansion/K^{-1}: 70.6×10^{-6}

Lattice structure (cell dimensions/pm), space group

α-Na hexagonal ($a = 376.7$, $c = 615.4$), P6$_3$/mmc
β-Na b.c.c. ($a = 429.06$), Im3m
T(b.c.c.→hexagonal) = 5 K

X-ray diffraction: mass absorption coefficients (μ/ρ)/cm^2 g^{-1}:
CuK$_\alpha$ 30.1   MoK$_\alpha$ 3.21

Sodium

Atomic number: 11
Thermal neutron capture cross-section/barns: 0.534 ± 0.005
Number of isotopes (including nuclear isomers): 7
Isotope mass range: $20 \rightarrow 26$

Key isotopes

Nuclide	Atomic mass	Natural abundance (%)	Half life $T_{1/2}$	Decay mode and energy (MeV)	Nuclear spin I	Nuclear magnetic moment μ	Uses
^{22}Na	21.9944	0	2.602y	β^+ (1.82); EC; γ	3	+1.746	tracer
^{23}Na	22.9898	100	stable		3/2	+2.217 40	NMR
^{24}Na	23.990 96	0	15.0h	β^- (4.17); γ	4	+1.690	tracer, medical

NMR

^{23}Na

Relative sensitivity (^{1}H = 1.00)	9.25×10^{-2}
Absolute sensitivity (^{1}H = 1.00)	9.25×10^{-2}
Receptivity (^{13}C = 1.00)	525
Magnetogyric ratio/rad T^{-1} s^{-1}	7.0761×10^7
Quadrupole moment/m^2	0.12×10^{-28}
Frequency (^{1}H = 100 MHz; 2.3488 T)/MHz	26.451

Reference: NaCl (aq)

Ground state electron configuration: [Ne]3s^1
Term symbol: $^2S_{1/2}$
Electron affinity (M→M$^-$)/kJ mol^{-1}: 52.9

Ionization energies/kJ mol^{-1}

1. M → M$^+$	495.8		6. M^{5+}→M^{6+}	16 610
2. M$^+$→M^{2+}	4 562.4		7. M^{6+}→M^{7+}	20 114
3. M^{2+}→M^{3+}	6 912		8. M^{7+}→M^{8+}	25 490
4. M^{3+}→M^{4+}	9 543		9. M^{8+}→M^{9+}	28 933
5. M^{4+}→M^{5+}	13 353		10. M^{9+}→M^{10+}	141 360

Principal lines in atomic spectrum

Wavelength/nm	Species	Sensitivity	Application
330.232	I	U3	AA, AE
330.299	I	U4	AA, AE
568.824	I		AE
588.995	I	U1	AA, AE
589.592*	I	U2	AE

* Lowest energy transition from ground state to nearest empty orbital

Abundance: Earth's crust 22 700 p.p.m.; seawater 10 600 p.p.m.
Biological role: Essential; non-toxic

<table>
<tr><td rowspan="2">**Sr**</td><td>Atomic number: **38**</td></tr>
<tr><td>Relative atomic mass ($^{12}C = 12.0000$): **87.62**</td></tr>
</table>

Chemical properties

Silvery white, soft metal obtained by high temperature reduction of SrO with Al. Protected as bulk metal by oxide film but will burn in air and reacts with water. Used in fireworks and flares to give red colour. Found as celestite, $SrSO_4$.

Radii/pm: Sr^{2+} 127; atomic 215.1 (α form); covalent 192

Electronegativity: 0.95 (Pauling); 0.99 (Allred-Rochow)

Effective nuclear charge: 2.85 (Slater); 6.07 (Clementi); 8.09 (Froese–Fischer)

Standard reduction potentials $E^{\ominus}/V$

Also $\quad \begin{array}{l} SrO_2^* \\ SrO_2 \end{array} \quad \begin{array}{l} \underline{2.333} \\ \underline{1.492} \end{array} \quad \begin{array}{l} Sr^{2+} \\ SrO \ (hyd) \end{array}$

*See oxidation states

Oxidation states

Sr^{II} ([Kr])	SrO, SrO_2 (peroxide), $Sr(OH)_2$, $[Sr(H_2O)_x]^{2+}$ (aq), Sr^{2+} salts, SrF_2, $SrCl_2$ etc., $SrCO_3$, some complexes SrH_2 is $Sr^{2+}2H^-$

Physical properties

Melting point/K: 1042
Boiling point/K: 1657
ΔH_{fusion}/kJ mol^{-1}: 9.16
ΔH_{vap}/kJ mol^{-1}: 154.4

Thermodynamic properties (298.15 K, 0.1 MPa)

State	$\Delta_f H^{\ominus}$/kJ mol^{-1}	$\Delta_f G^{\ominus}$/kJ mol^{-1}	$S^{\ominus}$/J K^{-1} mol^{-1}	C_p/J K^{-1} mol^{-1}
Solid	0	0	52.3	26.4
Gas	164.4	130.9	164.62	20.786

Density/kg m^{-3}: 2540 [293 K]; 2375 [liquid at m.p.]
Thermal conductivity/W m^{-1} K^{-1}: 35.3 [300 K]
Electrical resistivity/Ω m: 23.0×10^{-8} [293 K]
Mass magnetic susceptibility/kg^{-1} m^3: $+1.32 \times 10^{-8}$ (s)
Molar volume/cm^3: 34.50
Coefficient of linear thermal expansion/K^{-1}: 23×10^{-6}

Lattice structure (cell dimensions/pm), space group

α-Sr f.c.c. ($a = 608.49$), Fm3m
β-Sr h.c.p. ($a = 432$, $c = 706$), P6$_3$/mmc
γ-Sr b.c.c. ($a = 485$), Im3m
$T(\alpha \rightarrow \beta) = 506$ K
$T(\beta \rightarrow \gamma) = 813$ K

X-ray diffraction: mass absorption coefficients (μ/ρ)/cm^2 g^{-1}:
CuK$_\alpha$ 125   MoK$_\alpha$ 95.0

Recognized as an element in 1790 by A. Crawford at Edinburgh, Scotland. Isolated in 1808 by Sir Humphry Davy at London, UK

[Named after Strontian, Scotland]

Strontium

Atomic number: 38
Thermal neutron capture cross-section/barns: 1.21 ± 0.06
Number of isotopes (including nuclear isomers): 18
Isotope mass range: $80 \rightarrow 95$

Nuclear properties

Key isotopes

Nuclide	Atomic mass	Natural abundance (%)	Half life $T_{1/2}$	Decay mode and energy (MeV)	Nuclear spin I	Nuclear magnetic moment μ	Uses
^{82}Sr		0	25d	EC($c.$ 0.4); γ			tracer
^{84}Sr	83.9134	0.56	stable				
^{85}Sr	84.913	0	64d	EC(1.11); γ			tracer medical
^{86}Sr	85.9094	9.86	stable		0		
^{87}Sr	86.9089	7.00	stable		9/2	-1.093	NMR
^{88}Sr	87.9056	82.58	stable		0		
^{89}Sr	88.907	0	52.7d	β^-(1.463); γ			
^{90}Sr	89.907	0	28.1y	β^-(0.546); no γ			tracer, medical

NMR

^{87}Sr

Relative sensitivity (^{1}H = 1.00) 2.69×10^{-3}
Absolute sensitivity (^{1}H = 1.00) 1.88×10^{-4}
Receptivity (^{13}C = 1.00)   1.07
Magnetogyric ratio/rad T^{-1} s^{-1} -1.1593×10^7
Quadrupole moment/m^2 0.36×10^{-28}
Frequency (^{1}H = 100 MHz; 2.3488 T)/MHz 4.333
Reference: Sr^{2+} (aq)

Ground state electron configuration: [Kr]$5s^2$
Term symbol: $^1S_0$
Electron affinity (M$\rightarrow$M$^-$)/kJ mol^{-1}: -146

Electron shell properties

Ionization energies/kJ mol^{-1}

1. M $\rightarrow$M$^+$	549.5		6. M$^{5+}\rightarrow$M^{6+}	8 760
2. M$^+\rightarrow$M^{2+}	1064.2		7. M$^{6+}\rightarrow$M^{7+}	10 200
3. M$^{2+}\rightarrow$M^{3+}	4210		8. M$^{7+}\rightarrow$M^{8+}	11 800
4. M$^{3+}\rightarrow$M^{4+}	5500		9. M$^{8+}\rightarrow$M^{9+}	15 600
5. M$^{4+}\rightarrow$M^{5+}	6910		10. M$^{9+}\rightarrow$M^{10+}	17 100

Principal lines in atomic spectrum

Wavelength/nm	Species	Sensitivity	Application
407.771	II	V1	AE
460.733	I	U1	AA, AE
483.208	I	U2	AE
487.249	I	U3	AE
689.258*	I		

* Lowest energy transition from ground state to nearest empty orbital

Abundance: Earth's crust 384 p.p.m.; seawater 8 p.p.m.
Biological role: None; non-toxic; mimics calcium

S

Atomic number: 16
Relative atomic mass ($^{12}C = 12.0000$): 32.066

Chemical properties

Several allotropes, of which orthorhombic S_8 is most stable. Stable to air and water but burns if ignited. Attacked by oxidizing acids. Found as native sulfur and as metal sulfide ores. Recovered from H_2S in natural gas. Key industrial chemical.

Radii/pm: S^{4+} 37; S^{6+} 29; atomic 104 (S_8); covalent 104: van der Waals 185

Electronegativity: 2.58 (Pauling); 2.44 (Allred-Rochow)

Effective nuclear charge: 5.45 (Slater); 5.48 (Clementi); 6.04 (Froese–Fischer)

Standard reduction potentials $E^{\ominus}/V$

VI   V   IV   III   II*   0   −II

Acid solution SO_4^{2-} —-0.07— $S_2O_6^{2-}$ —0.57— H_2SO_3 —-0.07— $HS_2O_4^-$ —0.87— $S_2O_3^{2-}$ —0.60— S —0.14— H_2S (0.16; 0.40; 0.50)

Also $S_2O_8^{2-}$ —0.25— SO_4^-

Alkaline solution SO_4^{2-} —-0.94— SO_3^{2-} —-0.58— $S_2O_3^{2-}$ —-0.74— S —-0.45— S^{2-} (−0.66)

*Average oxidation state

Covalent bonds

	r/pm	E/kJ mol^{-1}
S—H	133.5	347
S—C	182	272
S=C	155	476
S—O	150	265
S=O	144	c. 525
S—F	156	328
S—Cl	199	255
S—S	208	226

Oxidation states

S^{-II}	H_2S, S^{2-}	} polysulfides S_n^{2-}
S^{-I}	H_2S_2, etc.	
S^0	S_6, S_8, etc.	
S^{I}	$S_2O?$, S_2F_2, S_2Cl_2	
S^{II}	SF_2, SCl_2	
S^{III}	$Na_2S_2O_4$	
S^{IV}	SO_2, SO_3^{2-} (aq), SF_4, SCl_4, $SOCl_2$	
S^{V}	$Na_2S_2O_6$, S_2F_{10}	
S^{VI}	SO_3, H_2SO_4, SO_4^{2-} (aq), etc., SF_6, HSO_3F, SO_2Cl_2	

Physical properties

Melting point/K: 386.0 (α) 392.2 (β) 380.0 (γ)
Boiling point/K: 717.824
ΔH_{fusion}/kJ mol^{-1}: 1.23 ΔH_{vap}/kJ mol^{-1}: 9.62

Thermodynamic properties (298.15 K, 0.1 MPa)

State	$\Delta_f H^{\ominus}$/kJ mol^{-1}	$\Delta_f G^{\ominus}$/kJ mol^{-1}	$S^{\ominus}$/J K^{-1} mol^{-1}	C_p/J K^{-1} mol^{-1}
Solid (α)	0	0	31.80	22.64
Solid (β)	0.33	n.a.	n.a.	n.a.
Gas	278.805	238.250	167.821	23.673

Density/kg m^{-3}: 2070 (α), 1957 (β) [293 K]; 1819 [liquid, 393 K]
Thermal conductivity/W m^{-1} K^{-1}: 0.269 (α) [300 K]
Electrical resistivity/Ω m: 2×10^{15} [293 K]
Mass magnetic susceptibility/kg^{-1} m^3: -6.09×10^{-9} (α); -5.83×10^{-9} (β)
Molar volume/cm^3: 15.49
Coefficient of linear thermal expansion/K^{-1}: 74.33×10^{-6}

Lattice structure (cell dimensions/pm), space group

α-S_8 orthorhombic ($a = 1046.46$, $b = 1286.60$, $c = 2448.60$), Fddd
β-S_8 monoclinic ($a = 1102$, $b = 1096$, $c = 1090$, $\alpha = 96.7°$), P2$_1$/c
γ-S_8 monoclinic ($a = 857$, $b = 1305$, $c = 823$, $\alpha = 112°54'$), P2/c
ε-S_6 rhombohedral ($a = 646$, $\alpha = 115°18'$), R$\bar{3}$
In addition to the above ring forms there are also S_7, S_{9-12}, S_{18} and S_{20} rings.
Plastic sulfur is long chains of Sn also known in several forms χ, ψ, φ, μ and ω.
$T(\alpha \rightarrow \beta) = 366.7$ K

X-ray diffraction: mass absorption coefficients (μ/ρ)/cm^2 g^{-1}:
CuK$_\alpha$ 89.1   MoK$_\alpha$ 9.55

Sulfur

Atomic number: 16
Thermal neutron capture cross-section/barns: 0.51
Number of isotopes (including nuclear isomers): 10
Isotope mass range: $29 \rightarrow 38$

Key isotopes

Nuclide	Atomic mass	Natural abundance (%)	Half life $T_{1/2}$	Decay mode and energy (MeV)	Nuclear spin I	Nuclear magnetic moment μ	Uses
^{32}S	31.972 07	95.02	stable		0		
^{33}S	32.971 46	0.75	stable		3/2	+0.6435	NMR
^{34}S	33.967 86	4.21	stable		0		
^{35}S	34.969 0	0	87.9d	β^-(0.1674); no γ	3/2	±1.00	tracer
^{36}S	35.967 09	0.02	stable		0		

NMR

^{33}S

Relative sensitivity (^{1}H = 1.00)	2.26×10^{-3}
Absolute sensitivity (^{1}H = 1.00)	1.72×10^{-5}
Receptivity (^{13}C = 1.00)	0.0973
Magnetogyric ratio/rad T^{-1} s^{-1}	2.0534×10^7
Quadrupole moment/m^2	-0.05×10^{-28}
Frequency (^{1}H = 100 MHz; 2.3488 T)/MHz	7.670

Reference: CS$_2$

Ground state electron configuration: [Ne]$3s^2 3p^4$
Term symbol: $^3P_2$
Electron affinity ($M \rightarrow M^-$)/kJ mol^{-1}: 200.4

Ionization energies/kJ mol^{-1}

1. $M \rightarrow M^+$	999.6	6. $M^{5+} \rightarrow M^{6+}$	8 495
2. $M^+ \rightarrow M^{2+}$	2251	7. $M^{6+} \rightarrow M^{7+}$	27 106
3. $M^{2+} \rightarrow M^{3+}$	3361	8. $M^{7+} \rightarrow M^{8+}$	31 669
4. $M^{3+} \rightarrow M^{4+}$	4564	9. $M^{8+} \rightarrow M^{9+}$	36 578
5. $M^{4+} \rightarrow M^{5+}$	7013	10. $M^{9+} \rightarrow M^{10+}$	43 138

Principal lines in atomic spectrum

Wavelength/nm	Species	Sensitivity	Application
191.468* (vac)	I		
469.413	I	U7	AE
921.291	I	U4	AE
922.811	I	U5	AE
923.749	I	U6	AE

* Lowest energy transition from ground state to nearest empty orbital

Abundance: Atmosphere SO$_2$ up to 1 p.p.m., H$_2$S 0.0001 p.p.m. (volume); Earth's crust 340 p.p.m.; seawater 900 p.p.m.
Biological role: Essential; non-toxic as element or sulfate

Ta

Atomic number: 73
Relative atomic mass ($^{12}C = 12.0000$): **180.9479**

Chemical properties

Shiny, silvery metal, soft when pure. Found as tantalite $(Fe,Mn)Ta_2O_6$ but mostly obtained as by-product of tin extraction. Very corrosion resistant due to oxide film; attacked by HF and fused alkalis. Used in chemical plants and surgery.

Radii/pm: Ta^{3+} 72; Ta^{4+} 68; Ta^{5+} 64; atomic 143; covalent 134
Electronegativity: 1.5 (Pauling); 1.33 (Allred-Rochow)
Effective nuclear charge: 3.30 (Slater); 9.53 (Clementi); 13.78 (Froese–Fischer)

Standard reduction potentials $E^{\ominus}/V$

V		0
Ta_2O_5	$\xrightarrow{-0.81}$	Ta

Oxidation states

Ta^{-III}	(d^8)	$[Ta(CO)_5]^{3-}$
Ta^{-I}	(d^6)	$[Ta(CO)_5]^{-}$
Ta^{I}	(d^4)	$[(C_5H_5)Ta(CO)_4]$
Ta^{II}	(d^3)	TaO?
Ta^{III}	(d^2)	TaF_3, $TaCl_3$, $TaBr_3$
Ta^{IV}	(d^1)	TaO_2, $TaCl_4$, $TaBr_4$, TaI_4
Ta^{V}	(d^0, f^{14})	Ta_2O_5, $[Ta_6O_{19}]^{8-}$ (aq), TaF_5, $TaCl_5$, TaF_6^-, TaF_7^{2-}, $TaOF_3$, TaO_2F

Physical properties

Melting point/K: 3269
Boiling point/K: 5698 ± 100
ΔH_{fusion}/kJ mol^{-1}: 31.4
ΔH_{vap}/kJ mol^{-1}: 758.22

Thermodynamic properties (298.15 K, 0.1 MPa)

State	$\Delta_f H^{\ominus}$/kJ mol^{-1}	$\Delta_f G^{\ominus}$/kJ mol^{-1}	$S^{\ominus}$/J K^{-1} mol^{-1}	C_p/J K^{-1} mol^{-1}
Solid	0	0	41.51	25.36
Gas	782.0	739.3	185.214	20.857

Density/kg m^{-3}: 16 654 [293 K]; 15 000 [liquid at m.p.]
Thermal conductivity/W m^{-1} K^{-1}: 57.5 [300 K]
Electrical resistivity/Ω m: 12.45×10^{-8} [298 K]
Mass magnetic susceptibility/kg^{-1} m^3: $+1.07 \times 10^{-8}$ (s)
Molar volume/cm^3: 10.87
Coefficient of linear thermal expansion/K^{-1}: 6.6×10^{-6}

Lattice structure (cell dimensions/pm), space group

b.c.c. ($a = 330.29$), Im3m

X-ray diffraction: mass absorption coefficients (μ/ρ)/cm^2 g^{-1}:
CuK$_\alpha$ 166   MoK$_\alpha$ 95.4

Tantalum

Atomic number: 73
Thermal neutron capture cross-section/barns: 22 ± 1
Number of isotopes (including nuclear isomers): 18
Isotope mass range: $172 \rightarrow 186$

Nuclear properties

Key isotopes

Nuclide	Atomic mass	Natural abundance (%)	Half life $T_{1/2}$	Decay mode and energy (MeV)	Nuclear spin I	Nuclear magnetic moment μ	Uses
^{180}Ta	179.9415	0.012	$>1 \times 10^{12}$y	EC, β^-			
^{181}Ta	180.9480	99.988	stable		7/2	+2.35	NMR
^{182}Ta		0	115.1d	$\beta^-(1.811)$; γ			tracer

NMR

	^{181}Ta
Relative sensitivity (^{1}H = 1.00)	3.60×10^{-2}
Absolute sensitivity (^{1}H = 1.00)	3.60×10^{-2}
Receptivity (^{13}C = 1.00)	204
Magnetogyric ratio/rad T^{-1} s^{-1}	3.2073×10^7
Quadrupole moment /m^2	3×10^{-28}
Frequency (^{1}H = 100 MHz; 2.3488 T)/MHz	11.970

Reference: TaF$_6^-$ (aq)

Ground state electron configuration: $[Xe]4f^{14}5d^36s^2$
Term symbol: $^4F_{3/2}$
Electron affinity (M$\rightarrow$M$^-$)/kJ mol^{-1}: 14

Electron shell properties

Ionization energies/kJ mol^{-1}

1. M $\rightarrow$M$^+$	761	6. M$^{5+}\rightarrow$M^{6+}	
2. M$^+\rightarrow$M^{2+}	(1500)	7. M$^{6+}\rightarrow$M^{7+}	
3. M$^{2+}\rightarrow$M^{3+}	(2100)	8. M$^{7+}\rightarrow$M^{8+}	
4. M$^{3+}\rightarrow$M^{4+}	(3200)	9. M$^{8+}\rightarrow$M^{9+}	
5. M$^{4+}\rightarrow$M^{5+}	(4300)	10. M$^{9+}\rightarrow$M^{10+}	

Principal lines in atomic spectrum

Wavelength/nm	Species	Sensitivity	Application
265.327	I		
260.863	I		AE
271.467	I		AA
331.116	I	U1	AE
540.251*	I		

* Lowest energy transition from ground state to nearest empty orbital

Abundance: Earth's crust 1.7 p.p.m.; seawater <0.0025 p.p.m.
Biological role: None; non-toxic

<table>
<tr><td>**Tc**</td><td>**Atomic number: 43**
Relative atomic mass ($^{12}C = 12.0000$): 98.9062</td></tr>
</table>

Chemical properties

Radioactive metal produced in tonne quantities from fission products of uranium nuclear fuel. Silvery as metal but usually obtained as grey powder. Resists oxidation, slowly tarnishes in moist air, burns in oxygen, dissolves in nitric and sulfuric acids.

Radii/pm: Tc^{2+} 95; Tc^{4+} 72; Tc^{7+} 56; atomic 135.8

Electronegativity: 1.9 (Pauling); 1.36 (Allred-Rochow)

Effective nuclear charge: 3.60 (Slater); 7.23 (Clementi); 10.28 (Froese–Fischer)

Standard reduction potentials $E^{\ominus}$/V

Oxidation states

Tc^{-I}	(d^8)	$[Tc(CO)_5]^-$
Tc^0	(d^7)	$Tc_2(CO)_{10}$
Tc^{IV}	(d^3)	TcO_2, TcO_3^{2-} (aq), $TcCl_4$, complexes
Tc^V	(d^2)	TcO_3^- (aq), $TcCl_5$, complexes
Tc^{VI}	(s^1)	TcO_3 ?, TcF_6, $TcOCl_4$, complexes
Tc^{VII}	$(d^0, [Kr])$	Tc_2O_7, TcO_4^- (aq), TcO_3Cl, complexes

Physical properties

Melting point/K: 2445
Boiling point/K: 5150
ΔH_{fusion}/kJ mol^{-1}: 23.81
ΔH_{vap}/kJ mol^{-1}: 585.22

Thermodynamic properties (298.15 K, 0.1 MPa)

State	$\Delta_f H^{\ominus}$/kJ mol^{-1}	$\Delta_f G^{\ominus}$/kJ mol^{-1}	$S^{\ominus}$/J K^{-1} mol^{-1}	C_p/J K^{-1} mol^{-1}
Solid	0	0	n.a.	25
Gas	678	n.a.	181.07	20.79

Density/kg m^{-3}: 11 500 (est.) [293 K]
Thermal conductivity/W m^{-1} K^{-1}: 50.6 [300 K]
Electrical resistivity/Ω m: 22.6×10^{-8} [393 K]
Mass magnetic susceptibility/kg^{-1} m^3: $+3.1 \times 10^{-8}$ (s)
Molar volume/cm^3: 8.6 (est.)
Coefficient of linear thermal expansion/K^{-1}: 8.06×10^{-6}

Lattice structure (cell dimensions/pm), space group

h.c.p. ($a = 274.3$, $c = 440.0$), P6$_3$/mmc

X-ray diffraction: mass absorption coefficients (μ/ρ)/cm^2 g^{-1}:
CuK$_\alpha$ 172   MoK$_\alpha$ 19.7

Technetium

Atomic number: 43
Thermal neutron capture cross-section/barns: 22 ± 3 (^{99}Tc)
Number of isotopes (including nuclear isomers): 23
Isotope mass range: $92 \rightarrow 107$

Nuclear properties

Key isotopes

Nuclide	Atomic mass	Natural abundance (%)	Half life $T_{1/2}$	Decay mode and energy (MeV)	Nuclear spin I	Nuclear magnetic moment μ	Uses
^{97}Tc		0	2.6×10^6y	EC(c. 0.3)			
^{98}Tc	97.9072	0	1.5×10^6y	β^-(1.7); γ			
^{99}Tc	98.903	0	2.12×10^5y	β^-(0.292); no γ	9/2	+5.681	NMR tracer, medical
^{99m}Tc		0	6.049h	IT(0.1427); γ			

NMR

^{99}Tc

Relative sensitivity (^{1}H = 1.00) —
Absolute sensitivity (^{1}H = 1.00) —
Receptivity (^{13}C = 1.00) —
Magnetogyric ratio/rad T^{-1} s^{-1} 6.0503×10^7
Quadrupole moment /m^2 -0.13×10^{-28}
Frequency (^{1}H = 100 MHz; 2.3488 T)/MHz   22.508
Reference: TcO_4^- (aq)

Ground state electron configuration: [Kr]$4d^5 5s^2$
Term symbol: $^6S_{5/2}$
Electron affinity (M$\rightarrow$M$^-$)/kJ mol^{-1}: 96

Electron shell properties

Ionization energies/kJ mol^{-1}

1. M $\rightarrow$ M$^+$	702		6. M^{5+} $\rightarrow$ M^{6+}	(7 300)
2. M$^+$ $\rightarrow$ M^{2+}	1472		7. M^{6+} $\rightarrow$ M^{7+}	(9 100)
3. M^{2+} $\rightarrow$ M^{3+}	2850		8. M^{7+} $\rightarrow$ M^{8+}	(15 600)
4. M^{3+} $\rightarrow$ M^{4+}	(4100)		9. M^{8+} $\rightarrow$ M^{9+}	(17 800)
5. M^{4+} $\rightarrow$ M^{5+}	(5700)		10. M^{9+} $\rightarrow$ M^{10+}	(19 900)

Principal lines in atomic spectrum

Wavelength/nm	Species	Sensitivity	Application
403.163	I		
426.227	I		
429.706	I		
485.359	I		
608.523*	I		

* Lowest energy transition from ground state to nearest empty orbital

Abundance: Earth's crust 0.0007 p.p.m.; seawater nil
Biological role: None; toxicity n.a., but radioactive

Te

Atomic number: 52
Relative atomic mass ($^{12}C = 12.0000$): 127.60

Chemical properties

Silvery white, metallic looking in bulk but usually obtained as dark grey powder. Semi-metal. Obtained from anode slime in Cu refining. Burns in air or oxygen. Unaffected by water or HCl but dissolves in HNO_3. Used in alloys to improve machinability.

Radii/pm: Te^{4+} 97; Te^{6+} 56; atomic 143.2; covalent 137; van der Waals 220; Te^{2-} 211

Electronegativity: 2.1 (Pauling); 2.01 (Allred-Rochow)

Effective nuclear charge: 6.95 (Slater); 10.81 (Clementi); 13.51 (Froese–Fischer)

Standard reduction potentials $E^{\ominus}/V$

	VI		IV		0		−I		−II
Acid solution	H_2TeO_4	$\underline{\quad 0.93 \quad}$	Te^{4+}	$\underline{\quad 0.57 \quad}$	Te	$\underline{\quad -0.74 \quad}$	Te_2^{2-}	$\underline{\quad -0.64 \quad}$	H_2Te
		$\underline{\quad 1.00 \quad}$	TeO_2	$\underline{\quad 0.53 \quad}$					
Alkaline solution	TeO_4^{2-}	$\underline{\quad 0.07 \quad}$	TeO_3^{2-}	$\underline{\quad -0.42 \quad}$	Te			$\underline{\quad -1.14 \quad}$	Te^{2-}

Covalent bonds r/pm E/kJ mol^{-1}

	r/pm	E/kJ mol^{-1}
Te–H	c. 170	c. 240
Te–C	205	
Te–O	200	268
Te–F	185	335
Te–Cl	231	251
Te–Te	286	235

Oxidation states

Te^{-II}	H_2Te, Te^{2-}
Te^{-I}	Te_2^{2-}
Te^{0-I}	cluster cations Te_4^{2+}, Te_6^{4+}
Te^{II}	TeO, $TeCl_2$, $TeBr_2$
Te^{IV}	TeO_2, H_2TeO_3, TeO_3^{2-} (aq), TeF_4, $TeCl_4$, TeF_5^-
Te^{V}	Te_2F_{10}
Te^{VI}	TeO_3, H_2TeO_4, TeO_4^{2-} (aq), H_6TeO_6, TeF_6, TeF_8^{2-}

Physical properties

Melting point/K: 722.7
Boiling point/K: 1263.0
ΔH_{fusion}/kJ mol^{-1}: 13.5 ΔH_{vap}/kJ mol^{-1}: 104.6

Thermodynamic properties (298.15 K, 0.1 MPa)

State	$\Delta_f H^{\ominus}$/kJ mol^{-1}	$\Delta_f G^{\ominus}$/kJ mol^{-1}	$S^{\ominus}$/J K^{-1} mol^{-1}	C_p/J K^{-1} mol^{-1}
Solid	0	0	49.71	25.73
Gas	196.73	157.08	182.74	20.786

Density/kg m^{-3}: 6240 [293 K]; 5797 [liquid at m.p.]
Thermal conductivity/W m^{-1} K^{-1}: 2.35 [300 K]
Electrical resistivity/Ω m: 4.36×10^{-3} [298 K]
Mass magnetic susceptibility/kg^{-1} m^3: -3.9×10^{-9} (s)
Molar volume/cm^3: 20.45
Coefficient of linear thermal expansion/K^{-1}: 16.75×10^{-6}

Lattice structure (cell dimensions/pm), space group

Hexagonal ($a = 445.65$, $c = 592.68$), P$3_1$21 or P$3_2$21
High pressure forms: ($a = 420.8$, $c = 1203.6$), R$\bar{3}$m
($a = 460.3$, $c = 382.2$), R$\bar{3}$m

X-ray diffraction: mass absorption coefficients (μ/ρ)/cm^2 g^{-1}:
CuK$_\alpha$ 282   MoK$_\alpha$ 35.0

Discovered in 1783 by Baron Franz Joseph Müller von
Reichenstein at Sibiu, Romania
[Latin, *tellus* = earth]

Tellurium

Atomic number: 52
Thermal neutron capture cross-section/barns: 4.7 ± 0.1
Number of isotopes (including nuclear isomers): 29
Isotope mass range: $115 \rightarrow 135$

Nuclear properties

Key isotopes

Nuclide	Atomic mass	Natural abundance (%)	Half life $T_{1/2}$	Decay mode and energy (MeV)	Nuclear spin I	Nuclear magnetic moment μ	Uses
^{120}Te	119.9045	0.096	stable				
^{122}Te	121.9030	2.60	stable				
^{123}Te	122.9042	0.908	1.2×10^{13}y	EC(c. 0.06); no γ	1/2	-0.7359	NMR
^{124}Te	123.9028	4.816	stable				
^{125m}Te		0	58d	IT; γ	3/2	$+0.60$	tracer
^{125}Te	124.9044	7.18	stable		1/2	-0.8872	NMR
^{126}Te	125.9032	18.95	stable		0		
^{127}Te		0	9.4h	β^- (0.69); γ			tracer
^{128}Te	127.9047	31.69	stable		0		
^{130}Te	129.9067	33.80	stable		0		

NMR

	$[^{123}\text{Te}]$	^{125}Te
Relative sensitivity (^{1}H = 1.00)	1.80×10^{-2}	3.15×10^{-2}
Absolute sensitivity (^{1}H = 1.00)	1.56×10^{-4}	2.20×10^{-3}
Receptivity (^{13}C = 1.00)	0.89	12.5
Magnetogyric ratio/rad T^{-1} s^{-1}	-7.0006×10^7	-8.4398×10^7
Frequency (^{1}H = 100 MHz; 2.3488 T)/MHz	26.207	31.596

Reference: Te(CH$_3$)$_2$

Ground state electron configuration: [Kr]$4d^{10}5s^25p^4$
Term symbol: $^3P_2$
Electron affinity (M→M$^-$)/kJ mol^{-1}: 190.2

Electron shell properties

Ionization energies/kJ mol^{-1}

1. M →M$^+$	869.2	6. M^{5+}→M^{6+}	6 822
2. M $^+$→M^{2+}	1795	7. M^{6+}→M^{7+}	13 200
3. M^{2+}→M^{3+}	2698	8. M^{7+}→M^{8+}	(15 800)
4. M^{3+}→M^{4+}	3610	9. M^{8+}→M^{9+}	(18 500)
5. M^{4+}→M^{5+}	5668	10. M^{9+}→M^{10+}	(21 200)

Principal lines in atomic spectrum

Wavelength/nm	Species	Sensitivity	Application
200.202	I		
214.281	I		AE
225.902*	I		
238.326	I	U3	AE
238.578	I	U2	AE

* Lowest energy transition from ground state to nearest empty orbital

Abundance: Earth's crust 0.001 p.p.m.; seawater n.a. $< 1 \times 10^{-10}$ p.p.m.
Biological role: None; very toxic; teratogenic

Atomic number: 65

Relative atomic mass ($^{12}C = 12.0000$): 158.92534

Chemical properties

Silvery white metal, rare member of the lanthanide (rare earth) group. Obtained from monazite. Slowly oxidized by air, reacts with cold water. Used in solid state devices and lasers.

Radii/pm: Tb^{3+} 93; Tb^{4+} 81; atomic 178.2; covalent 159

Electronegativity: n.a. (Pauling); 1.10 (Allred-Rochow)

Effective nuclear charge: 2.85 (Slater); 8.30 (Clementi); 11.39 (Froese–Fischer)

Standard reduction potentials $E^\ominus$/V

	IV		III		0
Acid solution	Tb^{4+}	—3.1—	Tb^{3+}	—−2.31—	Tb
Alkaline solution	TbO_2	—0.9—	$Tb(OH)_3$	—−2.82—	Tb

Oxidation states

Tb^{III} (f^8)	Tb_2O_3, $Tb(OH)_3$, $[Tb(H_2O)_x]^{3+}$ (aq), TbF_3, $TbCl_3$, etc., Tb^{3+} salts, complexes
Tb^{IV} (f^7)	TbO_2, TbF_4

Physical properties

Melting point/K: 1629

Boiling point/K: 3396

ΔH_{fusion}/kJ mol^{-1}: 16.3

ΔH_{vap}/kJ mol^{-1}: 391

Thermodynamic properties (298.15 K, 0.1 MPa)

State	$\Delta_f H^\ominus$/kJ mol^{-1}	$\Delta_f G^\ominus$/kJ mol^{-1}	$S^\ominus$/J K^{-1} mol^{-1}	C_p/J K^{-1} mol^{-1}
Solid	0	0	73.22	28.91
Gas	388.7	349.7	203.58	24.56

Density/kg m^{-3}: 8229 [293 K]

Thermal conductivity/W m^{-1} K^{-1}: 11.1 [300 K]

Electrical resistivity/Ω m: 114×10^{-8} [298 K]

Mass magnetic susceptibility/kg^{-1} m^3: $+1.15 \times 10^{-5}$ (s)

Molar volume/cm^3: 19.31

Coefficient of linear thermal expansion/K^{-1}: 7.0×10^{-6}

Lattice structure (cell dimensions/pm), space group

Tb rhombic ($a = 359.0$, $b = 626.0$, $c = 571.5$), Cmcm
α-Tb h.c.p. ($a = 360.10$, $c = 569.36$), P6$_3$/mmc
β-Tb b.c.c. ($a = 402$), Im3m
$T(\alpha \rightarrow rhombic) = 220$ K
$T(\alpha \rightarrow \beta) = 1590$ K

X-ray diffraction: mass absorption coefficients (μ/ρ)/cm^2 g^{-1}:
CuK$_\alpha$ 273   MoK$_\alpha$ 67.5

Terbium

Atomic number: 65
Thermal neutron capture cross-section/barns: 30 ± 10
Number of isotopes (including nuclear isomers): 24
Isotope mass range: $147 \rightarrow 164$

**Nuclear
properties**

Key isotopes

Nuclide	Atomic mass	Natural abundance (%)	Half life $T_{1/2}$	Decay mode and energy (MeV)	Nuclear spin I	Nuclear magnetic moment μ	Uses
^{159}Tb	159.9250	100	stable		3/2	+2.008	NMR
^{160}Tb		0	72.1d	β^- (1.72); γ	3	± 1.70	tracer

NMR

	^{159}Tb
Relative sensitivity (^{1}H $= 1.00$)	5.83×10^{-2}
Absolute sensitivity (^{1}H $= 1.00$)	5.83×10^{-2}
Receptivity (^{13}C $= 1.00$)	3.94×10^2
Magnetogyric ratio/rad T^{-1} s^{-1}	6.4306×10^7
Quadrupole moment/m^2	1.3×10^{-28}
Frequency (^{1}H $= 100$ MHz; 2.3488 T)/MHz	22.678

Ground state electron configuration: [Xe]$4f^9 6s^2$
Term symbol: $^6H_{15/2}$
Electron affinity (M$\rightarrow$M$^-$)/kJ mol^{-1}: $\leqslant 50$

**Electron
shell
properties**

Ionization energies/kJ mol^{-1}

1. $M \rightarrow M^+$ 564.6	6. $M^{5+} \rightarrow M^{6+}$
2. $M^+ \rightarrow M^{2+}$ 1112	7. $M^{6+} \rightarrow M^{7+}$
3. $M^{2+} \rightarrow M^{3+}$ 2114	8. $M^{7+} \rightarrow M^{8+}$
4. $M^{3+} \rightarrow M^{4+}$ 3839	9. $M^{8+} \rightarrow M^{9+}$
5. $M^{4+} \rightarrow M^{5+}$	10. $M^{9+} \rightarrow M^{10+}$

Principal lines in atomic spectrum

Wavelength/nm	Species	Sensitivity	Application
350.917	II		AE
356.174	II		AE
387.418	II		AE
431.883	I		AA
432.643	I		AA

Abundance: Earth's crust 1.2 p.p.m.; seawater 1.4×10^{-6} p.p.m.
Biological role: None; toxicity n.a., but probably low

TI

Atomic number: 81
Relative atomic mass ($^{12}C = 12.0000$): **204.3833**

Chemical properties

Soft, silvery-grey metal. Tarnishes readily in moist air and with steam reacts to form TlOH. Attacked by acids, rapidly by HNO_3. Little used because of its toxicity, but still employed in special glass.

Radii/pm: Tl^+ 149; Tl^{3+} 105; atomic 170.4 (α form); covalent 155

Electronegativity: 1.62 (Tl^I) 2.04 (Tl^{III}) (Pauling);
1.44 (Allred-Rochow)

Effective nuclear charge: 5.00 (Slater); 12.25 (Clementi);
13.50 (Froese–Fischer)

Standard reduction potentials $E^{\ominus}/V$

	III		I		0
Acid solution	Tl^{3+}	$\underline{\quad 1.25 \quad}$	Tl^+	$\underline{\quad -0.3363 \quad}$	Tl
		$\underline{\qquad\qquad 0.72 \qquad\qquad}$			

Covalent bonds r/pm E/kJ mol^{-1}

	r/pm	E/kJ mol^{-1}
Tl^I–H	187.0	185
Tl^{III}–C	230	125
Tl^{III}–O	226	375
Tl^{III}–F	195	460
Tl^{III}–Cl	248	368
Tl–Tl	340.8	c. 63

Oxidation states

Tl^{-1} (s^2) Tl_2O, TlOH, Tl_2CO_3, $[Tl(H_2O)_6]^+$ (aq), Tl^+ salts, TlF, TlCl, etc.

Tl^{III} (d^{10}) Tl_2O_3, $[Tl(H_2O)_6]^{3+}$ (aq), TlF_3, $TlCl_3$, $TlBr_3$, $(CH_3)_2Tl^+$ (aq), $TlCl_6^{3-}$

Physical properties

Melting point/K: 576.7
Boiling point/K: 1730
ΔH_{fusion}/kJ mol^{-1}: 4.31
ΔH_{vap}/kJ mol^{-1}: 166.1

Thermodynamic properties (298.15 K, 0.1 MPa)

State	$\Delta_f H^{\ominus}$/kJ mol^{-1}	$\Delta_f G^{\ominus}$/kJ mol^{-1}	$S^{\ominus}$/J K^{-1} mol^{-1}	C_p/J K^{-1} mol^{-1}
Solid	0	0	64.18	26.32
Gas	182.21	147.41	180.963	20.786

Density/kg m^{-3}: 11 850 [293 K]; 11 290 [liquid at m.p.]
Thermal conductivity/W m^{-1} K^{-1}: 46.1 [300 K]
Electrical resistivity/Ω m: 18.0×10^{-8} [273 K]
Mass magnetic susceptibility/kg^{-1} m^3: -3.13×10^{-9} (s)
Molar volume/cm^3: 17.24
Coefficient of linear thermal expansion/K^{-1}: 28×10^{-6}

Lattice structure (cell dimensions/pm), space group

α-Tl hexagonal ($a = 345.6$, $c = 552.5$), P6$_3$/mmc
β-Tl cubic ($a = 388.2$), Im3m
γ-Tl f.c.c. ($a = 485.1$), Fm3m
$T(\alpha \rightarrow \beta) = 503$ K

X-ray diffraction: mass absorption coefficients (μ/ρ)/cm^2 g^{-1}:
CuK$_\alpha$ 224   MoK$_\alpha$ 119

Discovered in 1861 by W. Crookes at London, UK; isolated in
1862 by C.-A. Lamy at Paris, France
[Greek, *thallos* = green twig]

Thallium

Atomic number: 81
Thermal neutron capture cross-section/barns: 3.4 ± 0.5
Number of isotopes (including nuclear isomers): 28
Isotope mass range: $191 \rightarrow 210$

**Nuclear
properties**

Key isotopes

Nuclide	Atomic mass	Natural abundance (%)	Half life $T_{1/2}$	Decay mode and energy (MeV)	Nuclear spin I	Nuclear magnetic moment μ	Uses
^{203}Tl	202.9723	29.524	stable		1/2	+1.6115	NMR
^{204}Tl		0	3.81y	β^-(0.763); EC	2	±0.089	tracer
^{205}Tl	204.9745	70.476	stable		1/2	+1.6274	NMR
^{208}Tl		trace	3.10m	β^-(4.994); γ			

NMR

	$[^{203}$Tl$]$	^{205}Tl
Relative sensitivity (^{1}H = 1.00)	0.18	0.19
Absolute sensitivity (^{1}H = 1.00)	5.51×10^{-2}	0.13
Receptivity (^{13}C = 1.00)	289	769
Magnetogyric ratio/rad T^{-1} s^{-1}	15.3078×10^7	15.4584×10^7
Frequency (^{1}H = 100 MHz; 2.3488 T)/MHz	57.149	57.708

Reference: TlNO$_3$ (aq)

Ground state electron configuration: [Xe]$4f^{14}5d^{10}6s^26p^1$
Term symbol: $^2P_{1/2}$
Electron affinity (M$\rightarrow$M$^-$)/kJ mol^{-1}: 30

**Electron
shell
properties**

Ionization energies/kJ mol^{-1}

1. M $\rightarrow$ M$^+$ 589.3	6. M$^{5+}\rightarrow$M^{6+} (8 300)	
2. M$^+\rightarrow$M^{2+} 1971.0	7. M$^{6+}\rightarrow$M^{7+} (9 500)	
3. M$^{2+}\rightarrow$M^{3+} 2878	8. M$^{7+}\rightarrow$M^{8+} (11 300)	
4. M$^{3+}\rightarrow$M^{4+} (4900)	9. M$^{8+}\rightarrow$M^{9+} (14 000)	
5. M$^{4+}\rightarrow$M^{5+} (6100)	10. M$^{9+}\rightarrow$M^{10+} (16 000)	

Principal lines in atomic spectrum

Wavelength/nm	Species	Sensitivity	Application
237.969	I		AA
276.787	I		AA
351.924	I	U3	AE
377.572*	I	U2	AA, AE
535.046	I	U1	AE

* Lowest energy transition from ground state to nearest empty orbital

Abundance: Earth's crust 0.7 p.p.m.; seawater 1×10^{-5} p.p.m.
Biological role: None; very toxic; teratogenic

Th

Atomic number: 90

Relative atomic mass ($^{12}C = 12.0000$): 232.0381

Chemical properties

Radioactive silvery metal, found in large deposits. Metal protected by oxide coating. Attacked by steam and slowly by acids. Metal itself is soft and ductile but alloys can be very strong. Used in incandescent gas mantles.

Radii/pm: Th^{3+} 101; Th^{4+} 99; atomic 179.8

Electronegativity: 1.3 (Pauling); 1.11 (Allred-Rochow)

Effective nuclear charge: 1.95 (Slater)

Standard reduction potentials $E^{\ominus}/V$

	IV		0
Acid solution	Th^{4+}	$\underline{\quad -1.83 \quad}$	Th
Alkaline solution	ThO_2	$\underline{\quad -2.56 \quad}$	Th

Oxidation states

Th^{II} (d^2)	ThO, ThH_2
Th^{III} (d^1)	ThI_3
Th^{IV} ([Rn])	ThO_2, $[Th(H_2O)_x]^{4+}$(aq), ThF_4, $ThCl_4$etc., ThF_7^{3-}, Th^{4+} salts, complexes

Physical properties

Melting point/K: 2023

Boiling point/K: c. 5060

ΔH_{fusion}/kJ mol^{-1}: < 19.2

ΔH_{vap}/kJ mol^{-1}: 513.67

Thermodynamic properties (298.15 K, 0.1 MPa)

State	$\Delta_f H^{\ominus}$/kJ mol^{-1}	$\Delta_f G^{\ominus}$/kJ mol^{-1}	$S^{\ominus}$/J K^{-1} mol^{-1}	C_p/J K^{-1} mol^{-1}
Solid	0	0	53.39	27.32
Gas	598.3	557.53	190.15	20.79

Density/kg m^{-3}: 11 720 [293 K]

Thermal conductivity/W m^{-1} K^{-1}: 54.0 [300 K]

Electrical resistivity/Ω m: 13.0×10^{-8} [273 K]

Mass magnetic susceptibility/kg^{-1} m^3: $+7.2 \times 10^{-9}$ (s)

Molar volume/cm^3: 19.80

Coefficient of linear thermal expansion/K^{-1}: 12.5×10^{-6}

Lattice structure (cell dimensions/pm), space group

α-Th f.c.c. ($a = 508.42$). Fm3m

β-Th b.c.c. ($a = 411$), Im3m

$T(\alpha \rightarrow \beta) = 1673$ K

X-ray diffraction: mass absorption coefficients (μ/ρ)/cm^2 g^{-1}:

CuK_α 327 MoK_α 143

Thorium

Atomic number: 90
Thermal neutron capture cross-section/barns: 7.4 ± 0.1
Number of isotopes (including nuclear isomers): 12
Isotope mass range: $223 \rightarrow 234$

Nuclear properties

Key isotopes

Nuclide	Atomic mass	Natural abundance (%)	Half life $T_{1/2}$	Decay mode and energy (MeV)	Nuclear spin I	Nuclear magnetic moment μ	Uses
^{228}Th	228.0287	trace	1.913y	$\alpha(5.521)$; γ			
^{229}Th	229.0316	0	7340y	$\alpha(5.167)$; γ			NMR
^{230}Th	230.0331	trace	8.0×10^4y	$\alpha(4.767)$; γ			
^{231}Th		trace	25.5h	$\beta^-(0.318)$; γ			
^{232}Th	232.0382	100	1.41×10^{10}y	$\alpha(4.08)$; γ			tracer
^{234}Th	234.0436	trace	24.1d	$\beta^-(0.263)$; γ			

NMR

	^{229}Th
Relative sensitivity (^{1}H $= 1.00$)	—
Absolute sensitivity (^{1}H $= 1.00$)	—
Receptivity (^{13}C $= 1.00$)	—
Magnetogyric ratio/rad T^{-1} s^{-1}	0.40×10^7
Quadrupole moment/m^2	4.4×10^{-28}
Frequency (^{1}H $= 100$ MHz; 2.3488 T)/MHz	1.5

Ground state electron configuration: $[Rn]6d^2 7s^2$
Term symbol: $^3F_2$
Electron affinity $(M \rightarrow M^-)$/kJ mol^{-1}: n.a.

Electron shell properties

Ionization energies/kJ mol^{-1}

1. $M \rightarrow M^+$ 587	6. $M^{5+} \rightarrow M^{6+}$
2. $M^+ \rightarrow M^{2+}$ 1110	7. $M^{6+} \rightarrow M^{7+}$
3. $M^{2+} \rightarrow M^{3+}$ 1978	8. $M^{7+} \rightarrow M^{8+}$
4. $M^{3+} \rightarrow M^{4+}$ 2780	9. $M^{8+} \rightarrow M^{9+}$
5. $M^{4+} \rightarrow M^{5+}$	10. $M^{9+} \rightarrow M^{10+}$

Principal lines in atomic spectrum

Wavelength/nm	Species	Sensitivity	Application
324.445	I		AA
324.576	II		AA
360.103	II		AE
401.914	II		AE
576.055	I		AA

Abundance: Earth's crust 8.1 p.p.m.; seawater 0.0007 p.p.m.
Biological role: None; low toxicity, but dangerous due to radioactivity

Tm

Atomic number: 69

Relative atomic mass ($^{12}C = 12.0000$): 168.93421

Silvery metal, rarest member of the lanthanide group (rare earths). Obtained from euxenite and monazite. Tarnishes in air and reacts with water. Few uses but some employed as radiation source in portable X-ray equipment.

Radii/pm: Tm^{3+} 87; Tm^{4+} 94; atomic 174.6; covalent 156

Electronegativity: 1.25 (Pauling); 1.11 (Allred-Rochow)

Effective nuclear charge: 2.85 (Slater); 8.58 (Clementi); 11.80 (Froese–Fischer)

Standard reduction potentials $E^{\ominus}/V$

	III	II	0
Acid solution	Tm^{3+} $\xrightarrow{-2.3}$	Tm^{2+} $\xrightarrow{-2.3}$	Tm
Alkaline solution	$Tm(OH)_3$	$\xrightarrow{\qquad -2.83 \qquad}$	Tm

-2.32 (III to 0, Acid solution)

Oxidation states

Tm^{II} (f^{13})	$TmCl_2$, $TmBr_2$, TmI_2
Tm^{III} (f^{12})	Tm_2O_3, $Tm(OH)_3$, $[Tm(H_2O)_x]^{3+}$ (aq), Tm^{3+} salts, TmF_3, $TmCl_3$ etc., complexes

Melting point/K: 1818

Boiling point/K: 2220

ΔH_{fusion}/kJ mol^{-1}: 18.4

ΔH_{vap}/kJ mol^{-1}: 247

Thermodynamic properties (298.15 K, 0.1 MPa)

State	$\Delta_f H^{\ominus}$/kJ mol^{-1}	$\Delta_f G^{\ominus}$/kJ mol^{-1}	$S^{\ominus}$/J K^{-1} mol^{-1}	C_p/J K^{-1} mol^{-1}
Solid	0	0	74.01	27.03
Gas	232.2	197.5	190.113	20.786

Density/kg m^{-3}: 9321 [293 K]

Thermal conductivity/W m^{-1} K^{-1}: 16.8 [300 K]

Electrical resistivity/Ω m: 79.0×10^{-8} [298 K]

Mass magnetic susceptibility/kg^{-1} m^3: $+1.90 \times 10^{-6}$ (s)

Molar volume/cm^3: 18.12

Coefficient of linear thermal expansion/K^{-1}: 13.3×10^{-6}

Lattice structure (cell dimensions/pm), space group

h.c.p. ($a = 353.75$; $c = 555.46$), P6$_3$/mmc

X-ray diffraction: mass absorption coefficients (μ/ρ)/cm^2 g^{-1}: CuK$_\alpha$ 140   MoK$_\alpha$ 80.8

Thulium

Atomic number: 69
Thermal neutron capture cross-section/barns: 115 ± 15
Number of isotopes (including nuclear isomers): 18
Isotope mass range: $161 \rightarrow 176$

Key isotopes

Nuclide	Atomic mass	Natural abundance (%)	Half life $T_{1/2}$	Decay mode and energy (MeV)	Nuclear spin I	Nuclear magnetic moment μ	Uses
^{169}Tm	168.9344	100	stable		1/2	-0.231	NMR
^{170}Tm		0	134d	$\beta^-(0.96); \gamma$	1	± 0.246	tracer

NMR

^{169}Tm

Relative sensitivity (^{1}H $= 1.00$)	5.66×10^{-4}
Absolute sensitivity (^{1}H $= 1.00$)	5.66×10^{-4}
Receptivity (^{13}C $= 1.00$)	3.21
Magnetogyric ratio/rad T^{-1} s^{-1}	-2.21×10^7
Frequency (^{1}H $= 100$ MHz; 2.3488 T)/MHz	8.271

Ground state electron configuration: $[Xe]4f^{13}6s^2$
Term symbol: $^2F_{7/2}$
Electron affinity $(M \rightarrow M^-)$/kJ mol^{-1}: $\leqslant 50$

Ionization energies/kJ mol^{-1}

1. $M \rightarrow M^+$	596.7	6. $M^{5+} \rightarrow M^{6+}$	
2. $M^+ \rightarrow M^{2+}$	1163	7. $M^{6+} \rightarrow M^{7+}$	
3. $M^{2+} \rightarrow M^{3+}$	2285	8. $M^{7+} \rightarrow M^{8+}$	
4. $M^{3+} \rightarrow M^{4+}$	4119	9. $M^{8+} \rightarrow M^{9+}$	
5. $M^{4+} \rightarrow M^{5+}$		10. $M^{9+} \rightarrow M^{10+}$	

Principal lines in atomic spectrum

Wavelength/nm	Species	Sensitivity	Application
371.791	I		AA
376.133	II		AE
376.192	II		AE
409.419	I		AA
410.584	I		AA

Abundance: Earth's crust 0.5 p.p.m.; seawater 4×10^{-8} p.p.m.
Biological role: None; non-toxic; stimulatory

Chemical properties

Soft, pliable, silvery-white metal. Chief ore is cassiteite (SnO_2). Unreactive to oxygen (protected by oxide film) and water but dissolves in acids and bases. Used in solders, alloys, tinplate, polymer additives and antifouling paints.

Radii/pm: Sn^{2+} 93; Sn^{4+} 74; atomic 140.5; covalent 140; Sn^{4+} 294

Electronegativity: 1.96 (Pauling); 1.72 (Allred-Rochow)

Effective nuclear charge: 5.65 (Slater); 9.10 (Clementi); 11.11 (Froese–Fischer)

Standard reduction potentials $E^{\ominus}$/V

	IV		II		0		−IV
Acid solution	SnO_2	$\underline{-0.088}$	SnO	$\underline{-0.104}$	Sn	$\underline{-1.071}$	SnH_4
	Sn^{4+}	$\underline{0.15}$	Sn^{2+}	$\underline{-0.137}$			

[Alkaline solutions contain many different forms]

Covalent bonds	r/pm	E/kJ mol^{-1}
Sn–H	170.1	<314
Sn–C	217	225
Sn^{II}–O	195	557
Sn^{IV}–F	188	322
Sn^{IV}–Cl	231	315
Sn–Sn (α)	281	195

Oxidation states

Sn^{II} SnO, SnF_2, $SnCl_2$, etc., $[Sn(OH)]^+$ (aq), $[Sn_3(OH)_4]^{2+}$ (aq), Sn^{2+} salts

Sn^{IV} SnO_2, SnH_4, SnF_4, $SnCl_4$, etc., $[SnCl_6]^{2-}$ (aq HCl), $[Sn(OH)_6]^{2-}$ (aqbase), organotin compounds

Physical properties

Melting point/K: 505.118
Boiling point/K: 2543
ΔH_{fusion}/kJ mol^{-1}: 7.20
ΔH_{vap}/kJ mol^{-1}: 296.2

Thermodynamic properties (298.15 K, 0.1 MPa)

State	$\Delta_f H^{\ominus}$/kJ mol^{-1}	$\Delta_f G^{\ominus}$/kJ mol^{-1}	$S^{\ominus}$/J K^{-1} mol^{-1}	C_p/J K^{-1} mol^{-1}
Solid (α)	−2.09	0.13	44.14	25.77
Solid (β)	0	0	51.55	26.99
Gas	302.1	267.3	168.486	21.259

Density/kg m^{-3}: 5750 (α); 7310 (β) [293 K]; 6973 [liquid at m.p.]

Thermal conductivity/W m^{-1} K^{-1}: 66.6 (α) [300 K]

Electrical resistivity/Ω m: 11.0×10^{-8} (α) [273 K]

Mass magnetic susceptibility/kg^{-1} m^3: -4.0×10^{-9} (α); $+3.3 \times 10^{-10}$ (β)

Molar volume/cm^3: 16.24 (β)

Coefficient of linear thermal expansion/K^{-1}: 5.3×10^{-6} (α); 21.2×10^{-6} (β)

Lattice structure (cell dimensions/pm), space group

α-Sn (grey) cubic ($a = 648.92$), Fd3m
β-Sn (white) tetragonal ($a = 583.16$, $c = 318.13$), I4$_2$/amd
$T(\alpha \rightarrow \beta) = 286.4$ K [β form at room temperatures]

X-ray diffraction: mass absorption coefficients (μ/ρ)/cm^2 g^{-1}:
CuK$_{\alpha}$ 256   MoK$_{\alpha}$ 31.1

Tin

Atomic number: 50
Thermal neutron capture cross-section/barns: 0.63 ± 0.01
Number of isotopes (including nuclear isomers): 28
Isotope mass range: $108 \rightarrow 128$

Key isotopes

Nuclide	Atomic mass	Natural abundance (%)	Half life $T_{1/2}$	Decay mode and energy (MeV)	Nuclear spin I	Nuclear magnetic moment μ	Uses
^{112}Sn	111.9040	1.0	stable				
^{113}Sn		0	115d	EC(1.02); γ	1/2	± 0.88	tracer
^{114}Sn	113.9030	0.7	stable				
^{115}Sn	114.9035	0.4	stable		1/2	-0.9178	NMR
^{116}Sn	115.9021	14.7	stable		0		
^{117}Sn	116.9031	7.7	stable		1/2	-1.000	NMR
^{118}Sn	117.9018	24.3	stable		0		
^{119}Sn	118.9034	8.6	stable		1/2	-1.0461	NMR
^{120}Sn		32.4	stable		0		
^{121}Sn		0	27.5h	β^-(0.383); no γ	3/2	± 0.699	tracer
^{122}Sn	121.9034	4.6	stable				
^{124}Sn	123.9052	5.6	stable				

NMR

	$[^{115}Sn]$	$[^{117}Sn]$	^{119}Sn
Relative sensitivity ($^1H = 1.00$)	3.5×10^{-2}	4.52×10^{-2}	5.18×10^{-2}
Absolute sensitivity ($^1H = 1.00$)	1.22×10^{-4}	3.44×10^{-3}	4.44×10^{-3}
Receptivity ($^{13}C = 1.00$)	0.693	19.54	25.2
Magnetogyric ratio/rad $T^{-1} s^{-1}$	-8.7475×10^7	-9.5319×10^7	-9.9756×10^7
Frequency ($^1H = 100$ MHz; 2.3488 T)/MHz	32.699	35.625	37.272

Reference: $Sn(CH_3)_4$

Ground state electron configuration: $[Kr]4d^{10}5s^25p^2$
Term symbol: $^3P_0$
Electron affinity $(M \rightarrow M^-)$/kJ mol^{-1}: 121

Ionization energies/kJ mol^{-1}

1. $M \rightarrow M^+$	708.6	6. $M^{5+} \rightarrow M^{6+}$	(9 900)
2. $M^+ \rightarrow M^{2+}$	1411.8	7. $M^{6+} \rightarrow M^{7+}$	(12 200)
3. $M^{2+} \rightarrow M^{3+}$	2943.0	8. $M^{7+} \rightarrow M^{8+}$	(14 600)
4. $M^{3+} \rightarrow M^{4+}$	3930.2	9. $M^{8+} \rightarrow M^{9+}$	(17 000)
5. $M^{4+} \rightarrow M^{5+}$	6974	10. $M^{9+} \rightarrow M^{10+}$	(20 600)

Principal lines in atomic spectrum

Wavelength/nm	Species	Sensitivity	Application
224.605	I		AA
283.999	I	U1	AE
286.333*	I	U2	AA, AE
303.412	I		AA, AE
326.233	I	U3	AE

* Lowest energy transition from ground state to nearest empty orbital

Abundance: Earth's crust 2.1 p.p.m.; seawater 0.003 p.p.m.
Biological role: Non-toxic but organotins used as biocides

Ti

Atomic number: 22
Relative atomic mass ($^{12}C = 12.0000$): **47.88**

Chemical properties

Hard, lustrous, silvery metal. Main ores are ilmenite, $FeTiO_3$, and rutile, TiO_2. Resists corrosion due to oxide layer, but powdered metal burns in air. Unaffected both by acids, (except HF and concentrated H_2SO_4), and alkalis. Used in paints and lightweight alloys.

Radii/pm: Ti^{2+} 80; Ti^{3+} 69; atomic 144.8; covalent 132
Electronegativity: 1.54 (Pauling); 1.32 (Allred-Rochow)
Effective nuclear charge: 3.15 (Slater); 4.82 (Clementi); 6.37 (Froese–Fischer)

Standard reduction potentials $E^{\ominus}/V$

	IV	III	II	0

Acid solution

$$\overset{-0.86}{\phantom{TiO^{2+}}}$$
$$\overset{-1.21}{\phantom{Ti^{3+}}}$$

Acid solution TiO^{2+} $\xrightarrow{0.1}$ Ti^{3+} $\xrightarrow{-0.37}$ Ti^{2+} $\xrightarrow{-1.63}$ Ti
TiO_2 $\xrightarrow{-0.56}$ Ti_2O_3 $\xrightarrow{-1.23}$ TiO $\xrightarrow{-1.31}$

Alkaline solution TiO_2 $\xrightarrow{-1.38}$ Ti_2O_3 $\xrightarrow{-1.95}$ TiO $\xrightarrow{-2.13}$ Ti

Oxidation states

Ti^{-I}	(d^5)	rare $[Ti(bipyridyl)_3]^-$
Ti^0	(d^4)	rare $[Ti(bipyridyl)_3]$
Ti^{II}	(d^2)	TiO, $TiCl_2$, $TiBr_2$, TiI_2, no solution chemistry (reduces H_2O); complexes
Ti^{III}	(d^1)	Ti_2O_3, $[Ti(H_2O)_6]^{3+}$ (aq), TiF_3, $TiCl_3$ etc., complexes
Ti^{IV}	$([Ar])$	TiO_2, TiO^{2+} (aq), $[Ti(OH)_3]^{2+}$ (aq), TiF_4, $TiCl_4$ etc., titanates (TiO_4^{4-}, TiO_3^{2-}), complexes

Physical properties

Melting point/K: 1933
Boiling point/K: 3560
ΔH_{fusion}/kJ mol^{-1}: 20.9
ΔH_{vap}/kJ mol^{-1}: 425.5

Thermodynamic properties (298.15 K, 0.1 MPa)

State	$\Delta_f H^{\ominus}$/kJ mol^{-1}	$\Delta_f G^{\ominus}$/kJ mol^{-1}	$S^{\ominus}$/J K^{-1} mol^{-1}	C_p/J K^{-1} mol^{-1}
Solid	0	0	30.63	25.02
Gas	469.9	425.1	180.298	24.430

Density/kg m^{-3}: 4540 [293 K]; 4110 [liquid at m.p.]
Thermal conductivity/W m^{-1} K^{-1}: 21.9 [300 K]
Electrical resistivity/Ω m: 42.0×10^{-8} [293 K]
Mass magnetic susceptibility/kg^{-1} m^3: $+4.01 \times 10^{-8}$ (s)
Molar volume/cm^3: 10.55
Coefficient of linear thermal expansion/K^{-1}: 8.35×10^{-6}

Lattice structure (cell dimensions/pm), space group

α-Ti h.c.p. ($a = 295.11$, $c = 468.43$), P6$_3$/mmc
β-Ti b.c.c. ($a = 330.65$), Im3m
$T(\alpha \rightarrow \beta) = 1155$ K
High pressure form: ($a = 462.5$; $c = 281.3$), P$\bar{3}$m1

X-ray diffraction: mass absorption coefficients (μ/ρ)/cm^2 g^{-1}:
CuK$_\alpha$ 208   MoK$_\alpha$ 24.2

Discovered in 1791 by Rev. W. Gregor at Creed, Cornwall, UK,
and independently by M. H. Klaproth in 1795 at Berlin, Germany

Titanium

[Called after the Titans, sons of the Earth goddess]

Atomic number: 22

Thermal neutron capture cross-section/barns: 6.1 ± 0.2

Number of isotopes (including nuclear isomers): 9

Isotope mass range: $43 \rightarrow 51$

Nuclear properties

Key isotopes

Nuclide	Atomic mass	Natural abundance (%)	Half life $T_{1/2}$	Decay mode and energy (MeV)	Nuclear spin I	Nuclear magnetic moment μ	Uses
^{44}Ti		0	48y	EC(0.16); γ			tracer
^{46}Ti	45.95263	8.2	stable				
^{47}Ti	45.9518	7.4	stable		5/2	-0.78846	NMR
^{48}Ti	45.948	73.8	stable				
^{49}Ti	45.94787	5.4	stable		7/2	-1.0414	NMR
^{50}Ti	49.9448	5.2	stable				

NMR

	^{47}Ti	^{49}Ti
Relative sensitivity (^{1}H $= 1.00$)	2.09×10^{-3}	3.76×10^{-3}
Absolute sensitivity (^{1}H $= 1.00$)	1.52×10^{-4}	2.07×10^{-4}
Receptivity (^{13}C $= 1.00$)	0.864	1.18
Magnetogyric ratio/rad T^{-1} s^{-1}	1.5084×10^7	1.5080×10^7
Quadrupole moment/m^2	$+0.29 \times 10^{-28}$	$+0.24 \times 10^{-28}$
Frequency (^{1}H $= 100$ MHz; 2.3488 T)/MHz	5.637	5.638

Reference: TiF_6^{2-} (conc. HF)

Ground state electron configuration: [Ar]$3d^2 4s^2$

Term symbol: $^3F_2$

Electron affinity (M$\rightarrow$M$^-$)/kJ mol^{-1}: -2

Electron shell properties

Ionization energies/kJ mol^{-1}

1. M $\rightarrow$ M$^+$	658	6. M$^{5+} \rightarrow$ M^{6+}	11516
2. M$^+ \rightarrow$ M^{2+}	1310	7. M$^{6+} \rightarrow$ M^{7+}	13590
3. M$^{2+} \rightarrow$ M^{3+}	2652	8. M$^{7+} \rightarrow$ M^{8+}	16260
4. M$^{3+} \rightarrow$ M^{4+}	4175	9. M$^{8+} \rightarrow$ M^{9+}	18640
5. M$^{4+} \rightarrow$ M^{5+}	9573	10. M$^{9+} \rightarrow$ M^{10+}	20830

Principal lines in atomic spectrum

Wavelength/nm	Species	Sensitivity	Application
319.992	I		AA
334.941	II		
364.268	I		AA, AE
365.350	I	U2	AA, AE
498.173	I	U1	

Abundance: Earth's crust 6320 p.p.m.; seawater 0.001 p.p.m.

Biological role: None; non-toxic; suspected carcinogen; stimulatory

Atomic number: 74

Relative atomic mass ($^{12}C = 12.0000$): 183.85

Chemical properties

Obtained as dull grey powder, difficult to melt. Metal is lustrous and silvery white. Ores are scheelite $CaWO_4$ and wolframite $(Fe,Mn)WO_4$. Resists attack by oxygen, acids, and alkalis. Used in alloys, light bulb filaments, and cutting tools.

Radii/pm: W^{4+} 68; W^{6+} 62; atomic 137.0; covalent 130

Electronegativity: 2.36 (Pauling); 1.40 (Allred-Rochow)

Effective nuclear charge: 4.35 (Slater); 9.85 (Clementi); 14.22 (Froese–Fischer)

Standard reduction potentials $E^{\ominus}/V$

	VI		V		IV		0

Acid solution

$$WO_3 \xrightarrow{-0.029} W_2O_5 \xrightarrow{-0.031} WO_2 \xrightarrow{-0.119} W$$

with $WO_3 \xrightarrow{-0.090} W_2O_5$

Alkaline solution

$$WO_4^{2-} \xrightarrow{-1.259} WO_2 \xrightarrow{-0.982} W$$

with $WO_4^{2-} \xrightarrow{-1.074} WO_2$

Oxidation states

W^{-IV}	(d^{10})	$[W(CO)_4]^{4-}$
W^{-II}	(d^8)	$[W(CO)_5]^{2-}$
W^{-I}	(d^7)	$[W_2(CO)_{10}]^{2-}$
W^0	(d^6)	$W(CO)_6$
W^{II}	(d^4)	WCl_2, WBr_2, WI_2, complexes
W^{III}	(d^3)	WCl_3, WBr_3, WI_3, complexes
W^{IV}	(d^2)	WO_2, WF_4, WCl_4, etc. WS_2, complexes
W^{V}	(d^1)	W_2O_5, WF_5, WCl_5, WF_6^-, complexes
W^{VI}	(d^0, f^{14})	WO_3, WO_4^{2-}, WF_6, WCl_6, $WOCl_4$, polytungstates, complexes

NB. There are no aqua ions of W in any oxidation state.

Physical properties

Melting point/K: 3680 ± 20

Boiling point/K: 5930

ΔH_{fusion}/kJ mol^{-1}: 35.2

ΔH_{vap}/kJ mol^{-1}: 824.2

Thermodynamic properties (298.15 K, 0.1 MPa)

State	$\Delta_f H^{\ominus}$/kJ mol^{-1}	$\Delta_f G^{\ominus}$/kJ mol^{-1}	$S^{\ominus}$/J K^{-1} mol^{-1}	C_p/J K^{-1} mol^{-1}
Solid	0	0	32.64	24.27
Gas	849.4	807.1	173.950	21.309

Density/kg m^{-3}: 19 300 [293 K]; 17 700 [liquid at m.p.]

Thermal conductivity/W m^{-1} K^{-1}: 174 [300 K]

Electrical resistivity/Ω m: 5.65×10^{-8} [300 K]

Mass magnetic susceptibility/kg^{-1} m^3: $+4.0 \times 10^{-9}$ (s)

Molar volume/cm^3: 9.53

Coefficient of linear thermal expansion/K^{-1}: 4.59×10^{-6}

Lattice structure (cell dimensions/pm), space group

b.c.c. ($a = 316.522$), Im3m

X-ray diffraction: mass absorption coefficients (μ/ρ)/cm^2 g^{-1}:
CuK_{α} 172 MoK_{α} 99.1

Tungsten (Wolfram)

Atomic number: 74
Thermal neutron capture cross-section/barns: 18.5 ± 0.5
Number of isotopes (including nuclear isomers): 22
Isotope mass range: $173 \rightarrow 189$

Key isotopes

Nuclide	Atomic mass	Natural abundance (%)	Half life $T_{1/2}$	Decay mode and energy (MeV)	Nuclear spin I	Nuclear magnetic moment μ	Uses
^{180}W	179.9470	0.10	stable				
^{182}W	181.9483	26.3	stable		0		
^{183}W	182.9503	14.3	stable		1/2	+0.1169	NMR
^{184}W	183.9510	30.7	stable		0		
^{185}W		0	75d	β^-(0.432); no γ	3/2		tracer
^{186}W	185.9543	28.6	stable		0		
^{187}W		0	23.9h	β^-(1.315); γ	3/2		tracer

NMR

	^{183}W
Relative sensitivity ($^1H = 1.00$)	7.20×10^{-4}
Absolute sensitivity ($^1H = 1.00$)	1.03×10^{-5}
Receptivity ($^{13}C = 1.00$)	0.0589
Magnetogyric ratio/rad T^{-1} s^{-1}	1.1145×10^7
Frequency ($^1H = 100$ MHz; 2.3488 T)/MHz	4.161

Reference: WF_6

Ground state electron configuration: $[Xe]4f^{14}5d^46s^2$
Term symbol: $^5D_0$
Electron affinity ($M \rightarrow M^-$)/kJ mol^{-1}: 119

Ionization energies/kJ mol^{-1}

1. $M \rightarrow M^+$	770	6. $M^{5+} \rightarrow M^{6+}$	(5900)	
2. $M^+ \rightarrow M^{2+}$	(1700)	7. $M^{6+} \rightarrow M^{7+}$		
3. $M^{2+} \rightarrow M^{3+}$	(2300)	8. $M^{7+} \rightarrow M^{8+}$		
4. $M^{3+} \rightarrow M^{4+}$	(3400)	9. $M^{8+} \rightarrow M^{9+}$		
5. $M^{4+} \rightarrow M^{5+}$	(4600)	10. $M^{9+} \rightarrow M^{10+}$		

Principal lines in atomic spectrum

Wavelength/nm	Species	Sensitivity	Application
202.998	II		
255.135	I		AA
268.142	I		AA
400.875	I	U3	AE
429.461	I	U2	AE
430.211	I	U1	AE
498.259*	I		

* Lowest energy transition from ground state to nearest empty orbital

Abundance: Earth's crust 1.2 p.p.m.; seawater $< 1 \times 10^{-6}$ p.p.m.
Biological role: None; probably of low toxicity

Atomic number: 92

Relative atomic mass ($^{12}C = 12.0000$): 238.0289

Chemical properties

Radioactive silvery metal; large deposits. Malleable, ductile, and tarnishes in air. Attacked by steam and acids but not by alkalis. Used a nuclear fuel and in nuclear weapons.

Radii/pm: U^{3+} 103; U^{4+} 97; U^{5+} 89; U^{6+} 80; atomic 138.5

Electronegativity: 1.38 (Pauling); 1.22 (Allred-Rochow)

Effective nuclear charge: 1.80 (Slater)

Standard reduction potentials $E^{\ominus}/V$

	VI		V		IV		III		0

$$-0.027 \qquad\qquad -1.38$$

Acid solution $\quad UO_2^{2+} \xrightarrow{0.16} UO_2^{+} \xrightarrow{-0.38} U^{4+} \xrightarrow{-0.52} U^{3+} \xrightarrow{-1.66} U$

Alkaline solution $UO_2(OH)_2 \xrightarrow{\qquad -0.3 \qquad} UO_2 \xrightarrow{-2.6} U(OH)_3 \xrightarrow{-2.10} U$

Oxidation states

U^{II} ($f^3 d^1$)	UO?
U^{III} (f^3)	$[U(H_2O)_x]^{3+}$ (aq) unstable, UF_3, UCl_3 etc., $[U(C_5H_5)_3]$
U^{IV} (f^2)	UO_2, $[U(H_2O)_x]^{4+}$ (aq), salts, UF_4, UCl_4 etc., $[UCl_6]^{2-}$
U^{V} (f^1)	U_2O_5, UO_2^{+} (aq) unstable, UF_5, UCl_5, UBr_5, UF_6^{-}, UF_7^{2-}, UF_8^{3-}
U^{VI} (f^0, [Rn])	UO_3 (U_3O_8), UO_2^{2+} (aq), salts, UF_6, UCl_6, complexes

Physical properties

Melting point/K: 1405.5

Boiling point/K: 4018

ΔH_{fusion}/kJ mol^{-1}: 15.5

ΔH_{vap}/kJ mol^{-1}: 417.1

Thermodynamic properties (298.15 K, 0.1 MPa)

State	$\Delta_f H^{\ominus}$/kJ mol^{-1}	$\Delta_f G^{\ominus}$/kJ mol^{-1}	$S^{\ominus}$/J K^{-1} mol^{-1}	C_p/J K^{-1} mol^{-1}
Solid	0	0	50.21	27.665
Gas	535.6	491.2	199.77	23.694

Density/kg m^{-3}: 18 950 [293 K]; 17 907 [liquid at m.p.]

Thermal conductivity/W m^{-1} K^{-1}: 27.6 [300 K]

Electrical resistivity/Ω m: 30.8×10^{-8} [273 K]

Mass magnetic susceptibility/kg^{-1} m^3: $+2.16 \times 10^{-8}$ (s)

Molar volume/cm^3: 12.56

Coefficient of linear thermal expansion/K^{-1}: 12.6×10^{-6}

Lattice structure (cell dimensions/pm), space group

α-U orthorhombic ($a = 284.785$, $b = 585.801$, $c = 494.553$), Cmcm
β-U tetragonal ($a = 1076.0$, $c = 565.2$), P4$_2$/mnm or P4$_2$nm
γ-U b.c.c. ($a = 352.4$), Im3m

$T(\alpha \rightarrow \beta) = 941$ K
$T(\beta \rightarrow \gamma) = 1047$ K

X-ray diffraction: mass absorption coefficients (μ/ρ)/cm^2 g^{-1}:
CuK$_\alpha$ 352   MoK$_\alpha$ 153

Discovered in 1789 by M. H. Klaproth at Berlin, Germany;
isolated in 1841 by E.-M. Peligot at Paris, France
[Named after the planet Uranus]

Uranium

Atomic number: 92

Thermal neutron capture cross-section/barns: 7.60 ± 0.07

Number of isotopes (including nuclear isomers): 15

Isotope mass range: $227 \rightarrow 240$

**Nuclear
properties**

Key isotopes

Nuclide	Atomic mass	Natural abundance (%)	Half life $T_{1/2}$	Decay mode and energy (MeV)	Nuclear spin I	Nuclear magnetic moment μ	Uses
^{234}U	234.0409	0.005	2.47×10^5y	$\alpha(4.856); \gamma$			
^{235}U	235.0439	0.720	7.00×10^8y	$\alpha(4.681);$ SF$; \gamma$	7/2	-0.43	NMR
^{236}U	236.0457	0	2.39×10^7y	$\alpha(4.573);$ SF$; \gamma$			
^{238}U	238.0508	99.275	4.51×10^9y	$\alpha(4.268);$ SF$: \gamma$			

NMR

^{235}U

Relative sensitivity ($^1H = 1.00$) $\qquad$ 1.21×10^{-4}

Absolute sensitivity ($^1H = 1.00$) $\qquad$ 8.71×10^{-7}

Receptivity ($^{13}C = 1.00$) $\qquad$ 5.4×10^{-3}

Magnetogyric ratio/rad T^{-1} s^{-1} $\qquad$ -0.4926×10^7

Quadrupole moment/m^2 $\qquad$ 4.55×10^{-28}

Frequency ($^1H = 100$ MHz; 2.3488 T)/MHz $\quad$ 1.790

Reference: UF_6

Ground state electron configuration: $[Rn]5f^36d^17s^2$

Term symbol: $^5L_6$

Electron affinity ($M \rightarrow M^-$)/kJ mol^{-1}: n.a.

**Electron
shell
properties**

Ionization energies/kJ mol^{-1}

1. $M \rightarrow M^+$	584	6. $M^{5+} \rightarrow M^{6+}$	
2. $M^+ \rightarrow M^{2+}$	1420	7. $M^{6+} \rightarrow M^{7+}$	
3. $M^{2+} \rightarrow M^{3+}$		8. $M^{7+} \rightarrow M^{8+}$	
4. $M^{3+} \rightarrow M^{4+}$		9. $M^{8+} \rightarrow M^{9+}$	
5. $M^{4+} \rightarrow M^{5+}$		10. $M^{9+} \rightarrow M^{10+}$	

Principal lines in atomic spectrum

Wavelength/nm	Species	Sensitivity	Application
356.659	I		AA
358.488	I		AA
367.007	II		AE
385.958	II		
591.540	I		AE

Abundance: Earth's crust 2.3 p.p.m.; seawater 0.003 p.p.m.

Biological role: None; toxic and dangerous due to radioactivity

V

Atomic number: 23
Relative atomic mass ($^{12}C = 12.0000$): **50.9415**

Chemical properties

Shiny, silvery metal, soft when pure. Found as patronite VS_4 but obtained as by-product from other ores and Venezuelan oils. Resists corrosion due to protective oxide film. Attacked by concentrated acids but not by fused alkalis. Used mainly as alloys and in steel.

Radii/pm: V^{2+} 72; V^{3+} 65; V^{4+} 61; V^{5+} 59; atomic 132.1

Electronegativity: 1.63 (Pauling); 1.45 (Allred-Rochow)

Effective nuclear charge: 3.30 (Slater); 4.98 (Clementi); 6.65 (Froese–Fischer)

Standard reduction potentials $E^{\ominus}$/V

Acid solution (pH < 3)

Alkaline solution (pH > 12)

Oxidation states

V^{-III}	(d^8)	rare $[V(CO)_5]^{3-}$
V^{-I}	(d^6)	$[V(CO)_6]^-$
V^0	(d^5)	$[V(CO)_6]$
V^I	(d^4)	$[V(bipyridyl)_3]^+$
V^{II}	(d^3)	VO, $[V(H_2O)_6]^{2+}$ (aq), VF_2, VCl_2, complexes
V^{III}	(d^2)	V_2O_3, $[V(H_2O)_6]^{3+}$ (aq), VF_3, VCl_3, $[VCl_4]^-$
V^{IV}	(d^1)	VO_2, VO^{2+} (aq), VF_4, VCl_4, complexes
V^V	$(d^0, [Ar])$	V_2O_5, VO_2^+ (aq), VO_4^{3-} (aq, alkali), VF_5, VF_6^-, complexes

Physical properties

Melting point/K: 2160
Boiling point/K: 3650
ΔH_{fusion}/kJ mol^{-1}: 17.6
ΔH_{vap}/kJ mol^{-1}: 459.70

Thermodynamic properties (298.15 K, 0.1 MPa)

State	$\Delta_f H^{\ominus}$/kJ mol^{-1}	$\Delta_f G^{\ominus}$/kJ mol^{-1}	$S^{\ominus}$/J K^{-1} mol^{-1}	C_p/J K^{-1} mol^{-1}
Solid	0	0	28.91	24.89
Gas	514.21	754.43	182.298	26.012

Density/kg m^{-3}: 6110 [292 K]; 5550 [liquid at m.p.]
Thermal conductivity/W m^{-1} K^{-1}: 30.7 [300 K]
Electrical resistivity/K: 24.8×10^{-8} [293 K]
Mass magnetic susceptibility/kg^{-1} m^3: $+6.28 \times 10^{-8}$ (s)
Molar volume/cm^3: 8.34
Coefficient of linear thermal expansion/K^{-1}: 8.3×10^{-6}

Lattice structure (cell dimensions/pm), space group

b.c.c. ($a = 302.40$), Im3m

X-ray diffraction: mass absorption coefficients (μ/ρ)/cm^2 g^{-1}:
CuK$_\alpha$ 233   MoK$_\alpha$ 27.5

Vanadium

Atomic number: 23
Thermal neutron capture cross-section/barns: 5.06 ± 0.06
Number of isotopes (including nuclear isomers): 9
Isotope mass range: $46 \rightarrow 54$

Nuclear properties

Key isotopes

Nuclide	Atomic mass	Natural abundance (%)	Half life $T_{1/2}$	Decay mode and energy (MeV)	Nuclear spin I	Nuclear magnetic moment μ	Uses
^{48}V	47.952	0	16.0d	β^+(4.013), EC; γ	4	± 1.6	tracer
^{49}V	48.948	0	330d	EC(0.601); no γ	7/2	± 4.5	tracer
^{50}V	49.9472	0.250	6×10^{15}y	EC	6	$+3.3470$	NMR
^{51}V	50.9440	99.750	stable		7/2	$+5.1485$	NMR

NMR

	$[^{50}\text{V}]$	^{51}V
Relative sensitivity (^{1}H = 1.00)	5.55×10^{-2}	0.38
Absolute sensitivity (^{1}H = 1.00)	1.33×10^{-4}	0.38
Receptivity (^{13}C = 1.00)	0.755	2150
Magnetogyric ratio/rad T^{-1} s^{-1}	2.6491×10^7	7.0362×10^7
Quadrupole moment/m^2	$+0.21 \times 10^{-28}$	-0.052×10^{-28}
Frequency (^{1}H = 100 MHz; 2.3488 T)/MHz	9.970	26.289

Reference: VOCl$_3$

Ground state electron configuration: [Ar]3d^{3}4s^2
Term symbol: ^{4}F$_{3/2}$
Electron affinity (M$\rightarrow$M$^-$)/kJ mol^{-1}: 61

Electron shell properties

Ionization energies/kJ mol^{-1}

1. M $\rightarrow$ M$^+$	650	6. M^{5+} $\rightarrow$ M^{6+}	12 362
2. M$^+$ $\rightarrow$ M^{2+}	1414	7. M^{6+} $\rightarrow$ M^{7+}	14 489
3. M^{2+} $\rightarrow$ M^{3+}	2828	8. M^{7+} $\rightarrow$ M^{8+}	16 760
4. M^{3+} $\rightarrow$ M^{4+}	4507	9. M^{8+} $\rightarrow$ M^{9+}	19 860
5. M^{4+} $\rightarrow$ M^{5+}	6294	10. M^{9+} $\rightarrow$ M^{10+}	22 240

Principal lines in atomic spectrum

Wavelength/nm	Species	Sensitivity	Application
309.311	II	V1	AE
318.398	I		AA, AE
318.540	I	U2	AA, AE
437.924	I	U1	AA, AE
485.148*	I		

* Lowest energy transition from ground state to nearest empty orbital

Abundance: Earth's crust 136 p.p.m.; seawater 0.002 p.p.m.
Biological role: Essential trace element; some compounds quite toxic; stimulatory

Xe

Atomic number: 54
Relative atomic mass (^{12}C = 12.0000): 131.29

Chemical properties

Colourless, odourless gas obtained from liquid air. Inert towards most other chemicals but reacts with fluorine gas to form xenon fluorides. Oxides, acids and salts known. Little used outside research.

Radii/pm: Xe^+ 190; atomic 218; covalent 209
Electronegativity: 2.6 (Pauling); n.a. (Allred-Rochow)
Effective nuclear charge: 8.25 (Slater); 12.42 (Clementi); 15.61 (Froese-Fischer)

Standard reduction potentials $E^\ominus$/V

| | VIII | IV | II | I | 0 |

Acid solution

$$H_4XeO_6 \xrightarrow{2.42} XeO_3 \xrightarrow{2.12} Xe$$

with overbar $\overline{2.18}$ from H_4XeO_6 to XeO_3

$$XeF_2 \xrightarrow{0.9} XeF \xrightarrow{3.4}$$ and $\overline{2.32}$

Alkaline solution $HXeO_6^{3-} \xrightarrow{0.99} HXeO_4^- \xrightarrow{1.24} Xe$ with $\overline{1.18}$

Covalent bonds

	r/pm	E/kJ mol^{-1}
Xe–O (XeO$_3$)	176	84
Xe–F	194	133

Oxidation states

Xe^0	clathrates: $Xe_8(H_2O)_{46}$, Xe (quinol)$_3$
Xe^{II}	XeF_2, $[XeF]^+[AsF_6]^-$
Xe^{IV}	XeF_4
Xe^{VI}	XeO_3, $XeOF_4$, XeO_2F_2, XeF_6, XeF_7^-, XeF_8^{2-}, $[XeF_5]^+[AsF_6]^-$
Xe^{VIII}	XeO_4, XeO_3F_2, Ba_2XeO_6, XeO_6^{4-} (aq)

Physical properties

Melting point/K: 161.3
Boiling point/K: 166.1
ΔH_{fusion}/kJ mol^{-1}: 3.10
ΔH_{vap}/kJ mol^{-1}: 12.65

Thermodynamic properties (298.15 K, 0.1 MPa)

State	$\Delta_f H^\ominus$/kJ mol^{-1}	$\Delta_f G^\ominus$/kJ mol^{-1}	$S^\ominus$/J K^{-1} mol^{-1}	C_p/J K^{-1} mol^{-1}
Gas	0	0	169.683	20.786

Density/kg m^{-3}: 3540 [solid, m.p.]; 2939 [liquid, b.p.]; 5.8971 [gas, 273 K]
Thermal conductivity/W m^{-1} K^{-1}: 0.005 69 [300 K]$_g$
Mass magnetic susceptibility/kg^{-1} m^3: -4.20×10^{-9} (g)
Molar volume/cm^3: 37.09 [161 K]

Lattice structure (cell dimensions/pm), space group

f.c.c. (88 K) ($a = 619.7$), Fm3m

X-ray diffraction: mass absorption coefficients (μ/ρ)/cm^2 g^{-1}:
CuK$_\alpha$ 306 MoK$_\alpha$ 39.2

Discovered in 1898 by Sir William Ramsay and M. W. Travers at London, UK

[Greek, *xenos* = stranger]

Xenon

Atomic number: 54
Thermal neutron capture cross-section/barns: 24.5
Number of isotopes (including nuclear isomers): 31
Isotope mass range: 118→142

Key isotopes

Nuclide	Atomic mass	Natural abundance (%)	Half life $T_{1/2}$	Decay mode and energy (MeV)	Nuclear spin I	Nuclear magnetic moment μ	Uses
^{127}Xe		0	36.4d	EC(0.44); γ			tracer
^{129}Xe	128.9048	26.4	stable		1/2	−0.7768	NMR
^{130}Xe	129.9035	4.1	stable				
^{131}Xe	130.9051	21.2	stable		3/2	+0.6908	NMR
^{132}Xe	131.9042	26.9	stable		0		
^{133}Xe		0	5.270d	β^-(0.427); γ			tracer, medical
^{134}Xe	133.9054	10.4	stable		0		
^{136}Xe	135.9072	8.9	stable		0		

NMR

	^{129}Xe	$[^{131}$Xe]
Relative sensitivity (^{1}H = 1.00)	2.12×10^{-2}	2.76×10^{-3}
Absolute sensitivity (^{1}H = 1.00)	5.60×10^{-3}	5.84×10^{-4}
Receptivity (^{13}C = 1.00)	31.8	3.31
Magnetogyric ratio/rad T^{-1} s^{-1}	-7.4003×10^7	2.1939×10^7
Quadrupole moment/m^2	—	-0.12×10^{-28}
Frequency (^{1}H = 100 MHz; 2.3488 T)/MHz	27.660	8.199

Reference: XeOF$_4$

Ground state electron configuration: [Kr]$4d^{10}5s^25p^6$
Term symbol: $^1S_0$
Electron affinity (M→M$^-$)/kJ mol^{-1}: −41 (calc.)

Ionization energies/kJ mol^{-1}

1. M →M$^+$	1170.4	6. M^{5+}→M^{6+}	(6 600)
2. M$^+$→M^{2+}	2046	7. M^{6+}→M^{7+}	(9 300)
3. M^{2+}→M^{3+}	3097	8. M^{7+}→M^{8+}	(10 600)
4. M^{3+}→M^{4+}	(4300)	9. M^{8+}→M^{9+}	(19 800)
5. M^{4+}→M^{5+}	(5500)	10. M^{9+}→M^{10+}	(23 000)

Principal lines in atomic spectrum

Wavelength/nm	Species	Sensitivity	Application
146.961* (vac)	I		
450.098	I	U4	AE
462.428	I	U3	AE
467.123	I	U2	AE
823.164	I		

* Lowest energy transition from ground state to nearest empty orbital

Abundance: Atmosphere 0.086 p.p.m. (by volume); Earth's crust nil; seawater 4.7×10^{-5} p.p.m.
Biological role: None; non-toxic as element

Yb

Atomic number: 70
Relative atomic mass ($^{12}C = 12.0000$): 173.04

Chemical properties

Soft, silvery-white metal of the lanthanide (rare earth) group. Obtained from euxenite. Slowly oxidized by air, reacts with water. Little used.

Radii/pm: Yb^{2+} 113; Yb^{3+} 86; atomic 194; covalent 170

Electronegativity: n.a. (Pauling); 1.06 (Allred–Rochow)

Effective nuclear charge: 2.85 (Slater); 8.59 (Clementi); 11.90 (Froese–Fischer)

Standard reduction potentials $E^{\ominus}/V$

	III		II		0
Acid solution	Yb^{3+}	$\xrightarrow{-1.05}$	Yb^{2+}	$\xrightarrow{-2.8}$	Yb
Alkaline solution	$Yb(OH)_3$		$\xrightarrow{\hspace{2cm}-2.74\hspace{2cm}}$		Yb

(Acid solution III→0: -2.22)

Oxidation states

Yb^{II} (f^{14})	YbO, YbS, YbF_2, $YbCl_2$ etc.
Yb^{III} (f^{13})	Yb_2O_3, $Yb(OH)_3$, $[Yb(H_2O)_x]^{3+}$ (aq), Yb^{3+} salts, YbF_3, $YbCl_3$, etc. $YbCl_6^{3-}$, complexes

Physical properties

Melting point/K: 1097
Boiling point/K: 1466
ΔH_{fusion}/kJ mol^{-1}: 9.20
ΔH_{vap}/kJ mol^{-1}: 159

Thermodynamic properties (298.15 K, 0.1 MPa)

State	$\Delta_f H^{\ominus}$/kJ mol^{-1}	$\Delta_f G^{\ominus}$/kJ mol^{-1}	$S^{\ominus}$/J K^{-1} mol^{-1}	C_p/J K^{-1} mol^{-1}
Solid	0	0	59.87	26.74
Gas	152.3	118.4	173.126	20.786

Density/kg m^{-3}: 6965 [293 K]
Thermal conductivity/W m^{-1} K^{-1}: 34.9 [300 K]
Electrical resistivity/Ω m: 29.0×10^{-8} [293 K]
Mass magnetic susceptibility/kg^{-1} m^3: $+1.81 \times 10^{-8}$ (s)
Molar volume/cm^3: 24.84
Coefficient of linear thermal expansion/K^{-1}: 25.0×10^{-6}

Lattice structure (cell dimensions/pm), space group

α-Yb f.c.c. ($a = 548.62$), Fm3m
β-Yb b.c.c. ($a = 444$), Im3m
$T(\alpha \rightarrow \beta) = 1073$ K

X-ray diffraction: mass absorption coefficients (μ/ρ)/cm^2 g^{-1}:
CuK$_\alpha$ 146   MoK$_\alpha$ 84.5

Ytterbium

Atomic number: 70

Thermal neutron capture cross-section/barns: 37 ± 3

Number of isotopes (including nuclear isomers): 16

Isotope mass range: $164 \rightarrow 177$

Nuclear properties

Key isotopes

Nuclide	Atomic mass	Natural abundance (%)	Half life $T_{1/2}$	Decay mode and energy (MeV)	Nuclear spin I	Nuclear magnetic moment μ	Uses
^{168}Yb	167.9339	0.14	stable				
^{169}Yb		0	31.8d	EC(c. 1.2); γ			tracer, medical
^{170}Yb	169.9349	3.06	stable				
^{171}Yb	170.9365	14.4	stable		1/2	+0.4919	NMR
^{172}Yb	171.9366	21.9	stable				
^{173}Yb	172.9383	16.1	stable		5/2	−0.6776	NMR
^{174}Yb	173.9390	31.8	stable				
^{175}Yb		0	101h	β^-(0.467); γ	(7/2)	± 0.3	tracer
^{176}Yb	175.9427	12.7	stable				

NMR (rarely studied)

	^{171}Yb	^{173}Yb
Relative sensitivity (^{1}H = 1.00)	5.46×10^{-3}	1.33×10^{-3}
Absolute sensitivity (^{1}H = 1.00)	7.81×10^{-4}	2.14×10^{-4}
Receptivity (^{13}C = 1.00)	4.05	1.14
Magnetogyric ratio/rad T^{-1} s^{-1}	4.718×10^7	1.310×10^7
Quadrupole moment/m^2	—	0.4–3.9×10^{-28}
Frequency (^{1}H = 100 MHz; 2.3488 T)/MHz	17.613	4.852

Ground state electron configuration: $[Xe]4f^{14}6s^2$

Term symbol: $^1S_0$

Electron affinity $(M \rightarrow M^-)$/kJ mol^{-1}: $\leqslant 50$

Electron shell properties

Ionization energies/kJ mol^{-1}

1. $M \rightarrow M^+$ 603.4	6. $M^{5+} \rightarrow M^{6+}$
2. $M^+ \rightarrow M^{2+}$ 1176	7. $M^{6+} \rightarrow M^{7+}$
3. $M^{2+} \rightarrow M^{3+}$ 2415	8. $M^{7+} \rightarrow M^{8+}$
4. $M^{3+} \rightarrow M^{4+}$ 4220	9. $M^{8+} \rightarrow M^{9+}$
5. $M^{4+} \rightarrow M^{5+}$	10. $M^{9+} \rightarrow M^{10+}$

Principal lines in atomic spectrum

Wavelength/nm	Species	Sensitivity	Application
246.450	I		AA
328.937	II		AE
346.437	I		AA
369.420	II		AE
398.799	I		AA, AE

Abundance: Earth's crust 3.1 p.p.m.; seawater 2×10^{-7} p.p.m.

Biological role: None; non-toxic

Y	Atomic number: 39
	Relative atomic mass ($^{12}C = 12.0000$): **88.90585**

Chemical properties

Soft, silvery-white metal. Stable in air due to formation of oxide film. Burns easily. Reacts with water to form hydrogen, H_2. Used as yttrium phosphors to give red colours in television screens.

Radii/pm: Y^{3+} 106; atomic 181; covalent 162

Electronegativity: 1.22 (Pauling); 1.11 (Allred–Rochow)

Effective nuclear charge: 3.00 (Slater); 6.26 (Clementi); 8.72 (Froese–Fischer)

Standard reduction potentials $E^{\ominus}/V$

	III		0
Acid solution	Y^{3+}	$\underline{-2.37}$	Y
Alkaline solution	$Y(OH)_3$	$\underline{-2.85}$	Y

Oxidation states

Y^{III} ([Kr]) Y_2O_3, $Y(OH)_3$, $[Y(H_2O)_x]^{3+}$ (aq), Y^{3+} salts, $Y_2(CO_3)_3$, YF_3, YCl_3 etc., YOCl, some complexes

YH_2–YH_3 consists of $Y^{3+}H^-$, complex bonding

Physical properties

Melting point/K: 1795

Boiling point/K: 3611

ΔH_{fusion}/kJ mol^{-1}: 17.2

ΔH_{vap}/kJ mol^{-1}: 367.4

Thermodynamic properties (298.15 K, 0.1 MPa)

State	$\Delta_f H^{\ominus}$/kJ mol^{-1}	$\Delta_f G^{\ominus}$/kJ mol^{-1}	$S^{\ominus}$/J K^{-1} mol^{-1}	C_p/J K^{-1} mol^{-1}
Solid	0	0	44.43	26.53
Gas	421.3	381.1	1179.48	25.86

Density/kg m^{-3}: 4469 [293 K]

Thermal conductivity/W m^{-1} K^{-1}: 17.2 [300 K]

Electrical resistivity/Ω m: 57.0×10^{-8} [298 K]

Mass magnetic susceptibility/kg^{-1} m^3: $+2.70 \times 10^{-8}$ (s)

Molar volume/cm^3: 19.89

Coefficient of linear thermal expansion/K^{-1}: 10.6×10^{-6}

Lattice structure (cell dimensions/pm), space group

α-Y h.c.p. (a = 364.74; c = 573.06), P6$_3$/mmc
β-Y b.c.c. (a = 411), Im3m
$T(\alpha \rightarrow \beta) = 1763$ K

X-ray diffraction: mass absorption coefficients (μ/ρ)/cm^2 g^{-1}:
CuK$_\alpha$ 134   MoK$_\alpha$ 100

Yttrium

Atomic number: 39
Thermal neutron capture cross-section/barns: 1.3 ± 0.1
Number of isotopes (including nuclear isomers): 21
Isotope mass range: $82 \rightarrow 96$

Nuclear properties

Key isotopes

Nuclide	Atomic mass	Natural abundance (%)	Half life $T_{1/2}$	Decay mode and energy (MeV)	Nuclear spin I	Nuclear magnetic moment μ	Uses
^{88}Y	87.910	0	106.6d	β^+(3.621); EC, γ			tracer
^{89}Y	88.9054	100	stable		1/2	-0.1373	NMR
^{90}Y		0	64h	β^-(2.27) ; no γ	2	-1.63	tracer

NMR

^{89}Y

Relative sensitivity (^{1}H = 1.00)	1.18×10^{-4}
Absolute sensitivity (^{1}H = 1.00)	1.18×10^{-4}
Receptivity (^{13}C = 1.00)	0.668
Magnetogyric ratio/rad T^{-1} s^{-1}	-1.3108×10^7
Frequency (^{1}H = 100 MHz; 2.3488 T)/MHz	4.899

Reference: $Y(NO_3)_3$ (aq)

Ground state electron configuration: $[Kr]4d^1 5s^2$
Term symbol: $^2D_{3/2}$
Electron affinity $(M \rightarrow M^-)$/kJ mol^{-1}: -39

Electron shell properties

Ionization energies/kJ mol^{-1}

1. $M \rightarrow M^+$	616	6. $M^{5+} \rightarrow M^{6+}$	8 970
2. $M^+ \rightarrow M^{2+}$	1181	7. $M^{6+} \rightarrow M^{7+}$	11 200
3. $M^{2+} \rightarrow M^{3+}$	1980	8. $M^{7+} \rightarrow M^{8+}$	12 400
4. $M^{3+} \rightarrow M^{4+}$	5963	9. $M^{8+} \rightarrow M^{9+}$	14 137
5. $M^{4+} \rightarrow M^{5+}$	7430	10. $M^{9+} \rightarrow M^{10+}$	18 400

Principal lines in atomic spectrum

Wavelength/nm	Species	Sensitivity	Application
371.030	II	V1	AE
407.738	I		AA
410.238	I		AA
464.370	I	U2	AE
467.484	I	U1	AE
668.758*	I		

* Lowest energy transition from ground state to nearest empty orbital

Abundance: Earth's crust 31 p.p.m.; seawater 0.0003 p.p.m.
Biological role: None; probably of low toxicity; suspected carcinogen

Zn

Atomic number: 30

Relative atomic mass ($^{12}C = 12.0000$): 65.39

Chemical properties

Bluish-white metal, brittle when cast. Major ores are zinc blende, ZnS, and calamine (smithsonite) $ZnCO_3$. Tarnishes in air, reacts with acids and alkalis. Used ion galvanizing iron, in alloys, e.g. brass, in batteries, as ZnO in rubber, and as polymer stabilizer.

Radii/pm: Zn^{2+} 83; atomic 133.2; covalent 125

Electronegativity: 1.65 (Pauling); 1.66 (Allred-Rochow)

Effective nuclear charge: 4.35 (Slater); 5.97 (Clementi); 8.28 (Froese–Fischer)

Standard reduction potentials $E^{\ominus}/V$

	II		0
Acid solution	Zn^{2+}	$\underline{\quad -0.7626 \quad}$	Zn
Alkaline solution	$Zn(OH)_4^{2-}$	$\underline{\quad -1.285 \quad}$	Zn
	$Zn(OH)_2$	$\underline{\quad -1.246 \quad}$	Zn

Oxidation states

Zn^I ($d^{10} s^1$)	rare Zn_2^{2+} in $Zn/ZnCl_2$ glass
Zn^{II} (d^{10})	ZnO, ZnS, $Zn(OH)_2$, $[Zn(H_2O)_6]^{2+}$ (aq), $[Zn(OH)_4]^{2-}$ (aq alkali) Zn^{2+} salts, ZnF_2, $ZnCl_2$ etc., many complexes

Physical properties

Melting point/K: 692.73

Boiling point/K: 1180

ΔH_{fusion}/kJ mol^{-1}: 6.67

ΔH_{vap}/kJ mol^{-1}: 114.2

Thermodynamic properties (298.15 K, 0.1 MPa)

State	$\Delta_f H^{\ominus}$/kJ mol^{-1}	$\Delta_f G^{\ominus}$/kJ mol^{-1}	$S^{\ominus}$/J K^{-1} mol^{-1}	C_p/J K^{-1} mol^{-1}
Solid	0	0	41.63	25.40
Gas	130.729	95.145	160.984	20.786

Density/kg m^{-3}: 7133 [293 K]; 6577 [liquid at m.p.]

Thermal conductivity/W m^{-1} K^{-1}: 116 [300 K]

Electrical resistivity/Ω m: 5.916×10^{-8} [293 K]

Mass magnetic susceptibility/kg^{-1} m^3: -2.20×10^{-9} (s)

Molar volume/cm^3: 9.17

Coefficient of linear thermal expansion/K^{-1}: 25.0×10^{-6}

Lattice structure (cell dimensions/pm), space group

h.c.p. ($a = 266.47$; $c = 494.69$), $P6_3/mmc$

X-ray diffraction: mass absorption coefficients (μ/ρ)/cm^2 g^{-1}:
CuK$_\alpha$ 60.3   MoK$_\alpha$ 55.4

Zinc

Atomic number: 30
Thermal neutron capture cross-section/barns: 1.10 ± 0.04
Number of isotopes (including nuclear isomers): 15
Isotope mass range: $60 \rightarrow 72$

Key isotopes

Nuclide	Atomic mass	Natural abundance (%)	Half life $T_{1/2}$	Decay mode and energy (MeV)	Nuclear spin I	Nuclear magnetic moment μ	Uses
^{64}Zn	63.9291	48.6	$>8 \times 10^{15}$y		0		
^{65}Zn	64.926	0	243.6d	$\beta^+(1.353)$; EC; γ	5/2	$+0.7692$	tracer
^{66}Zn	65.9260	27.9	stable		0		
^{67}Zn	66.9271	4.1	stable		5/2	$+0.8752$	NMR
^{68}Zn	67.9249	18.8	stable		0		
^{69m}Zn	69.9253	0	13.9h	IT(0.4389); γ			tracer
^{70}Zn	69.9253	0.6	stable				

NMR

^{67}Zn

Relative sensitivity (^{1}H = 1.00) 2.85×10^{-3}
Absolute sensitivity (^{1}H = 1.00) 1.17×10^{-4}
Receptivity (^{13}C = 1.00) 0.665
Magnetogyric ratio/rad T^{-1} s^{-1} 1.6737×10^7
Quadrupole moment/m^2 0.15×10^{-28}
Frequency (^{1}H = 100 MHz; 2.3488 T)/MHz 6.254
References: $Zn(ClO_4)_2$ (aq)

Ground state electron configuration: $[Ar]3d^{10}4s^2$
Term symbol: $^1S_0$
Electron affinity $(M \rightarrow M^-)$/kJ mol^{-1}: 9

Ionization energies/kJ mol^{-1}

1.	M $\rightarrow$ M$^+$	906.4	6.	M$^{5+} \rightarrow$ M^{6+}	10 400
2.	M$^+ \rightarrow$ M^{2+}	1733.3	7.	M$^{6+} \rightarrow$ M^{7+}	12 900
3.	M$^{2+} \rightarrow$ M^{3+}	3832.6	8.	M$^{7+} \rightarrow$ M^{8+}	16 800
4.	M$^{3+} \rightarrow$ M^{4+}	5730	9.	M$^{8+} \rightarrow$ M^{9+}	19 600
5.	M$^{4+} \rightarrow$ M^{5+}	7970	10.	M$^{9+} \rightarrow$ M^{10+}	23 000

Principal lines in atomic spectrum

Wavelength/nm	Species	Sensitivity	Application
213.856	I	U1	AA, AE
250.200	II	V4	AE
255.795	II	V3	AE
307.590*	I		AA
330.259	I	U3	AE
334.502	I	U2	AE

* Lowest energy transition from ground state to nearest empty orbital

Abundance: Earth's crust 76 p.p.m.; seawater 0.01 p.p.m.
Biological role: Essential trace element; non-toxic except in large excess; carcinogenic

Zr

Atomic number: 40

Relative atomic mass ($^{12}C = 12.0000$): **91.224**

Chemical properties

Hard, lustrous, silvery metal. Major minerals baddeleyite, ZrO_2, and zircon, $ZrSiO_4$. Very corrosion-resistant due to oxide layer, but will burn in air. Unaffected by acids, (except HF), and alkalis. Used in alloys, coloured glazes, and nuclear reactors.

Radii/pm: Zr^{2+} 109; Zr^{4+} 87; atomic 160; covalent 145

Electronegativity: 1.33 (Pauling); 1.22 (Allred-Rochow)

Effective nuclear charge: 3.15 (Slater); 6.45 (Clementi); 9.20 (Froese–Fischer)

Standard reduction potentials $E^{\ominus}$/V

IV		0
Zr^{4+}	$\underline{-1.55}$	Zr

Oxidation states

Zr^0	(d^4)	rare $[Zr(bipyridyl)_3]$
Zr^I	(d^3)	ZrCl?
Zr^{II}	(d^2)	ZrO? $ZrCl_2$
Zr^{III}	(d^1)	$ZrCl_3$, $ZrBr_3$, ZrI_3, Zr^{3+} reduces H_2O
Zr^{IV}	([Kr])	ZrO_2, $Zr(OH)^{3+}$ (aq), ZrF_4, $ZrCl_4$ etc., ZrF_6^{2-}, ZrF_7^{3-}, ZrF_8^{4-}, zirconates, complexes

Physical properties

Melting point/K: 2125

Boiling point/K: 4650

ΔH_{fusion}/kJ mol^{-1}: 23.0

ΔH_{vap}/kJ mol^{-1}: 566.7

Thermodynamic properties (298.15 K, 0.1 MPa)

State	$\Delta_f H^{\ominus}$/kJ mol^{-1}	$\Delta_f G^{\ominus}$/kJ mol^{-1}	$S^{\ominus}$/J K^{-1} mol^{-1}	C_p/J K^{-1} mol^{-1}
Solid	0	0	38.99	25.36
Gas	608.8	566.5	181.36	26.65

Density/kg m^{-3}: 6506 [293 K]; 5800 [liquid at m.p.]

Thermal conductivity/W m^{-1} K^{-1}: 22.7 [300 K]

Electrical resistivity/Ω m: 40.0×10^{-8} [293 K]

Mass magnetic susceptibility/kg^{-1} m^3: $+1.68 \times 10^{-8}$ (s)

Molar volume/cm^3: 14.02

Coefficient of linear thermal expansion/K^{-1}: 5.78×10^{-6}

Lattice structure (cell dimensions/pm), space group

α-Zr h.c.p. ($a = 323.21$; $b = 514.77$), P6$_3$/mmc

β-Zr b.c.c. ($a = 361.6$), Im3m

$T(\alpha \rightarrow \beta) = 1135$ K

High pressure form: ($a = 503.6$; $c = 310.9$), P$\bar{3}$m1

X-ray diffraction: mass absorption coefficients (μ/ρ)/cm^2 g^{-1}:
CuK$_\alpha$ 143   MoK$_\alpha$ 15.9

Discovered by M. H. Klaproth in 1789 at University of Berlin, Germany; isolated in 1824 by J. J. Berzelius at Stockholm, Sweden

[Arabic, *zargun*, = gold colour]

Zirconium

Atomic number: 40
Thermal neutron capture cross-section/barns: 0.182 ± 0.005
Number of isotopes (including nuclear isomers): 20
Isotope mass range: $81 \rightarrow 98$

Nuclear properties

Key isotopes

Nuclide	Atomic mass	Natural abundance (%)	Half life $T_{1/2}$	Decay mode and energy (MeV)	Nuclear spin I	Nuclear magnetic moment μ	Uses
^{90}Zr	89.9043	51.45	stable				
^{91}Zr	90.9053	11.32	stable		5/2	-1.303	NMR
^{92}Zr	91.9046	17.19	stable				
^{94}Zr	93.9061	17.28	stable				
^{95}Zr	94.9077	0	65d	$\beta^-(1.121)$; γ			tracer
^{96}Zr	95.9082	2.76	$>3.6 \times 10^{17}$y				
^{97}Zr		0	17h	$\beta^-(2.67)$; γ			tracer

NMR

^{91}Zr

Relative sensitivity (^{1}H $= 1.00$)	9.48×10^{-3}
Absolute sensitivity (^{1}H $= 1.00$)	1.06×10^{-3}
Receptivity (^{13}C $= 1.00$)	6.04
Magnetogyric ratio/rad T^{-1} s^{-1}	-2.4868×10^7
Quadrupole moment/m^2	-0.21×10^{-28}
Frequency (^{1}H $= 100$ MHz; 2.3488 T)/MHz	9.330

Reference: no agreed standard

Ground state electron configuration: $[\text{Kr}]4d^2 5s^2$
Term symbol: $^3F_2$
Electron affinity $(M \rightarrow M^-)$/kJ mol^{-1}: 43

Electron shell properties

Ionization energies/kJ mol^{-1}

1. $M \rightarrow M^+$ 660	6. $M^{5+} \rightarrow M^{6+}$ (9 500)
2. $M^+ \rightarrow M^{2+}$ 1267	7. $M^{6+} \rightarrow M^{7+}$ (11 200)
3. $M^{2+} \rightarrow M^{3+}$ 2218	8. $M^{7+} \rightarrow M^{8+}$ (13 800)
4. $M^{3+} \rightarrow M^{4+}$ 3313	9. $M^{8+} \rightarrow M^{9+}$ (15 700)
5. $M^{4+} \rightarrow M^{5+}$ 7860	10. $M^{9+} \rightarrow M^{10+}$ (17 500)

Principal lines in atomic spectrum

Wavelength/nm	Species	Sensitivity	Application
303.092	II		AA
333.198	II	V1	AE
351.960	I	U3	AA, AE
354.768	I	U2	AA, AE
360.119	I	U1	AA, AE
676.238*	I		

* Lowest energy transition from ground state to nearest empty orbital

Abundance: Earth's crust 162 p.p.m.; seawater 2.6×10^{-5} p.p.m.
Biological role: None; non-toxic

Tables of properties of the elements in order of element (A tables)
and in ranking order of property (B tables)

Table 1A Abundance of elements in the Earth's crust
p.p.m. (= g per tonne)

1	Hydrogen	1520	53	Iodine	0.46
2	Helium	0.003	54	Xenon	nil
3	Lithium	18	55	Caesium	2.6
4	Beryllium	2	56	Barium	390
5	Boron	9	57	Lanthanum	35
6	Carbon	180	58	Cerium	66
7	Nitrogen	19	59	Praseodymium	9.1
8	Oxygen	455 000	60	Neodymium	40
9	Fluorine	544	61	Promethium	trace
10	Neon	0.04	62	Samarium	7.0
11	Sodium	22 700	63	Europium	2.1
12	Magnesium	27 640	64	Gadolinium	6.1
13	Aluminium	83 000	65	Terbium	1.2
14	Silicon	272 000	66	Dysprosium	4.5
15	Phosphorus	1120	67	Holmium	1.3
16	Sulfur	340	68	Erbium	3.5
17	Chlorine	126	69	Thulium	0.5
18	Argon	0.04	70	Ytterbium	3.1
19	Potassium	18 400	71	Lutetium	0.8
20	Calcium	46 600	72	Hafnium	2.8
21	Scandium	25	73	Tantalum	1.7
22	Titanium	6320	74	Tungsten	1.2
23	Vanadium	136	75	Rhenium	0.0007
24	Chromium	122	76	Osmium	0.005
25	Manganese	1060	77	Iridium	0.001
26	Iron	62 000	78	Platinum	0.01
27	Cobalt	29	79	Gold	0.004
28	Nickel	99	80	Mercury	0.08
29	Copper	68	81	Thallium	0.7
30	Zinc	76	82	Lead	13
31	Gallium	19	83	Bismuth	0.008
32	Germanium	1.5	84	Polonium	trace
33	Arsenic	1.8	85	Astatine	trace
34	Selenium	0.05	86	Radon	1.7×10^{-10}
35	Bromine	2.5	87	Francium	trace
36	Krypton	nil	88	Radium	trace
37	Rubidium	78	89	Actinium	trace
38	Strontium	384	90	Thorium	8.1
39	Yttrium	31	91	Protoactinium	trace
40	Zirconium	162	92	Uranium	2.3
41	Niobium	20	93	Neptunium	trace
42	Molybdenum	1.2	94	Plutonium	trace
43	Technetium	trace	95	Americium	nil
44	Ruthenium	0.0001	96	Curium	nil
45	Rhodium	0.0001	97	Berkelium	nil
46	Palladium	0.015	98	Californium	nil
47	Silver	0.08	99	Einsteinium	nil
48	Cadmium	0.16	100	Fermium	nil
49	Indium	0.24	101	Mendelevium	nil
50	Tin	2.1	102	Nobelium	nil
51	Antimony	0.2	103	Lawrencium	nil
52	Tellurium	0.001			

Table 1B Abundance of elements in the Earth's crust in order of abundance p.p.m. (= g per tonne)

8	Oxygen	455 000	73	Tantalum	1.7
14	Silicon	272 000	32	Germanium	1.5
13	Aluminium	83 000	67	Holmium	1.3
26	Iron	62 000	42	Molybdenum	1.2
20	Calcium	46 600	74	Tungsten	1.2
12	Magnesium	27 640	65	Terbium	1.2
11	Sodium	22 700	71	Lutetium	0.8
19	Potassium	18 400	81	Thallium	0.7
22	Titanium	6320	69	Thulium	0.5
1	Hydrogen	1520	53	Iodine	0.46
15	Phosphorus	1120	49	Indium	0.24
25	Manganese	1060	51	Antimony	0.2
9	Fluorine	544	48	Cadmium	0.16
56	Barium	390	47	Silver	0.08
38	Strontium	384	80	Mercury	0.08
16	Sulfur	340	34	Selenium	0.05
6	Carbon	180	10	Neon	0.04
40	Zirconium	162	18	Argon	0.04
23	Vanadium	136	46	Palladium	0.015
17	Chlorine	126	78	Platinum	0.01
24	Chromium	122	83	Bismuth	0.008
28	Nickel	99	76	Osmium	0.005
37	Rubidium	78	79	Gold	0.004
30	Zinc	76	2	Helium	0.003
29	Copper	68	77	Iridium	0.001
58	Cerium	66	52	Tellurium	0.001
60	Neodymium	40	75	Rhenium	0.0007
57	Lanthanum	35	44	Ruthenium	0.0001
39	Yttrium	31	45	Rhodium	0.0001
27	Cobalt	29	86	Radon	1.7×10^{-10}
21	Scandium	25	43	Technetium	trace
41	Niobium	20	61	Promethium	trace
7	Nitrogen	19	84	Polonium	trace
31	Gallium	19	85	Astatine	trace
3	Lithium	18	87	Francium	trace
82	Lead	13	88	Radium	trace
59	Praseodymium	9.1	89	Actinium	trace
5	Boron	9	91	Protoactinium	trace
90	Thorium	8.1	93	Neptunium	trace
62	Samarium	7.0	94	Plutonium	trace
64	Gadolinium	6.1	36	Krypton	nil
66	Dysprosium	4.5	54	Xenon	nil
68	Erbium	3.5	95	Americium	nil
70	Ytterbium	3.1	96	Curium	nil
72	Hafnium	2.8	97	Berkelium	nil
55	Caesium	2.6	98	Californium	nil
35	Bromine	2.5	99	Einsteinium	nil
92	Uranium	2.3	100	Fermium	nil
50	Tin	2.1	101	Mendelevium	nil
63	Europium	2.1	102	Nobelium	nil
4	Beryllium	2	103	Lawrencium	nil
33	Arsenic	1.8			

Table 2A Boiling points/K

1	Hydrogen	20.28	53	Iodine		457.50
2	Helium	4.216	54	Xenon		166.1
3	Lithium	1620	55	Caesium		951.6
4	Beryllium	3243	56	Barium		1910
5	Boron	3931	57	Lanthanum		3730
6	Carbon	5100 (subl.)	58	Cerium		3699
7	Nitrogen	77.4	59	Praseodymium		3785
8	Oxygen	90.19	60	Neodymium		3341
9	Fluorine	85.01	61	Promethium	*c.*	3000
10	Neon	27.10	62	Samarium		2064
11	Sodium	1156.1	63	Europium		1870
12	Magnesium	1363	64	Gadolinium		3539
13	Aluminium	2740	65	Terbium		3396
14	Silicon	2628	66	Dysprosium		2835
15	Phosphorus (P_4)	553	67	Holmium		2968
16	Sulfur	717.824	68	Erbium		3136
17	Chlorine	238.6	69	Thulium		2220
18	Argon	87.3	70	Ytterbium		1466
19	Potassium	1047	71	Lutetium		3668
20	Calcium	1757	72	Hafnium		5470
21	Scandium	3104	73	Tantalum		5698
22	Titanium	3560	74	Tungsten		5930
23	Vanadium	3650	75	Rhenium		5900
24	Chromium	2945	76	Osmium		5300
25	Manganese	2235	77	Iridium		4403
26	Iron	3023	78	Platinum		4100
27	Cobalt	3143	79	Gold		3080
28	Nickel	3005	80	Mercury		629.73
29	Copper	2840	81	Thallium		1730
30	Zinc	1180	82	Lead		2013
31	Gallium	2676	83	Bismuth		1883
32	Germanium	3103	84	Polonium		1235
33	Arsenic	889 (subl.)	85	Astatine		610 (est.)
34	Selenium	958.1	86	Radon		211.4
35	Bromine	331.9	87	Francium		950
36	Krypton	120.85	88	Radium		1413
37	Rubidium	961	89	Actinium		3470
38	Strontium	1657	90	Thorium		5060
39	Yttrium	3611	91	Protactinium	*c.*	4300
40	Zirconium	4650	92	Uranium		4018
41	Niobium	5015	93	Neptunium		4175
42	Molybdenum	4885	94	Plutonium		3505
43	Technetium	5150	95	Americium		2880
44	Ruthenium	4173	96	Curium		n.a.
45	Rhodium	4000	97	Berkelium		n.a.
46	Palladium	3413	98	Californium		n.a.
47	Silver	2485	99	Einsteinium		n.a.
48	Cadmium	1038	100	Fermium		n.a.
49	Indium	2353	101	Mendelevium		n.a.
50	Tin	2543	102	Nobelium		n.a.
51	Antimony	1908	103	Lawrencium		n.a.
52	Tellurium	1263				

Table 2B Boiling points in order of temperature/K

74	Tungsten	5930		69	Thulium	2220
75	Rhenium	5900		62	Samarium	2064
73	Tantalum	5698		82	Lead	2013
72	Hafnium	5470		56	Barium	1910
76	Osmium	5300		51	Antimony	1908
43	Technetium	5150		83	Bismuth	1883
6	Carbon	5100 (subl.)		63	Europium	1870
90	Thorium	5060		20	Calcium	1757
41	Niobium	5015		81	Thallium	1730
42	Molybdenum	4885		38	Strontium	1657
40	Zirconium	4650		3	Lithium	1620
77	Iridium	4403		70	Ytterbium	1466
91	Protactinium	c. 4300		88	Radium	1413
93	Neptunium	4175		12	Magnesium	1363
44	Ruthenium	4173		52	Tellurium	1263
78	Platinum	4100		84	Polonium	1235
92	Uranium	4018		30	Zinc	1180
45	Rhodium	4000		11	Sodium	1156.1
5	Boron	3931		19	Potassium	1047
59	Praseodymium	3785		48	Cadmium	1038
57	Lanthanum	3730		37	Rubidium	961
58	Cerium	3699		34	Selenium	958.1
71	Lutetium	3668		55	Caesium	951.6
23	Vanadium	3650		87	Francium	950
39	Yttrium	3611		33	Arsenic	889 (subl.)
22	Titanium	3560		16	Sulfur	717.824
64	Gadolinium	3539		80	Mercury	629.73
94	Plutonium	3505		85	Astatine	610 (est)
89	Actinium	3470		15	Phosphorus (P_4)	553
46	Palladium	3413		53	Iodine	457.50
65	Terbium	3396		35	Bromine	331.9
60	Neodymium	3341		17	Chlorine	238.6
4	Beryllium	3243		86	Radon	211.4
27	Cobalt	3143		54	Xenon	166.1
68	Erbium	3136		36	Krypton	120.85
21	Scandium	3104		8	Oxygen	90.19
32	Germanium	3103		18	Argon	87.3
79	Gold	3080		9	Fluorine	85.01
26	Iron	3023		7	Nitrogen	77.4
28	Nickel	3005		10	Neon	27.10
61	Promethium	c. 3000		1	Hydrogen	20.28
67	Holmium	2968		2	Helium	4.216
24	Chromium	2945				
95	Americium	2880		96	Curium	n.a.
29	Copper	2840		97	Berkelium	n.a.
66	Dysprosium	2835		98	Californium	n.a.
13	Aluminium	2740		99	Einsteinium	n.a.
31	Gallium	2676		100	Fermium	n.a.
14	Silicon	2628		101	Mendelevium	n.a.
50	Tin	2543		102	Nobelium	n.a.
47	Silver	2485		103	Lawrencium	n.a.
49	Indium	2353				
25	Manganese	2235				

Table 3A Coefficient of linear thermal expansion/10^{-6} K^{-1}

1	Hydrogen	n.a./gas	53	Iodine	n.a.
2	Helium	n.a./gas	54	Xenon	n.a./gas
3	Lithium	56	55	Caesium	97
4	Beryllium	11.5	56	Barium	18.1
5	Boron	5	57	Lanthanum	4.9
6	Carbon (diam.)	1.19	58	Cerium	8.5
7	Nitrogen	n.a./gas	59	Praseodymium	6.79
8	Oxygen	n.a./gas	60	Neodymium	6.7
9	Fluorine	n.a./gas	61	Promethium	n.a.
10	Neon	n.a./gas	62	Samarium	10.4
11	Sodium	70.6	63	Europium	32
12	Magnesium	26.1	64	Gadolinium	8.6
13	Aluminium	23.03	65	Terbium	7.0
14	Silicon	4.2	66	Dysprosium	10.0
15	Phosphorus	124.5	67	Holmium	9.5
16	Sulfur	74.33	68	Erbium	9.2
17	Chlorine	n.a./gas	69	Thulium	13.3
18	Argon	n.a./gas	70	Ytterbium	25.0
19	Potassium	83	71	Lutetium	8.12
20	Calcium	22	72	Hafnium	5.9
21	Scandium	10.0	73	Tantalum	6.6
22	Titanium	8.35	74	Tungsten	4.59
23	Vanadium	8.3	75	Rhenium	6.63
24	Chromium	6.2	76	Osmium	6.1‡
25	Manganese	22	77	Iridium	6.8
26	Iron	12.3	78	Platinum	9.0
27	Cobalt	13.36	79	Gold	14.16
28	Nickel	13.3	80	Mercury	n.a./liquid
29	Copper	16.5	81	Thallium	28
30	Zinc	25.0	82	Lead	29.1
31	Gallium	31.5*	83	Bismuth	13.4
32	Germanium	5.57	84	Polonium	23.0
33	Arsenic	4.7	85	Astatine	n.a.
34	Selenium	36.9	86	Radon	n.a./gas
35	Bromine	n.a./liquid	87	Francium	n.a.
36	Krypton	n.a./gas	88	Radium	20.2
37	Rubidium	90	89	Actinium	14.9
38	Strontium	23	90	Thorium	12.5
39	Yttrium	10.6	91	Protactinium	7.3
40	Zirconium	5.78	92	Uranium	12.6
41	Niobium	7.06	93	Neptunium	27.5
42	Molybdenum	5.43	94	Plutonium	55
43	Technetium	8.06	95	Americium	n.a.
44	Ruthenium	9.1	96	Curium	n.a.
45	Rhodium	8.40	97	Berkelium	n.a.
46	Palladium	11.2	98	Californium	n.a.
47	Silver	19.2	99	Einsteinium	n.a.
48	Cadmium	29.8	100	Fermium	n.a.
49	Indium	33	101	Mendelevium	n.a.
50	Tin (α)	5.3†	102	Nobelium	n.a.
51	Antimony	8.5	103	Lawrencium	n.a.
52	Tellurium	16.75			

*This value is along the b axis; it is 16.5 along c and 11.5 along a.
†21.2 for β tin.
‡This value is along the b axis; it is 6.8 along c and 4.3 along a.

Table 3B Coefficient of linear thermal expansion in order of expansion/10^{-6} K^{-1}

15	Phosphorus	124.5	22	Titanium	8.35
55	Caesium	97	23	Vanadium	8.3
37	Rubidium	90	71	Lutetium	8.12
19	Potassium	83	43	Technetium	8.06
16	Sulfur	74.33	91	Protactinium	7.3
11	Sodium	70.6	41	Niobium	7.06
3	Lithium	56	65	Terbium	7.0
94	Plutonium	55	77	Iridium	6.8
34	Selenium	36.9	59	Praseodymium	6.79
49	Indium	33	60	Neodymium	6.7
63	Europium	32	75	Rhenium	6.63
31	Gallium	31.5*	73	Tantalum	6.6
48	Cadmium	29.8	24	Chromium	6.2
82	Lead	29.1	76	Osmium	6.1‡
81	Thallium	28	72	Hafnium	5.9
93	Neptunium	27.5	40	Zirconium	5.57
12	Magnesium	26.1	32	Germanium	5.78
30	Zinc	25.0	42	Molybdenum	5.43
70	Ytterbium	25.0	50	Tin (α)	5.3†
13	Aluminium	23.03	5	Boron	5
38	Strontium	23	57	Lanthanum	4.9
84	Polonium	23.0	33	Arsenic	4.7
20	Calcium	22	74	Tungsten	4.59
25	Manganese	22	14	Silicon	4.2
88	Radium	20.2	6	Carbon (diam.)	1.19
47	Silver	19.2			
56	Barium	18.1	1	Hydrogen	n.a./gas
52	Tellurium	16.75	2	Helium	n.a./gas
29	Copper	16.5	7	Nitrogen	n.a./gas
89	Actinium	14.9	8	Oxygen	n.a./gas
79	Gold	14.16	9	Fluorine	n.a./gas
83	Bismuth	13.4	10	Neon	n.a./gas
27	Cobalt	13.36	17	Chlorine	n.a./gas
28	Nickel	13.3	18	Argon	n.a./gas
69	Thulium	13.3	35	Bromine	n.a./liquid
92	Uranium	12.6	36	Krypton	n.a./gas
90	Thorium	12.5	53	Iodine	n.a.
26	Iron	12.3	54	Xenon	n.a./gas
4	Beryllium	11.5	61	Promethium	n.a.
46	Palladium	11.2	80	Mercury	n.a./liquid
39	Yttrium	10.6	85	Astatine	n.a.
62	Samarium	10.4	86	Radon	n.a./gas
21	Scandium	10.0	87	Francium	n.a.
66	Dysprosium	10.0	95	Americium	n.a.
67	Holmium	9.5	96	Curium	n.a.
68	Erbium	9.2	97	Berkelium	n.a.
44	Ruthenium	9.1	98	Californium	n.a.
78	Platinum	9.0	99	Einsteinium	n.a.
64	Gadolinium	8.6	100	Fermium	n.a.
51	Antimony	8.5	101	Mendelevium	n.a.
58	Cerium	8.5	102	Nobelium	n.a.
45	Rhodium	8.40	103	Lawrencium	n.a.

*This value is along the b axis; it is 16.5 along c and 11.5 along a.
†21.2 for β tin.
‡This value is along the b axis; it is 6.8 along c and 4.3 along a.

Table 4A Densities ρ for solids at 298 K, unless otherwise specified/kg m^{-3}

1	Hydrogen [11 K]	76.0	53	Iodine	4930
2	Helium (liq., 4 K)	124.8	54	Xenon	3540
3	Lithium	534	55	Caesium	1873
4	Beryllium	1847.7	56	Barium	3594
5	Boron	2340	57	Lanthanum	6145
6	Carbon (diam.)	3513	58	Cerium (α)	8240
7	Nitrogen [21 K]	1026	59	Praseodymium	6773
8	Oxygen [55 K]	2000	60	Neodymium	7007
9	Fluorine (liq., 85 K)	1516	61	Promethium	7220
10	Neon [24 K]	1444	62	Samarium	7520
11	Sodium	971	63	Europium	5243
12	Magnesium	1738	64	Gadolinium	7900.4
13	Aluminium	2698	65	Terbium	8229
14	Silicon	2329	66	Dysprosium	8550
15	Phosphorus (P_4)	1820	67	Holmium	8795
16	Sulfur (α)	2070	68	Erbium	9066
17	Chlorine [113 K]	2030	69	Thulium	9321
18	Argon [40 K]	1656	70	Ytterbium	6965
19	Potassium	862	71	Lutetium	9840
20	Calcium	1550	72	Hafnium	13 310
21	Scandium	2989	73	Tantalum	16 654
22	Titanium	4540	74	Tungsten	19 300
23	Vanadium	6110	75	Rhenium	21 020
24	Chromium	7190	76	Osmium	22 590
25	Manganese	7440	77	Iridium	22 420
26	Iron	7874	78	Platinum	21 450
27	Cobalt	8900	79	Gold	19 320
28	Nickel	8902	80	Mercury (liq.)	13 546
29	Copper	8960	81	Thallium	11 850
30	Zinc	7133	82	Lead	11 350
31	Gallium	5907	83	Bismuth	9747
32	Germanium	5323	84	Polonium	9320
33	Arsenic (α)	5780	85	Astatine	n.a.
34	Selenium	4790	86	Radon (liq., 211 K)	4400
35	Bromine [123 K]	4050			
36	Krypton [117 K]	2823	87	Francium	n.a.
37	Rubidium	1532	88	Radium *c.*	5000
38	Strontium	2540	89	Actinium	10 060
39	Yttrium	4469	90	Thorium	11 720
40	Zirconium	6506	91	Protactinium	15 370 (est.)
41	Niobium	8570	92	Uranium	18 950
42	Molybdenum	10 220	93	Neptunium	20 250
43	Technetium	11 500	94	Plutonium	19 840
44	Ruthenium	12 370	95	Americium	13 670
45	Rhodium	12 410	96	Curium	13 300
46	Palladium	12 020	97	Berkelium	14 790
47	Silver	10 500	98	Californium	n.a.
48	Cadmium	8650	99	Einsteinium	n.a.
49	Indium	7310	100	Fermium	n.a.
50	Tin (β)	7310	101	Mendelevium	n.a.
51	Antimony	6691	102	Nobelium	n.a.
52	Tellurium	6240	103	Lawrencium	n.a.

Table 4B Densities ρ for solids at 298 K, unless otherwise specified, in order of density/kg m^{-3}

76	Osmium	22 590
77	Iridium	22 420
78	Platinum	21 450
75	Rhenium	21 020
93	Neptunium	20 250
94	Plutonium	19 840
79	Gold	19 320
74	Tungsten	19 300
92	Uranium	18 950
73	Tantalum	16 654
91	Protactinium	15 370 (est)
97	Berkelium	14 790
95	Americium	13 670
80	Mercury (liq.)	13 546
72	Hafnium	13 310
96	Curium	13 300
45	Rhodium	12 410
44	Ruthenium	12 370
46	Palladium	12 020
81	Thallium	11 850
90	Thorium	11 720
43	Technetium	11 500
82	Lead	11 350
47	Silver	10 500
42	Molybdenum	10 220
89	Actinium	10 060
71	Lutetium	9840
83	Bismuth	9747
69	Thulium	9321
84	Polonium	9320
68	Erbium	9066
29	Copper	8960
28	Nickel	8902
27	Cobalt	8900
67	Holmium	8795
48	Cadmium	8650
41	Niobium	8570
66	Dysprosium	8550
58	Cerium (α)	8240
65	Terbium	8229
64	Gadolinium	7900.4
26	Iron	7874
62	Samarium	7520
25	Manganese	7440
49	Indium	7310
50	Tin (β)	7310
61	Promethium	7220
24	Chromium	7190
30	Zinc	7133
60	Neodymium	7007
70	Ytterbium	6965
59	Praseodymium	6773

51	Antimony	6691
40	Zirconium	6506
52	Tellurium	6240
57	Lanthanum	6145
23	Vanadium	6110
31	Gallium	5907
33	Arsenic (α)	5780
32	Germanium	5323
63	Europium	5243
88	Radium	c. 5000
53	Iodine	4930
34	Selenium	4790
22	Titanium	4540
39	Yttrium	4469
86	Radon (liq., 211 K)	4400
35	Bromine [123 K]	4050
56	Barium	3594
54	Xenon	3540
6	Carbon (diam.)	3513
21	Scandium	2989
36	Krypton [117 K]	2823
13	Aluminium	2698
38	Strontium	2540
5	Boron	2340
14	Silicon	2329
16	Sulfur (α)	2070
17	Chlorine [113 K]	2030
8	Oxygen [55 K]	2000
55	Caesium	1873
4	Beryllium	1847.7
15	Phosphorus (P_4)	1820
12	Magnesium	1738
18	Argon [40 K]	1656
20	Calcium	1550
37	Rubidium	1532
9	Fluorine (liq., 85 K)	1516
10	Neon [24 K]	1444
7	Nitrogen [21 K]	1026
11	Sodium	971
19	Potassium	862
3	Lithium	534
2	Helium (liq., 4 K)	124.8
1	Hydrogen [11 K]	76.0

85	Astatine	n.a.
87	Francium	n.a.
98	Californium	n.a.
99	Einsteinium	n.a.
100	Fermium	n.a.
101	Mendelevium	n.a.
102	Nobelium	n.a.
103	Lawrencium	n.a.

Table 5A Discovery

1	Hydrogen	1766	53	Iodine		1811
2	Helium	1895	54	Xenon		1898
3	Lithium	1817	55	Caesium		1860
4	Beryllium	1797	56	Barium		1808
5	Boron	1808	58	Lanthanum		1839
6	Carbon	pre-history	58	Cerium		1803
7	Nitrogen	1772	59	Praseodymium		1885
8	Oxygen	1774	60	Neodymium		1885
9	Fluorine	1886	61	Promethium		1945
10	Neon	1898	62	Samarium		1879
11	Sodium	1807	63	Europium		1901
12	Magnesium	1755	64	Gadolinium		1880
13	Aluminium	1825	65	Terbium		1843
14	Silicon	1824	66	Dysprosium		1886
15	Phosphorus	1669	67	Holmium		1878
16	Sulfur	pre-history	68	Erbium		1842
17	Chlorine	1774	69	Thulium		1879
18	Argon	1894	70	Ytterbium		1878
19	Potassium	1807	71	Lutetium		1907
20	Calcium	1808	72	Hafnium		1923
21	Scandium	1879	73	Tantalum		1802
22	Titanium	1791	74	Tungsten		1783
23	Vanadium	1801	75	Rhenium		1925
24	Chromium	1780	76	Osmium		1803
25	Manganese	1774	77	Iridium		1803
26	Iron	*c.* 2500 BC	78	Platinum	pre	1700
27	Cobalt	1735	79	Gold	*c.*	3000 BC
28	Nickel	1751	80	Mercury	*c.*	1500 BC
29	Copper	*c.* 5000 BC	81	Thallium		1861
30	Zinc	pre 1500*	82	Lead	*c.*	1000 BC
31	Gallium	1875	83	Bismuth	*c.*	1500
32	Germanium	1896	84	Polonium		1898
33	Arsenic	*c.* 1250	85	Astatine		1940
34	Selenium	1817	86	Radon		1900
35	Bromine	1826	87	Francium		1939
36	Krypton	1898	88	Radium		1898
37	Rubidium	1861	89	Actinium		1899
38	Strontium	1808	90	Thorium		1815
39	Yttrium	1794	91	Protactinium		1917
40	Zirconium	1789	92	Uranium		1789
41	Niobium	1801	93	Neptunium		1940
42	Molybdenum	1781	94	Plutonium		1940
43	Technetium	1937	95	Americium		1944
44	Ruthenium	1808	96	Curium		1944
45	Rhodium	1803	97	Berkelium		1949
46	Palladium	1803	98	Californium		1950
47	Silver	*c.* 3000 BC	99	Einsteinium		1952
48	Cadmium	1817	100	Fermium		1952
49	Indium	1863	101	Mendelevium		1955
50	Tin	*c.* 2100 BC	102	Nobelium		1958
51	Antimony	*c.* 1600 BC	103	Lawrencium		1961
52	Tellurium	1783				

*Zinc was known as the copper–zinc alloy, brass, around 20 BC.

Table 5B Discovery in chronological order

6	Carbon	pre-history		48	Cadmium	1817
16	Sulfur	pre-history		14	Silicon	1824
29	Copper	*c.* 5000 BC		13	Aluminium	1825
47	Silver	*c.* 3000 BC		35	Bromine	1826
79	Gold	*c.* 3000 BC		57	Lanthanum	1839
26	Iron	*c.* 2500 BC		68	Erbium	1842
50	Tin	*c.* 2100 BC		65	Terbium	1843
51	Antimony	*c.* 1600 BC		55	Caesium	1860
80	Mercury	*c.* 1500 BC		37	Rubidium	1861
82	Lead	*c.* 1000 BC		81	Thallium	1861
33	Arsenic	*c.* 1250		49	Indium	1863
30	Zinc	pre 1500*		31	Gallium	1875
83	Bismuth	*c.* 1500		67	Holmium	1878
15	Phosphorus	1669		70	Ytterbium	1878
78	Platinum	pre 1700		21	Scandium	1879
				62	Samarium	1879
27	Cobalt	1735		69	Thulium	1879
28	Nickel	1751		64	Gadolinium	1880
12	Magnesium	1755		59	Praseodymium	1885
1	Hydrogen	1766		60	Neodymium	1885
7	Nitrogen	1772		32	Germanium	1886
8	Oxygen	1774		9	Fluorine	1886
17	Chlorine	1774		66	Dysprosium	1886
25	Manganese	1774		18	Argon	1894
24	Chromium	1780		2	Helium	1895
42	Molybdenum	1781		36	Krypton	1898
52	Tellurium	1783		10	Neon	1898
74	Tungsten	1783		54	Xenon	1898
40	Zirconium	1789		84	Polonium	1898
92	Uranium	1789		88	Radium	1898
22	Titanium	1791		89	Actinium	1899
39	Yttrium	1794				
4	Beryllium	1797		86	Radon	1900
				63	Europium	1901
23	Vanadium	1801		71	Lutetium	1907
41	Niobium	1801		91	Protactinium	1917
73	Tantalum	1802		72	Hafnium	1923
45	Rhodium	1803		75	Rhenium	1925
46	Palladium	1803		43	Technetium	1937
76	Osmium	1803		87	Francium	1939
77	Iridium	1803		93	Neptunium	1940
58	Cerium	1803		85	Astatine	1940
19	Potassium	1807		94	Plutonium	1940
11	Sodium	1807		95	Americium	1944
5	Boron	1808		96	Curium	1944
20	Calcium	1808		61	Promethium	1945
38	Strontium	1808		97	Berkelium	1949
44	Ruthenium	1808		98	Californium	1950
56	Barium	1808		99	Einsteinium	1952
53	Iodine	1811		100	Fermium	1952
90	Thorium	1815		101	Mendelevium	1955
3	Lithium	1817		102	Nobelium	1958
34	Selenium	1817		103	Lawrencium	1961

*Zinc was known as the copper–zinc alloy, brass, around 20 BC.

Table 6A Electrical resistivities at 298 K/10^{-8} Ωm

1	Hydrogen	n.a.		53	Iodine	1.3×10^{15}
2	Helium	n.a.		54	Xenon	n.a.
3	Lithium	8.55		55	Caesium	20.0
4	Beryllium	4.0		56	Barium	50
5	Boron	1.8×10^{12}		57	Lanthanum	57
6	Carbon (graph.)	1.375×10^3		58	Cerium	73
7	Nitrogen	n.a.		59	Praseodymium	68
8	Oxygen	n.a.		60	Neodymium	64.0
9	Fluorine	n.a.		61	Promethium	50 (est.)
10	Neon	n.a.		62	Samarium	88.0
11	Sodium	4.2		63	Europium	90.0
12	Magnesium	4.45		64	Gadolinium	134.0
13	Aluminium	2.6548		65	Terbium	114
14	Silicon	1×10^5		66	Dysprosium	57.0
15	Phosphorus (P_4)	1×10^{17}		67	Holmium	87.0
16	Sulfur	2×10^{23}		68	Erbium	87
17	Chlorine	n.a.		69	Thulium	79.0
18	Argon	n.a.		70	Ytterbium	29.0
19	Potassium	6.15		71	Lutetium	79.0
20	Calcium	3.43		72	Hafnium	35.1
21	Scandium	61.0		73	Tantalum	12.45
22	Titanium	42.0		74	Tungsten	5.65
23	Vanadium	24.8		75	Rhenium	19.3
24	Chromium	12.7		76	Osmium	8.12
25	Manganese	185.0		77	Iridium	5.3
26	Iron	9.71		78	Platinum	10.6
27	Cobalt	6.24		79	Gold	2.35
28	Nickel	6.84		80	Mercury	94.1
29	Copper	1.6730		81	Thallium	18.0
30	Zinc	5.916		82	Lead	20.648
31	Gallium	27		83	Bismuth	106.8
32	Germanium	4.6×10^5		84	Polonium	140
33	Arsenic	26		85	Astatine	n.a.
34	Selenium	1×10^6		86	Radon	n.a.
35	Bromine	n.a.		87	Francium	n.a.
36	Krypton	n.a.		88	Radium	100
37	Rubidium	12.5		89	Actinium	n.a.
38	Strontium	23.0		90	Thorium	13.0
39	Yttrium	57.0		91	Protactinium	17.7
40	Zirconium	40.0		92	Uranium	30.8
41	Niobium	12.5		93	Neptunium	122
42	Molybdenum	5.2		94	Plutonium	146
43	Technetium	22.6 (373 K)		95	Americium	68
44	Ruthenium	7.6		96	Curium	n.a.
45	Rhodium	4.51		97	Berkelium	n.a.
46	Palladium	10.8		98	Californium	n.a.
47	Silver	1.59		99	Einsteinium	n.a.
48	Cadmium	6.83		100	Fermium	n.a.
49	Indium	8.37		101	Mendelevium	n.a.
50	Tin (α)	11.0		102	Nobelium	n.a.
51	Antimony	39.0		103	Lawrencium	n.a.
52	Tellurium	4.36×10^5				

Table 6B Electrical resistivities at 298 K in order of resistivity/10^{-8} Ωm

16	Sulfur	2×10^{23}	37	Rubidium	12.5
15	Phosphorus (P_4)	1×10^{17}	41	Niobium	12.5
53	Iodine	1.3×10^{15}	73	Tantalum	12.45
5	Boron	1.8×10^{12}	50	Tin (α)	11.0
34	Selenium	1×10^6	46	Palladium	10.8
32	Germanium	4.6×10^5	78	Platinum	10.6
52	Tellurium	4.36×10^5	26	Iron	9.71
14	Silicon	1×10^5	3	Lithium	8.55
6	Carbon (graph.)	1375	49	Indium	8.37
25	Manganese	185.0	76	Osmium	8.12
94	Plutonium	146	44	Ruthenium	7.6
84	Palonium	140	28	Nickel	6.84
64	Gadolinium	134.0	48	Cadmium	6.83
93	Neptunium	122	27	Cobalt	6.24
65	Terbium	114	19	Potassium	6.15
83	Bismuth	106.8	30	Zinc	5.916
88	Radium	100	74	Tungsten	5.65
80	Mercury	94.1	77	Iridium	5.3
63	Europium	90.0	42	Molybdenum	5.2
62	Samarium	88.0	45	Rhodium	4.51
67	Holmium	87.0	12	Magnesium	4.45
68	Erbium	87	11	Sodium	4.2
69	Thulium	79.0	4	Beryllium	4.0
71	Lutetium	79.0	20	Calcium	3.43
58	Cerium	73	13	Aluminium	2.6548
59	Praseodymium	68	79	Gold	2.35
95	Americium	68	29	Copper	1.6730
60	Neodymium	64.0	47	Silver	1.59
21	Scandium	61.0			
66	Dysprosium	57.0	1	Hydrogen	n.a.
39	Yttrium	57.0	2	Helium	n.a.
57	Lanthanum	57	7	Nitrogen	n.a.
56	Barium	50	8	Oxygen	n.a.
61	Promethium	50 (est.)	9	Fluorine	n.a.
22	Titanium	42.0	10	Neon	n.a.
40	Zirconium	40.0	17	Chlorine	n.a.
51	Antimony	39.0	18	Argon	n.a.
72	Hafnium	35.1	35	Bromine	n.a.
92	Uranium	30.8	36	Krypton	n.a.
70	Ytterbium	29.0	54	Xenon	n.a.
31	Gallium	27	85	Astatine	n.a.
33	Arsenic	26	86	Radon	n.a.
23	Vanadium	24.8	87	Francium	n.a.
38	Strontium	23.0	89	Actinium	n.a.
43	Technetium	22.6 (373 K)	96	Curium	n.a.
82	Lead	20.648	97	Berkelium	n.a.
55	Caesium	20.0	98	Californium	n.a.
75	Rhenium	19.3	99	Einsteinium	n.a.
81	Thallium	18.0	100	Fermium	n.a.
91	Protactinium	17.7	101	Mendelevium	n.a.
90	Thorium	13.0	102	Nobelium	n.a.
24	Chromium	12.7	103	Lawrencium	n.a.

Table 7A Electron affinities $\Delta E(M \rightarrow M^-)^*$/kJ mol^{-1}

1	Hydrogen	72.8		53	Iodine	295.3
2	Helium	−21 (calc.)		54	Xenon	−41 (calc.)
3	Lithium	59.8		55	Caesium	45.5
4	Beryllium	−18		56	Barium	−46
5	Boron	23		57	Lanthanum	53
6	Carbon	122.5		58	Cerium	
7	Nitrogen	−7		59	Praseodymium	
8	Oxygen	141		60	Neodymium	
9	Fluorine	322		61	Promethium	
10	Neon	−29 (calc.)		62	Samarium	
11	Sodium	52.9		63	Europium	
12	Magnesium	−21		64	Gadolinium	
13	Aluminium	44		65	Terbium	≤50
14	Silicon	133.6		66	Dysprosium	
15	Phosphorus	71.7		67	Holmium	
16	Sulfur	200.4		68	Erbium	
17	Chlorine	348.7		69	Thulium	
18	Argon	−35 (calc.)		70	Ytterbium	
19	Potassium	43.8		71	Lutetium	
20	Calcium	−186		72	Hafnium	61
21	Scandium	−70		73	Tantalum	14
22	Titanium	−2		74	Tungsten	119
23	Vanadium	61		75	Rhenium	37
24	Chromium	94		76	Osmium	139
25	Manganese	−94		77	Iridium	190
26	Iron	44		78	Platinum	247
27	Cobalt	102		79	Gold	223
28	Nickel	156		80	Mercury	−18
29	Copper	118.3		81	Thallium	30
30	Zinc	9		82	Lead	35.2
31	Gallium	36 (calc.)		83	Bismuth	101
32	Germanium	116		84	Polonium	186
33	Arsenic	77		85	Astatine	270
34	Selenium	195.0		86	Radon	−41 (calc.)
35	Bromine	324.5		87	Francium	44 (calc.)
36	Krypton	−39 (calc.)		88	Radium	n.a.
37	Rubidium	46.9		89	Actinium	n.a.
38	Strontium	−146		90	Thorium	n.a.
39	Yttrium	−39		91	Protactinium	n.a.
40	Zirconium	43		92	Uranium	n.a.
41	Niobium	109		93	Neptunium	n.a.
42	Molybdenum	114		94	Plutonium	n.a.
43	Technetium	96		95	Americium	n.a.
44	Ruthenium	146		96	Curium	n.a.
45	Rhodium	162		97	Berkelium	n.a.
46	Palladium	98.4		98	Californium	n.a.
47	Silver	125.7		99	Einsteinium	n.a.
48	Cadmium	−26		100	Fermium	n.a.
49	Indium	34		101	Mendelevium	n.a.
50	Tin	121		102	Nobelium	n.a.
51	Antimony	101		103	Lawrencium	n.a.
52	Tellurium	190.2				

*By convention electron affinities are positive for energy released.

Table 7B Electron affinities $\Delta E(M \rightarrow M^-)$ in order of energies*/kJ mol^{-1}

17	Chlorine	348.7		71	Lutetium	$\leqslant 50$
35	Bromine	324.5		37	Rubidium	46.9
9	Fluorine	322		55	Caesium	45.5
53	Iodine	295.3		13	Aluminium	44
85	Astatine	270		26	Iron	44
78	Platinum	247		87	Francium	44 (calc.)
79	Gold	223		19	Potassium	43.8
16	Sulfur	200.4		40	Zirconium	43
34	Selenium	195.0		75	Rhenium	37
52	Tellurium	190.2		31	Gallium	36 (calc.)
77	Iridium	190		82	Lead	35.2
84	Polonium	186		49	Indium	34
45	Rhodium	162		81	Thallium	30
28	Nickel	156		5	Boron	23
44	Ruthenium	146		73	Tantalum	14
8	Oxygen	141		30	Zinc	9
76	Osmium	139		22	Titanium	-2
14	Silicon	133.6		7	Nitrogen	-7
47	Silver	125.7		4	Beryllium	-18
6	Carbon	122.5		80	Mercury	-18
50	Tin	121		2	Helium	-21 (calc.)
74	Tungsten	119		12	Magnesium	-21
29	Copper	118.3		48	Cadmium	-26
32	Germanium	116		10	Neon	-29 (calc.)
42	Molybdenum	114		18	Argon	-35 (calc.)
41	Niobium	109		36	Krypton	-39 (calc.)
27	Cobalt	102		39	Yttrium	-39
51	Antimony	101		54	Xenon	-41 (calc.)
83	Bismuth	101		86	Radon	-41 (calc.)
46	Palladium	98.4		56	Barium	-46
43	Technetium	96		72	Hafnium	-61
24	Chromium	94		21	Scandium	-70
33	Arsenic	77		25	Manganese	-94
1	Hydrogen	72.8		38	Strontium	-146
15	Phosphorus	71.7		20	Calcium	-186
23	Vanadium	61				
3	Lithium	59.8		88	Radium	n.a.
57	Lanthanum	53		89	Actinium	n.a.
11	Sodium	52.9		90	Thorium	n.a.
58	Cerium	⎫		91	Protactinium	n.a.
59	Praseodymium	⎪		92	Uranium	n.a.
60	Neodymium	⎪		93	Neptunium	n.a.
61	Promethium	⎪		94	Plutonium	n.a.
62	Samarium	⎪		95	Americium	n.a.
63	Europium	⎪		96	Curium	n.a.
64	Gadolinium	⎬ ≤ 50		97	Berkelium	n.a.
65	Terbium	⎪		98	Californium	n.a.
66	Dysprosium	⎪		99	Einsteinium	n.a.
67	Holmium	⎪		100	Fermium	n.a.
68	Erbium	⎪		101	Mendelevium	n.a.
69	Thulium	⎪		102	Nobelium	n.a.
70	Ytterbium	⎭		103	Lawrencium	n.a.

*By convention electron affinities are positive for energy released.

Table 8A Electronegativities (Allred–Rochow values)

1	Hydrogen	2.20*	53	Iodine	2.21
2	Helium	n.a.	54	Xenon	2.6*
3	Lithium	0.97	55	Caesium	0.86
4	Beryllium	1.47	56	Barium	0.97
5	Boron	2.01	57	Lanthanum	1.08
6	Carbon	2.50	58	Cerium	1.06
7	Nitrogen	3.07	59	Praseodymium	1.07
8	Oxygen	3.50	60	Neodymium	1.07
9	Fluorine	4.10	61	Promethium	1.07
10	Neon	n.a.	62	Samarium	1.07
11	Sodium	1.08	63	Europium	1.01
12	Magnesium	1.23	64	Gadolinium	1.11
13	Aluminium	1.47	65	Terbium	1.10
14	Silicon	1.74	66	Dysprosium	1.10
15	Phosphorus	2.06	67	Holmium	1.10
16	Sulfur	2.44	68	Erbium	1.11
17	Chlorine	2.83	69	Thulium	1.11
18	Argon	n.a.	70	Ytterbium	1.06
19	Potassium	0.91	71	Lutetium	1.14
20	Calcium	1.04	72	Hafnium	1.23
21	Scandium	1.20	73	Tantalum	1.33
22	Titanium	1.32	74	Tungsten	1.40
23	Vanadium	1.45	75	Rhenium	1.46
24	Chromium	1.56	76	Osmium	1.52
25	Manganese	1.60	77	Iridium	1.55
26	Iron	1.64	78	Platinum	1.44
27	Cobalt	1.70	79	Gold	1.42
28	Nickel	1.75	80	Mercury	1.44
29	Copper	1.75	81	Thallium	1.44
30	Zinc	1.66	82	Lead	1.55
31	Gallium	1.82	83	Bismuth	1.67
32	Germanium	2.02	84	Polonium	1.76
33	Arsenic	2.20	85	Astatine	1.96
34	Selenium	2.48	86	Radon	c. 2.0 (est.)
35	Bromine	2.74	87	Francium	0.86
36	Krypton	2.9*	88	Radium	0.97
37	Rubidium	0.89	89	Actinium	1.00
38	Strontium	0.99	90	Thorium	1.11
39	Yttrium	1.11	91	Protactinium	1.14
40	Zirconium	1.22	92	Uranium	1.22
41	Niobium	1.23	93	Neptunium	1.22
42	Molybdenum	1.30	94	Plutonium	1.22
43	Technetium	1.36	95	Americium	n.a.
44	Ruthenium	1.42	96	Curium	n.a.
45	Rhodium	1.45	97	Berkelium	n.a.
46	Palladium	1.35	98	Californium	n.a.
47	Silver	1.42	99	Einsteinium	n.a.
48	Cadmium	1.46	100	Fermium	n.a.
49	Indium	1.49	101	Mendelevium	n.a.
50	Tin	1.72	102	Nobelium	n.a.
51	Antimony	1.82	103	Lawrencium	n.a.
52	Tellurium	2.01			

*Pauling electronegativity value.

Table 8B Electronegativities (Allred–Rochow values) in ranking order

9	Fluorine	4.10		22	Titanium	1.32
8	Oxygen	3.50		42	Molybdenum	1.30
7	Nitrogen	3.07		12	Magnesium	1.23
36	Krypton	2.9*		41	Niobium	1.23
17	Chlorine	2.83		72	Hafnium	1.23
35	Bromine	2.74		40	Zirconium	1.22
54	Xenon	2.6*		92	Uranium	1.22
6	Carbon	2.50		93	Neptunium	1.22
34	Selenium	2.48		94	Plutonium	1.22
16	Sulfur	2.44		21	Scandium	1.20
53	Iodine	2.21		71	Lutetium	1.14
1	Hydrogen	2.20*		91	Protactinium	1.14
33	Arsenic	2.20		39	Yttrium	1.11
15	Phosphorus	2.06		64	Gadolinium	1.11
32	Germanium	2.02		68	Erbium	1.11
5	Boron	2.01		69	Thulium	1.11
52	Tellurium	2.01		90	Thorium	1.11
86	Radon	c. 2.0 (est.)		65	Terbium	1.10
85	Astatine	1.96		66	Dysprosium	1.10
31	Gallium	1.82		67	Holmium	1.10
51	Antimony	1.82		57	Lanthanum	1.08
84	Polonium	1.76		11	Sodium	1.08
28	Nickel	1.75		59	Praseodymium	1.07
29	Copper	1.75		60	Neodymium	1.07
14	Silicon	1.74		61	Promethium	1.07
50	Tin	1.72		62	Samarium	1.07
27	Cobalt	1.70		58	Cerium	1.06
83	Bismuth	1.67		70	Ytterbium	1.06
30	Zinc	1.66		20	Calcium	1.04
26	Iron	1.64		63	Europium	1.01
25	Manganese	1.60		89	Actinium	1.00
24	Chromium	1.56		38	Strontium	0.99
77	Iridium	1.55		3	Lithium	0.97
82	Lead	1.55		56	Barium	0.97
76	Osmium	1.52		88	Radium	0.97
49	Indium	1.49		19	Potassium	0.91
4	Beryllium	1.47		37	Rubidium	0.89
13	Aluminium	1.47		37	Caesium	0.86
48	Cadmium	1.46		87	Francium	0.86
75	Rhenium	1.46				
45	Rhodium	1.45		2	Helium	n.a.
23	Vanadium	1.45		10	Neon	n.a.
78	Platinum	1.44		18	Argon	n.a.
80	Mercury	1.44		95	Americium	n.a.
81	Thallium	1.44		96	Curium	n.a.
47	Silver	1.42		97	Berkelium	n.a.
44	Ruthenium	1.42		98	Californium	n.a.
79	Gold	1.42		99	Einsteinium	n.a.
74	Tungsten	1.40		100	Fermium	n.a.
43	Technetium	1.36		101	Mendelevium	n.a.
46	Palladium	1.35		102	Nobelium	n.a.
73	Tantalum	1.33		103	Lawrencium	n.a.

*Pauling electronegativity value.

Table 9A Enthalpy of formation of gaseous atoms $\Delta_f H^{\ominus}/\text{kJ mol}^{-1}$ at 298.15 K and 0.1 MPa

1	Hydrogen	218.0	53	Iodine	106.8
2	Helium	0	54	Xenon	0
3	Lithium	159.4	55	Caesium	76.1
4	Beryllium	324.3	56	Barium	180
5	Boron	562.7	57	Lanthanum	431.0
6	Carbon	716.7	58	Cerium	423
7	Nitrogen	472.7	59	Praseodymium	355.6
8	Oxygen	249.2	60	Neodymium	327.6
9	Fluorine	79.0	61	Promethium	n.a.
10	Neon	0	62	Samarium	206.7
11	Sodium	107.3	63	Europium	175.3
12	Magnesium	147.7	64	Gadolinium	397.5
13	Aluminium	326.4	65	Terbium	388.7
14	Silicon	455.6	66	Dysprosium	290.4
15	Phosphorus	314.6	67	Holmium	300.8
16	Sulfur	278.8	68	Erbium	317.1
17	Chlorine	121.7	69	Thulium	232.2
18	Argon	0	70	Ytterbium	152.3
19	Potassium	89.2	71	Lutetium	427.6
20	Calcium	178.2	72	Hafnium	619.2
21	Scandium	377.8	73	Tantalum	782.0
22	Titanium	469.9	74	Tungsten	849.4
23	Vanadium	514.2	75	Rhenium	769.9
24	Chromium	396.6	76	Osmium	791
25	Manganese	280.7	77	Iridium	665.3
26	Iron	416.3	78	Platinum	565.3
27	Cobalt	424.7	79	Gold	366.1
28	Nickel	429.7	80	Mercury	61.3
29	Copper	338.3	81	Thallium	182.2
30	Zinc	130.7	82	Lead	195.0
31	Gallium	277.0	83	Bismuth	207.1
32	Germanium	376.6	84	Polonium	146
33	Arsenic	302.5	85	Astatine	n.a.
34	Selenium	227.1	86	Radon	0
35	Bromine	111.9	87	Francium	72.8
36	Krypton	0	88	Radium	159
37	Rubidium	80.9	89	Actinium	406
38	Strontium	164.4	90	Thorium	598.3
39	Yttrium	421.3	91	Protactinium	607
40	Zirconium	608.8	92	Uranium	535.6
41	Niobium	725.9	93	Neptunium	n.a.
42	Molybdenum	658.1	94	Plutonium	n.a.
43	Technetium	678	95	Americium	n.a.
44	Ruthenium	642.7	96	Curium	n.a.
45	Rhodium	556.9	97	Berkelium	n.a.
46	Palladium	378.2	98	Californium	n.a.
47	Silver	284.6	99	Einsteinium	n.a.
48	Cadmium	112.0	100	Fermium	n.a.
49	Indium	243.3	101	Mendelevium	n.a.
50	Tin	302.1	102	Nobelium	n.a.
51	Antimony	262.3	103	Lawrencium	n.a.
52	Tellurium	196.7			

Table 9B Enthalpy of formation of gaseous atoms of the elements $\Delta_f H^\ominus$ at 298.15 K and 0.1 MPa in order of $\Delta_f H^\ominus / \text{kJ mol}^{-1}$

74	Tungsten	849.4	51	Antimony	262.3
76	Osmium	791	8	Oxygen	249.2
73	Tantalum	782.0	49	Indium	243.3
75	Rhenium	769.9	69	Thulium	232.2
41	Niobium	725.9	34	Selenium	227.1
6	Carbon	716.7	1	Hydrogen	218.0
43	Technetium	678	83	Bismuth	207.1
77	Iridium	665.3	62	Samarium	206.7
42	Molybdenum	658.1	52	Tellurium	196.7
44	Ruthenium	642.7	82	Lead	195.0
72	Hafnium	619.2	81	Thallium	182.2
40	Zirconium	608.8	56	Barium	180
91	Protactinium	607	20	Calcium	178.2
90	Thorium	598.3	63	Europium	175.3
78	Platinum	565.3	38	Strontium	164.4
5	Boron	562.7	3	Lithium	159.4
45	Rhodium	556.9	88	Radium	159
92	Uranium	535.6	70	Ytterbium	152.3
23	Vanadium	514.2	12	Magnesium	147.7
7	Nitrogen	472.7	84	Polonium	146
22	Titanium	469.9	30	Zinc	130.7
14	Silicon	455.6	17	Chlorine	121.7
57	Lanthanum	431.0	48	Cadmium	112.0
28	Nickel	429.7	35	Bromine	111.9
71	Lutetium	427.6	11	Sodium	107.3
27	Cobalt	424.7	53	Iodine	106.8
58	Cerium	423	19	Potassium	89.2
39	Yttrium	421.3	37	Rubidium	80.9
26	Iron	416.3	9	Fluorine	79.0
89	Actinium	406	55	Caesium	76.1
64	Gadolinium	397.5	87	Francium	72.8
24	Chromium	396.6	80	Mercury	61.3
65	Terbium	388.7	2	Helium	0
46	Palladium	378.2	10	Neon	0
21	Scandium	377.8	18	Argon	0
32	Germanium	376.6	36	Krypton	0
79	Gold	366.1	54	Xenon	0
59	Praseodymium	355.6	86	Radon	0
29	Copper	338.3			
60	Neodymium	327.6	61	Promethium	n.a.
13	Aluminium	326.4	85	Astatine	n.a.
4	Beryllium	324.3	93	Neptunium	n.a.
68	Erbium	317.1	94	Plutonium	n.a.
15	Phosphorus	314.6	95	Americium	n.a.
33	Arsenic	302.5	96	Curium	n.a.
50	Tin	302.1	97	Berkelium	n.a.
67	Holmium	300.8	98	Californium	n.a.
66	Dysprosium	290.4	99	Einsteinium	n.a.
47	Silver	284.6	100	Fermium	n.a.
25	Manganese	280.7	101	Mendelevium	n.a.
16	Sulfur	278.8	102	Nobelium	n.a.
31	Gallium	277.0	103	Lawrencium	n.a.

Table 10A Enthalpies of vaporization $\Delta H_{vap}/kJ\ mol^{-1}$

1	Hydrogen	0.46	53	Iodine	41.67
2	Helium	0.082	54	Xenon	12.65
3	Lithium	147.7	55	Caesium	66.5
4	Beryllium	308.8	56	Barium	150.9
5	Boron	504.5	57	Lanthanum	402.1
6	Carbon	710.9	58	Cerium	398
7	Nitrogen	5.58	59	Praseodymium	357
8	Oxygen	6.82	60	Neodymium	328
9	Fluorine	3.26	61	Promethium	n.a.
10	Neon	1.736	62	Samarium	164.8
11	Sodium	99.2	63	Europium	176
12	Magnesium	127.6	64	Gadolinium	301
13	Aluminium	290.8	65	Terbium	391
14	Silicon	383.3	66	Dysprosium	293
15	Phosphorus (P_4)	51.9	67	Holmium	303
16	Sulfur	9.62	68	Erbium	280
17	Chlorine	20.42	69	Thulium	247
18	Argon	6.53	70	Ytterbium	159
19	Potassium	79.1	71	Lutetium	428
20	Calcium	150.6	72	Hafnium	570.7
21	Scandium	376.1	73	Tantalum	758.22
22	Titanium	425.5	74	Tungsten	824.2
23	Vanadium	459.7	75	Rhenium	704.25
24	Chromium	341.8	76	Osmium	738.06
25	Manganese	220.5	77	Iridium	612.1
26	Iron	340.2	78	Platinum	469
27	Cobalt	382.4	79	Gold	343.1
28	Nickel	374.8	80	Mercury	59.11
29	Copper	306.7	81	Thallium	166.1
30	Zinc	114.2	82	Lead	177.8
31	Gallium	270.3	83	Bismuth	179.1
32	Germanium	327.6	84	Polonium	100.8
33	Arsenic	31.9	85	Astatine	n.a.
34	Selenium	90	86	Radon	18.1
35	Bromine	30.5	87	Francium	n.a.
36	Krypton	9.05	88	Radium	136.7
37	Rubidium	75.7	89	Actinium	293
38	Strontium	154.4	90	Thorium	513.7
39	Yttrium	367.4	91	Protactinium	481
40	Zirconium	566.7	92	Uranium	417.1
41	Niobium	680.19	93	Neptunium	336.6
42	Molybdenum	589.9	94	Plutonium	343.5
43	Technetium	585.2	95	Americium	238.5
44	Ruthenium	567	96	Curium	n.a.
45	Rhodium	494.3	97	Berkelium	n.a.
46	Palladium	361.5	98	Californium	n.a.
47	Silver	257.7	99	Einsteinium	n.a.
48	Cadmium	100.0	100	Fermium	n.a.
49	Indium	231.8	101	Mendelevium	n.a.
50	Tin	296.2	102	Nobelium	n.a.
51	Antimony	165.8	103	Lawrencium	n.a.
52	Tellurium	104.6			

Table 10B Enthalpies of vaporization ΔH_{vap} in order of ΔH_{vap}/kJ mol^{-1}

74	Tungsten	824.2	25	Manganese	220.5
73	Tantalum	758.22	83	Bismuth	179.1
76	Osmium	738.06	82	Lead	177.8
6	Carbon	710.9	63	Europium	176
75	Rhenium	704.25	81	Thallium	166.1
41	Niobium	680.19	51	Antimony	165.8
77	Iridium	612.1	62	Samarium	164.8
42	Molybdenum	589.9	70	Ytterbium	159
43	Technetium	585.2	38	Strontium	154.4
72	Hafnium	570.7	56	Barium	150.9
44	Ruthenium	567	20	Calcium	150.6
40	Zirconium	566.7	3	Lithium	147.7
90	Thorium	513.7	88	Radium	136.7
5	Boron	504.5	12	Magnesium	127.6
45	Rhodium	494.3	30	Zinc	114.2
91	Protactinium	481	52	Tellurium	104.6
78	Platinum	469	84	Polonium	100.8
23	Vanadium	459.7	48	Cadmium	100.0
71	Lutetium	428	11	Sodium	99.2
22	Titanium	425.5	34	Selenium	90
92	Uranium	417.1	19	Potassium	79.1
57	Lanthanum	402.1	37	Rubidium	75.7
58	Cerium	398	55	Caesium	66.5
65	Terbium	391	80	Mercury	59.11
14	Silicon	383.3	15	Phosphorus (P_4)	51.9
27	Cobalt	382.4	53	Iodine	41.67
21	Scandium	376.1	33	Arsenic	31.9
28	Nickel	374.8	35	Bromine	30.5
39	Yttrium	367.4	17	Chlorine	20.42
46	Palladium	361.5	86	Radon	18.1
59	Praseodymium	357	54	Xenon	12.65
94	Plutonium	343.5	16	Sulfur	9.62
79	Gold	343.1	36	Krypton	9.05
24	Chromium	341.8	8	Oxygen	6.82
26	Iron	340.2	18	Argon	6.53
93	Neptunium	336.6	7	Nitrogen	5.58
60	Neodymium	328	9	Fluorine	3.26
32	Germanium	327.6	10	Neon	1.736
4	Beryllium	308.8	1	Hydrogen	0.46
29	Copper	306.7	2	Helium	0.082
67	Holmium	303			
64	Gadolinium	301	61	Promethium	n.a.
50	Tin	296.2	85	Astatine	n.a.
66	Dysprosium	293	87	Francium	n.a.
89	Actinium	293	96	Curium	n.a.
13	Aluminium	290.8	97	Berkelium	n.a.
68	Erbium	280	98	Californium	n.a.
31	Gallium	270.3	99	Einsteinium	n.a.
47	Silver	257.7	100	Fermium	n.a.
69	Thulium	247	101	Mendelevium	n.a.
95	Americium	238.5	102	Nobelium	n.a.
49	Indium	231.8	103	Lawrencium	n.a.

Table 11A Ionization energies of neutral atoms $\Delta E(M \rightarrow M^+)$ kJ mol^{-1}

1	Hydrogen	1312.0	53	Iodine	1008.4
2	Helium	2372.3	54	Xenon	1170.4
3	Lithium	513.3	55	Caesium	375.7
4	Beryllium	899.4	56	Barium	502.8
5	Boron	800.6	57	Lanthanum	538.1
6	Carbon	1086.2	58	Cerium	527.4
7	Nitrogen	1402.3	59	Praseodymium	523.1
8	Oxygen	1313.9	60	Neodymium	529.6
9	Fluorine	1681	61	Promethium	535.9
10	Neon	2080.6	62	Samarium	543.3
11	Sodium	495.8	63	Europium	546.7
12	Magnesium	737.7	64	Gadolinium	592.5
13	Aluminium	577.4	65	Terbium	564.6
14	Silicon	786.5	66	Dysprosium	571.9
15	Phosphorus	1011.7	67	Holmium	580.7
16	Sulfur	999.6	68	Erbium	588.7
17	Chlorine	1251.1	69	Thulium	596.7
18	Argon	1520.4	70	Ytterbium	603.4
19	Potassium	418.8	71	Lutetium	523.5
20	Calcium	589.7	72	Hafnium	642
21	Scandium	631	73	Tantalum	761
22	Titanium	658	74	Tungsten	770
23	Vanadium	650	75	Rhenium	760
24	Chromium	652.7	76	Osmium	840
25	Manganese	717.4	77	Iridium	880
26	Iron	759.3	78	Platinum	870
27	Cobalt	760.0	79	Gold	890.1
28	Nickel	736.7	80	Mercury	1007.0
29	Copper	745.4	81	Thallium	589.3
30	Zinc	906.4	82	Lead	715.5
31	Gallium	578.8	83	Bismuth	703.2
32	Germanium	762.1	84	Polonium	812
33	Arsenic	947.0	85	Astatine	930
34	Selenium	940.9	86	Radon	1037
35	Bromine	1139.9	87	Francium	400
36	Krypton	1350.7	88	Radium	509.3
37	Rubidium	403.0	89	Actinium	499
38	Strontium	549.5	90	Thorium	587
39	Yttrium	616	91	Protactinium	568
40	Zirconium	660	92	Uranium	584
41	Niobium	664	93	Neptunium	597
42	Molybdenum	685.0	94	Plutonium	585
43	Technetium	702	95	Americium	578.2
44	Ruthenium	711	96	Curium	581
45	Rhodium	720	97	Berkelium	601
46	Palladium	805	98	Californium	608
47	Silver	731.0	99	Einsteinium	619
48	Cadmium	867.6	100	Fermium	627
49	Indium	558.3	101	Mendelevium	635
50	Tin	708.6	102	Nobelium	642
51	Antimony	833.7	103	Lawrencium	n.a.
52	Tellurium	869.2			

Table 11B Ionization energies of neutral atoms $\Delta E(M{\rightarrow}M^+)$ in order of energy/kJ mol^{-1}

2	Helium	2372.3	40	Zirconium	660
10	Neon	2080.6	22	Titanium	658
9	Fluorine	1681	24	Chromium	652.7
18	Argon	1520.4	23	Vanadium	650
7	Nitrogen	1402.3	72	Hafnium	642
36	Krypton	1350.7	102	Nobelium	642
8	Oxygen	1313.9	101	Mendelevium	635
1	Hydrogen	1312.0	21	Scandium	631
17	Chlorine	1251.1	100	Fermium	630
54	Xenon	1170.4	99	Einsteinium	619
35	Bromine	1139.9	39	Yttrium	616
6	Carbon	1086.2	98	Californium	608
86	Radon	1037	70	Ytterbium	603.4
15	Phosphorus	1011.7	97	Berkelium	601
53	Iodine	1008.4	93	Neptunium	597
80	Mercury	1007.0	69	Thulium	596.7
16	Sulfur	999.6	64	Gadolinium	592.5
33	Arsenic	947.0	20	Calcium	589.7
34	Selenium	940.9	81	Thallium	589.3
85	Astatine	930	68	Erbium	588.7
30	Zinc	906.4	90	Thorium	587
4	Beryllium	899.4	94	Plutonium	585
79	Gold	890.1	92	Uranium	584
77	Iridium	880	96	Curium	581
78	Platinum	870	67	Holmium	580.7
52	Tellurium	869.2	31	Gallium	578.8
48	Cadmium	867.6	95	Americium	578.2
76	Osmium	840	13	Aluminium	577.4
51	Antimony	833.7	66	Dysprosium	571.9
84	Polonium	812	91	Protactinium	568
46	Palladium	805	65	Terbium	564.6
5	Boron	800.6	49	Indium	558.3
14	Silicon	786.5	38	Strontium	549.5
74	Tungsten	770	63	Europium	546.7
32	Germanium	762.1	62	Samarium	543.3
73	Tantalum	761	57	Lanthanum	538.1
27	Cobalt	760.0	61	Promethium	535.9
75	Rhenium	760	60	Neodymium	529.6
26	Iron	759.3	58	Cerium	527.4
29	Copper	745.4	71	Lutetium	523.5
12	Magnesium	737.7	59	Praseodymium	523.1
28	Nickel	736.7	3	Lithium	513.3
47	Silver	731.0	88	Radium	509.3
45	Rhodium	720	56	Barium	502.8
25	Manganese	717.4	89	Actinium	499
82	Lead	715.5	11	Sodium	495.8
44	Ruthenium	711	19	Potassium	418.8
50	Tin	708.6	37	Rubidium	403.0
83	Bismuth	703.2	87	Francium	400
43	Technetium	702	55	Caesium	375.7
42	Molybdenum	685.0			
41	Niobium	664	103	Lawrencium	n.a.

Table 12A Mass magnetic susceptibilities χ at 298 K/10^{-9} kg^{-1} m^3

1	Hydrogen	−25.0	53	Iodine	−4.40
2	Helium	−5.9	54	Xenon	−4.20
3	Lithium	+25.6	55	Caesium	+2.8
4	Beryllium	−13	56	Barium	+1.9
5	Boron	−7.8	57	Lanthanum	+11
6	Carbon (gra.)	−6.3	58	Cerium	+217
7	Nitrogen	−5.4	59	Praseodymium	+447
8	Oxygen	+1355	60	Neodymium	+490.2
9	Fluorine	n.a.	61	Promethium	n.a.
10	Neon	−4.20	62	Samarium	+152
11	Sodium	+8.8	63	Europium	+2810
12	Magnesium	+6.8	64	Gadolinium	+60 300
13	Aluminium	+7.7	65	Terbium	+11 500
14	Silicon	−1.8	66	Dysprosium	+8000
15	Phosphorus (red)	−8.4	67	Holmium	+5490
16	Sulfur (α)	−6.09	68	Erbium	+3330
17	Chlorine	−7.2	69	Thulium	+1900
18	Argon	−6.16	70	Ytterbium	+18.1
19	Potassium	+6.7	71	Lutetium	+1.3
20	Calcium	+14	72	Hafnium	+5.3
21	Scandium	+88	73	Tantalum	+10.7
22	Titanium	+40.1	74	Tungsten	+4.0
23	Vanadium	+62.8	75	Rhenium	+4.56
24	Chromium	+44.5	76	Osmium	+0.65
25	Manganese (α)	+121	77	Iridium	+1.67
26	Iron	ferromagnetic	78	Platinum	+13.01
27	Cobalt	ferromagnetic	79	Gold	−1.78
28	Nickel	ferromagnetic	80	Mercury	−2.095
29	Copper	−1.081	81	Thallium	−3.13
30	Zinc	−2.20	82	Lead	−1.39
31	Gallium	−3.9	83	Bismuth	−16.84
32	Germanium	−1.328	84	Polonium	n.a.
33	Arsenic (α)	−0.917	85	Astatine	n.a.
34	Selenium	−4.0	86	Radon	n.a.
35	Bromine	−4.44	87	Francium	n.a.
36	Krypton	−4.32	88	Radium	n.a.
37	Rubidium	+2.49	89	Actinium	n.a.
38	Strontium	+13.2	90	Thorium	+7.2
39	Yttrium	+27.0	91	Protactinium	n.a.
40	Zirconium	+16.8	92	Uranium	+21.6
41	Niobium	+27.6	93	Neptunium	n.a.
42	Molybdenum	+12	94	Plutonium	+31.7
43	Technetium	+31	95	Americium	+50
44	Ruthenium	+5.37	96	Curium	n.a.
45	Rhodium	+13.6	97	Berkelium	n.a.
46	Palladium	+67.02	98	Californium	n.a.
47	Silver	−2.27	99	Einsteinium	n.a.
48	Cadmium	−2.21	100	Fermium	n.a.
49	Indium	−7.0	101	Mendelevium	n.a.
50	Tin (α)	−4.0	102	Nobelium	n.a.
51	Antimony	−10	103	Lawrencium	n.a.
52	Tellurium	−3.9			

Table 12B Mass magnetic susceptibilities χ in order of $\chi/10^{-9}\ \text{kg}^{-1}\ \text{m}^3$

26	Iron	ferromagnetic	33	Arsenic (α)	-0.917
27	Cobalt	ferromagnetic	29	Copper	-1.081
28	Nickel	ferromagnetic	32	Germanium	-1.328
			82	Lead	-1.39
64	Gadolinium	$+60\,300$	79	Gold	-1.78
65	Terbium	$+11\,500$	14	Silicon	-1.8
66	Dysprosium	$+8000$	80	Mercury	-2.095
67	Holmium	$+5490$	30	Zinc	-2.20
68	Erbium	$+3330$	48	Cadmium	-2.21
63	Europium	$+2810$	47	Silver	-2.27
69	Thulium	$+1900$	81	Thallium	-3.13
8	Oxygen	$+1355$	31	Gallium	-3.9
60	Neodymium	$+490.2$	52	Tellurium	-3.9
59	Praseodymium	$+447$	34	Selenium	-4.0
58	Cerium	$+217$	50	Tin (α)	-4.0
62	Samarium	$+152$	54	Xenon	-4.20
25	Manganese (α)	$+121$	10	Neon	-4.20
21	Scandium	$+88$	36	Krypton	-4.32
46	Palladium	$+67.02$	35	Iodine	-4.40
23	Vanadium	$+62.8$	35	Bromine	-4.44
95	Americium	$+50$	7	Nitrogen	-5.4
24	Chromium	$+44.5$	2	Helium	-5.9
22	Titanium	$+40.1$	16	Sulfur (α)	-6.09
94	Plutonium	$+31.7$	18	Argon	-6.16
43	Technetium	$+31$	6	Carbon (graph.)	-6.3
41	Niobium	$+27.6$	49	Indium	-7.0
39	Yttrium	$+27.0$	17	Chlorine	-7.2
3	Lithium	$+25.6$	5	Boron	-7.8
92	Uranium	$+21.6$	15	Phosphorus (red)	-8.4
70	Ytterbium	$+18.1$	51	Antimony	-10
40	Zirconium	$+16.8$	4	Beryllium	-13
21	Calcium	$+14$	83	Bismuth	-16.84
45	Rhodium	$+13.6$	1	Hydrogen	-25.0
38	Strontium	$+13.2$			
78	Platinum	$+13.01$	9	Fluorine	n.a.
42	Molybdenum	$+12$	61	Promethium	n.a.
57	Lanthanum	$+11$	84	Polonium	n.a.
73	Tantalum	$+10.7$	85	Astatine	n.a.
11	Sodium	$+8.8$	86	Radon	n.a.
13	Aluminium	$+7.7$	87	Francium	n.a.
90	Thorium	$+7.2$	88	Radium	n.a.
12	Magnesium	$+6.8$	89	Actinium	n.a.
19	Potassium	$+6.7$	91	Protactinium	n.a.
44	Ruthenium	$+5.37$	93	Neptunium	n.a.
72	Hafnium	$+5.3$	96	Curium	n.a.
75	Rhenium	$+4.56$	97	Berkelium	n.a.
74	Tungsten	$+4.0$	98	Californium	n.a.
55	Caesium	$+2.8$	99	Einsteinium	n.a.
37	Rubidium	$+2.49$	100	Fermium	n.a.
56	Barium	$+1.9$	101	Mendelevium	n.a.
77	Iridium	$+1.67$	102	Nobelium	n.a.
71	Lutetium	$+1.3$	103	Lawrencium	n.a.
76	Osmium	$+0.65$			

Table 13A Melting points/K

1	Hydrogen	14.01	53	Iodine	386.7
2	Helium	0.95	54	Xenon	161.3
3	Lithium	453.69	55	Caesium	301.6
4	Beryllium	1551	56	Barium	1002
5	Boron	2573	57	Lanthanum	1194
6	Carbon (diam.)	3820	58	Cerium	1072
7	Nitrogen	63.29	59	Praseodymium	1204
8	Oxygen	54.8	60	Neodymium	1294
9	Fluorine	53.53	61	Promethium	1441
10	Neon	24.48	62	Samarium	1350
11	Sodium	370.96	63	Europium	1095
12	Magnesium	922.0	64	Gadolinium	1586
13	Aluminium	933.5	65	Terbium	1629
14	Silicon	1683	66	Dysprosium	1685
15	Phosphorus (P_4)	317.3	67	Holmium	1747
16	Sulfur (α)	386.0	68	Erbium	1802
17	Chlorine	172.2	69	Thulium	1818
18	Argon	83.8	70	Ytterbium	1097
19	Potassium	336.8	71	Lutetium	1936
20	Calcium	1112	72	Hafnium	2503
21	Scandium	1814	73	Tantalum	3269
22	Titanium	1933	74	Tungsten	3680
23	Vanadium	2160	75	Rhenium	3453
24	Chromium	2130	76	Osmium	3327
25	Manganese	1517	77	Iridium	2683
26	Iron	1808	78	Platinum	2045
27	Cobalt	1768	79	Gold	1337.58
28	Nickel	1726	80	Mercury	234.28
29	Copper	1356.6	81	Thallium	576.6
30	Zinc	692.73	82	Lead	600.65
31	Gallium	302.93	83	Bismuth	544.5
32	Germanium	1210.6	84	Polonium	527
33	Arsenic	1090	85	Astatine	575 (est.)
34	Selenium	490	86	Radon	202
35	Bromine	265.9	87	Francium	300
36	Krypton	116.6	88	Radium	973
37	Rubidium	312.2	89	Actinium	1320
38	Strontium	1042	90	Thorium	2023
39	Yttrium	1795	91	Protactinium	2113
40	Zirconium	2125	92	Uranium	1405.5
41	Niobium	2741	93	Neptunium	913
42	Molybdenum	2890	94	Plutonium	914
43	Technetium	2445	95	Americium	1267
44	Ruthenium	2583	96	Curium	n.a.
45	Rhodium	2239	97	Berkelium	n.a.
46	Palladium	1825	98	Californium	n.a.
47	Silver	1235.1	99	Einsteinium	n.a.
48	Cadmium	594.1	100	Fermium	n.a.
49	Indium	429.32	101	Mendelevium	n.a.
50	Tin (β)	505.118	102	Nobelium	n.a.
51	Antimony	903.9	103	Lawrencium	n.a.
52	Tellurium	722.7			

Table 13B Melting points in order of temperature/K

6	Carbon (diam.)	3820	38	Strontium	1042
74	Tungsten	3680	56	Barium	1002
75	Rhenium	3453	88	Radium	973
76	Osmium	3327	13	Aluminium	933.5
73	Tantalum	3269	12	Magnesium	922.0
42	Molybdenum	2890	94	Plutonium	914
41	Niobium	2741	93	Neptunium	913
77	Iridium	2683	51	Antimony	903.9
44	Ruthenium	2583	52	Tellurium	722.7
5	Boron	2573	30	Zinc	692.73
72	Hafnium	2503	82	Lead	600.65
43	Technetium	2445	48	Cadmium	594.1
45	Rhodium	2239	81	Thallium	576.6
23	Vanadium	2160	85	Astatine	575 (est.)
24	Chromium	2130	83	Bismuth	544.5
40	Zirconium	2125	84	Polonium	527
91	Protactinium	2113	50	Tin (β)	505.118
78	Platinum	2045	34	Selenium	490
90	Thorium	2023	3	Lithium	453.69
71	Lutetium	1936	49	Indium	429.32
22	Titanium	1933	53	Iodine	386.7
46	Palladium	1825	16	Sulfur (α)	386.0
69	Thulium	1818	11	Sodium	370.96
21	Scandium	1814	19	Potassium	336.8
26	Iron	1808	15	Phosphorus (P_4)	317.31
68	Erbium	1802	37	Rubidium	312.2
39	Yttrium	1795	31	Gallium	302.93
27	Cobalt	1768	55	Caesium	301.6
67	Holmium	1747	87	Francium	300
28	Nickel	1726	35	Bromine	265.9
66	Dysprosium	1685	80	Mercury	234.28
14	Silicon	1683	86	Radon	202
65	Terbium	1629	17	Chlorine	172.2
64	Gadolinium	1586	54	Xenon	161.3
4	Beryllium	1551	36	Krypton	116.6
25	Manganese	1517	18	Argon	83.8
61	Promethium	1441	7	Nitrogen	63.29
92	Uranium	1405.5	8	Oxygen	54.8
29	Copper	1356.6	9	Fluorine	53.53
62	Samarium	1350	10	Neon	24.48
79	Gold	1337.58	1	Hydrogen	14.01
89	Actinium	1320	2	Helium	0.95
60	Neodymium	1294			
95	Americium	1267	96	Curium	n.a.
47	Silver	1235.1	97	Berkelium	n.a.
32	Germanium	1210.6	98	Californium	n.a.
59	Praseodymium	1204	99	Einsteinium	n.a.
57	Lanthanum	1194	100	Fermium	n.a.
20	Calcium	1112	101	Mendelevium	n.a.
70	Ytterbium	1097	102	Nobelium	n.a.
63	Europium	1095	103	Lawrencium	n.a.
33	Arsenic	1090			
58	Cerium	1072			

Table 14A Neutron (thermal) capture cross-section/barns

1	Hydrogen	0.332	53	Iodine	6.2
2	Helium	c. 0.007	54	Xenon	24.5
3	Lithium	71	55	Caesium	30.0
4	Beryllium	0.0092	56	Barium	1.2
5	Boron	3837	57	Lanthanum	8.9
6	Carbon	0.0034	58	Cerium	0.73
7	Nitrogen	0.075	59	Praseodymium	11.5
8	Oxygen	0.0002	60	Neodymium	49
9	Fluorine	0.098	61	Promethium	8400
10	Neon	0.038	62	Samarium	5820
11	Sodium	0.534	63	Europium	4100
12	Magnesium	0.064	64	Gadolinium	49 000
13	Aluminium	0.232	65	Terbium	30
14	Silicon	0.160	66	Dysprosium	90
15	Phosphorus	0.19	67	Holmium	65
16	Sulfur	0.51	68	Erbium	0.16
17	Chlorine	44	69	Thulium	115
18	Argon	0.65	70	Ytterbium	37
19	Potassium	2.2	71	Lutetium	75
20	Calcium	0.44	72	Hafnium	103
21	Scandium	25	73	Tantalum	22
22	Titanium	6.1	74	Tungsten	18.5
23	Vanadium	5.06	75	Rhenium	85
24	Chromium	3.1	76	Osmium	15.3
25	Manganese	13.3	77	Iridium	425
26	Iron	2.56	78	Platinum	9
27	Cobalt	37.5	79	Gold	98.8
28	Nickel	4.51	80	Mercury	375
29	Copper	3.8	81	Thallium	3.4
30	Zinc	1.10	82	Lead	0.18
31	Gallium	3.1	83	Bismuth	0.034
32	Germanium	2.3	84	Polonium	<0.5
33	Arsenic	4.30	85	Astatine	n.a.
34	Selenium	12.2	86	Radon	0.72
35	Bromine	6.8	87	Francium	n.a.
36	Krypton	24.1	88	Radium	20
37	Rubidium	0.5	89	Actinium	810
38	Strontium	1.21	90	Thorium	7.4
39	Yttrium	1.3	91	Protactinium	200
40	Zirconium	0.182	92	Uranium	7.6
41	Niobium	1.15	93	Neptunium	170
42	Molybdenum	2.65	94	Plutonium	1.8
43	Technetium	22	95	Americium	180
44	Ruthenium	3.0	96	Curium	180
45	Rhodium	150	97	Berkelium	1000
46	Palladium	6.0	98	Californium	2100
47	Silver	63.8	99	Einsteinium	<40
48	Cadmium	2450	100	Fermium	26
49	Indium	194	101	Mendelevium	n.a.
50	Tin	0.63	102	Nobelium	n.a.
51	Antimony	5	103	Lawrencium	n.a.
52	Tellurium	4.7			

64	Gadolinium	49 000	23	Vanadium	5.06
61	Promethium	8400	51	Antimony	5
62	Samarium	5820	52	Tellurium	4.7
63	Europium	4100	28	Nickel	4.51
5	Boron	3837	33	Arsenic	4.30
48	Cadmium	2450	29	Copper	3.8
98	Californium	2100	81	Thallium	3.4
97	Berkelium	1000	24	Chromium	3.1
89	Actinium	810	31	Gallium	3.1
77	Iridium	425	44	Ruthenium	3.0
80	Mercury	375	42	Molybdenum	2.65
91	Protactinium	200	26	Iron	2.56
49	Indium	194	32	Germanium	2.3
95	Americium	180	19	Potassium	2.2
96	Curium	180	94	Plutonium	1.8
93	Neptunium	170	39	Yttrium	1.3
45	Rhodium	150	38	Strontium	1.21
69	Thulium	115	56	Barium	1.2
72	Hafnium	103	41	Niobium	1.15
79	Gold	98.8	30	Zinc	1.10
66	Dysprosium	90	74	Tungsten	1
75	Rhenium	85	58	Cerium	0.73
71	Lutetium	75	86	Radon	0.72
3	Lithium	71	18	Argon	0.65
67	Holmium	65	50	Tin	0.63
47	Silver	63.8	11	Sodium	0.534
60	Neodymium	49	16	Sulfur	0.51
17	Chlorine	44	37	Rubidium	0.5
99	Einsteinium	40	84	Polonium	0.5
27	Cobalt	37.5	20	Calcium	0.44
70	Ytterbium	37	1	Hydrogen	0.332
55	Caesium	30.0	13	Aluminium	0.232
65	Terbium	30	15	Phosphorus	0.19
100	Fermium	26	40	Zirconium	0.182
21	Scandium	25	82	Lead	0.18
54	Xenon	24.5	14	Silicon	0.160
36	Krypton	24.1	68	Erbium	0.16
73	Tantalum	22	9	Fluorine	0.098
43	Technetium	22	7	Nitrogen	0.075
88	Radium	20	12	Magnesium	0.064
76	Osmium	15.3	10	Neon	0.038
25	Manganese	13.3	83	Bismuth	0.034
34	Selenium	12.2	4	Beryllium	0.0092
59	Praseodymium	11.5	2	Helium	c. 0.007
78	Platinum	9	6	Carbon	0.0034
57	Lanthanum	8.9	8	Oxygen	0.0002
92	Uranium	7.6			
90	Thorium	7.4	85	Astatine	n.a.
35	Bromine	6.8	87	Francium	n.a.
53	Iodine	6.2	101	Mendelevium	n.a.
22	Titanium	6.1	102	Nobelium	n.a.
46	Palladium	6.0	103	Lawrencium	n.a.

Table 15A NMR frequencies of nuclei* at a field of 2.344 T
(^{1}H = 100 MHz)/MHz

1	Hydrogen	100.000	53	Iodine-127	20.007
2	Helium-3	76.178	54	Xenon-129	27.660
3	Lithium-7	38.863	55	Caesium-133	13.117
4	Beryllium-9	14.053	56	Barium-137	11.113
5	Boron-11	32.084	57	Lanthanum-139	14.126
6	Carbon-13	25.144	58	Cerium-139	10.862
7	Nitrogen-15	10.133	59	Praseodymium-141	29.291
8	Oxygen-17	13.557	60	Neodymium-143	5.437
9	Fluorine-19	94.077	61	Promethium-147	13.51
10	Neon-21	7.894	62	Samarium-147	4.128
11	Sodium-23	26.451	63	Europium-151	24.801
12	Magnesium-25	6.120	64	Gadolinium-155	3.819
13	Aluminium-27	26.057	65	Terbium-159	22.678
14	Silicon-29	19.865	66	Dysprosium-163	4.583
15	Phosphorus-31	40.481	67	Holmium-165	20.513
16	Sulfur-33	7.670	68	Erbium-167	2.890
17	Chlorine-35	9.798	69	Thulium-169	8.271
18	Argon-39	6.6	70	Ytterbium-171	17.613
19	Potassium-39	4.667	71	Lutetium-175	11.407
20	Calcium-43	6.728	72	Hafnium-177	3.120
21	Scandium-45	24.290	73	Tantalum-181	11.970
22	Titanium-49	5.638	74	Tungsten-183	4.161
23	Vanadium-51	26.289	75	Rhenium-187	22.513
24	Chromium-53	5.652	76	Osmium-187	2.282
25	Manganese-55	24.664	77	Iridium-191	1.718
26	Iron-57	3.231	78	Platinum-195	21.499
27	Cobalt-59	23.614	79	Gold-197	1.712
28	Nickel-61	8.936	80	Mercury-199	17.827
29	Copper-63	26.505	81	Thallium-205	57.708
30	Zinc-67	6.254	82	Lead-207	20.921
31	Gallium-71	30.495	83	Bismuth-209	16.069
32	Germanium-73	3.488	84	Polonium-209	28
33	Arsenic-75	17.126	85	Astatine	n.a.
34	Selenium-77	19.092	86	Radon	n.a.
35	Bromine-81	27.006	87	Francium	n.a.
36	Krypton-83	3.847	88	Radium	n.a.
37	Rubidium-87	32.721	89	Actinium-227	13.1
38	Strontium-87	4.333	90	Thorium-229	1.5
39	Yttrium-89	4.899	91	Protactinium-231	12.0
40	Zirconium-91	9.330	92	Uranium-235	1.790
41	Niobium-93	24.442	93	Neptunium-237	11.25
42	Molybdenum-95	6.514	94	Plutonium-239	3.63
43	Technetium-99	22.508	95	Americium-241	5.76
44	Ruthenium-101	4.941	96	Curium-247	0.75
45	Rhodium-103	3.172	97	Berkelium	n.a.
46	Palladium-105	4.576	98	Californium	n.a.
47	Silver-109	4.652	99	Einsteinium	n.a.
48	Cadmium-113	22.182	100	Fermium	n.a.
49	Indium-115	21.914	101	Mendelevium	n.a.
50	Tin-119	37.272	102	Nobelium	n.a.
51	Antimony-121	23.930	103	Lawrencium	n.a.
52	Tellurium-125	31.596			

*Where there are two or more isotopes that can be observed by NMR spectroscopy, only the one that is commonly used is included in this table.

Table 15B NMR frequencies of nuclei* at a field of 2.344 T (^{1}H = 100 MHz) in order of frequency/MHz

1	Hydrogen-1	100.000	58	Cerium-139	10.862
9	Fluorine-19	94.077	7	Nitrogen-15	10.133
2	Helium-3	76.178	17	Chlorine-35	9.798
81	Thallium-205	57.708	40	Zirconium-91	9.330
15	Phosphorus-31	40.481	28	Nickel-61	8.936
3	Lithium-7	38.863	69	Thulium-169	8.271
50	Tin-119	37.272	10	Neon-21	7.894
37	Rubidium-87	32.721	16	Sulfur-33	7.670
5	Boron-11	32.084	20	Calcium-43	6.728
52	Tellurium-125	31.596	18	Argon-39	6.6
31	Gallium-71	30.495	42	Molybdenum-95	6.514
59	Praseodymium-141	29.291	30	Zinc-67	6.254
84	Polonium-209	28	12	Magnesium-25	6.120
54	Xenon-129	27.660	95	Americium-241	5.76
35	Bromine-81	27.006	24	Chromium-53	5.652
29	Copper-63	26.505	22	Titanium-49	5.638
11	Sodium-23	26.451	60	Neodymium-143	5.437
23	Vanadium-51	26.289	44	Ruthenium-101	4.941
13	Aluminium-27	26.057	39	Yttrium-89	4.899
6	Carbon-13	25.144	19	Potassium-39	4.667
63	Europium-151	24.801	47	Silver-109	4.652
25	Manganese-55	24.664	66	Dysprosium-163	4.583
41	Niobium-93	24.442	46	Palladium-105	4.576
21	Scandium-45	24.290	38	Strontium-87	4.333
51	Antimony-121	23.930	74	Tungsten-183	4.161
27	Cobalt-59	23.614	62	Samarium-147	4.128
65	Terbium-159	22.678	36	Krypton-83	3.847
75	Rhenium-187	22.513	64	Gadolinium-155	3.819
43	Technetium-99	22.508	94	Plutonium-239	3.63
48	Cadmium-113	22.182	32	Germanium-73	3.488
49	Indium-115	21.914	26	Iron-57	3.231
78	Platinum-195	21.499	45	Rhodium-103	3.172
82	Lead-207	20.921	72	Hafnium-177	3.120
67	Holmium-165	20.513	68	Erbium-167	2.890
53	Iodine-127	20.007	76	Osmium-187	2.282
14	Silicon-29	19.865	92	Uranium-235	1.790
34	Selenium-77	19.092	77	Iridium-191	1.718
80	Mercury-199	17.827	79	Gold-197	1.712
70	Ytterbium-171	17.613	90	Thorium-229	1.5
33	Arsenic-75	17.126	96	Curium-247	0.75
83	Bismuth-209	16.069			
57	Lanthanum-139	14.126	85	Astatine	n.a.
4	Beryllium-9	14.053	86	Radon	n.a.
8	Oxygen-17	13.557	87	Francium	n.a.
61	Promethium-147	13.51	88	Radium	n.a.
55	Caesium-133	13.117	97	Berkelium	n.a.
89	Actinium-227	13.1	98	Californium	n.a.
91	Protactinium-231	12.0	99	Einsteinium	n.a.
73	Tantalum-181	11.970	100	Fermium	n.a.
71	Lutetium-175	11.407	101	Mendelevium	n.a.
93	Neptunium-237	11.25	102	Nobelium	n.a.
56	Barium-137	11.113	103	Lawrencium	n.a.

*Where there are two or more isotopes that can be observed by NMR spectroscopy, only the one that is commonly used is included in this table.

Table 16A Thermal conductivities at 300 K/W m^{-1} K^{-1}

1	Hydrogen (g)	0.1815	53	Iodine	0.449
2	Helium (g)	0.152	54	Xenon (g)	0.005 69
3	Lithium	84.7	55	Caesium	35.9
4	Beryllium	200	56	Barium	18.4
5	Boron	27.0	57	Lanthanum	13.5
6	Carbon (graph.)	1960	58	Cerium	11.4
7	Nitrogen (g)	0.025 98	59	Praseodymium	12.5
8	Oxygen (g)	0.026 74	60	Neodymium	16.5
9	Fluorine (g)	0.0279	62	Promethium	17.9 (est.)
10	Neon (g)	0.0493	62	Samarium	13.3
11	Sodium	141	63	Europium	13.9
12	Magnesium	156	64	Gadolinium	10.6
13	Aluminium	237	65	Terbium	11.1
14	Silicon	148	66	Dysprosium	10.7
15	Phosphorus (P_4)	0.235	67	Holmium	16.2
16	Sulfur (α)	0.269	68	Erbium	14.3
17	Chlorine (g)	0.0089	69	Thulium	16.8
18	Argon (g)	0.017 72	70	Ytterbium	34.9
19	Potassium	102.4	71	Lutetium	164
20	Calcium	200	72	Hafnium	23.0
21	Scandium	15.8	73	Tantalum	57.5
22	Titanium	21.9	74	Tungsten	174
23	Vanadium	30.7	75	Rhenium	47.9
24	Chromium	93.7	76	Osmium	87.6
25	Manganese	7.82	77	Iridium	147
26	Iron	80.2	78	Platinum	71.6
27	Cobalt	100	79	Gold	317
28	Nickel	90.7	80	Mercury (liq.)	8.34
29	Copper	401	81	Thallium	46.1
30	Zinc	116	82	Lead	35.3
31	Gallium	40.6	83	Bismuth	7.87
32	Germanium	59.9	84	Polonium	20
33	Arsenic (α)	50.0	85	Astatine	1.7
34	Selenium	2.04	86	Radon (g)	0.003 64 (est.)
35	Bromine (liq.)	0.122	87	Francium	15 (est.)
36	Krypton (g)	0.009 49	88	Radium	18.6 (est.)
37	Rubidium	58.2	89	Actinium	12
38	Strontium	35.3	90	Thorium	54.0
39	Yttrium	17.2	91	Protactinium	47 (est.)
40	Zirconium	22.7	92	Uranium	27.6
41	Niobium	53.7	93	Neptunium	6.3
42	Molybdenum	138	94	Plutonium	6.74
43	Technetium	50.6	95	Americium	10 (est.)
44	Ruthenium	117	96	Curium	10 (est.)
45	Rhodium	150	97	Berkelium	10 (est.)
46	Palladium	71.8	98	Californium	10 (est.)
47	Silver	429	99	Einsteinium	10 (est.)
48	Cadmium	96.8	100	Fermium	10 (est.)
49	Indium	81.6	101	Mendelevium	10 (est.)
50	Tin (α)	66.6	102	Nobelium	10 (est.)
51	Antimony	243	103	Lawrencium	10 (est.)
52	Tellurium	2.35			

Table 16B Thermal conductivities at 300 K in order of conductivity/W m^{-1} K^{-1}

6	Carbon (graph.)	1960	88	Radium	18.6 (est.)
47	Silver	429	56	Barium	18.4
29	Copper	401	61	Promethium	17.9 (est.)
79	Gold	317	39	Yttrium	17.2
51	Antimony	243	69	Thulium	16.8
13	Aluminium	237	60	Neodymium	16.5
4	Beryllium	200	67	Holmium	16.2
20	Calcium	200	21	Scandium	15.8
74	Tungsten	174	87	Francium	15 (est.)
71	Lutetium	164	68	Erbium	14.3
12	Magnesium	156	63	Europium	13.9
45	Rhodium	150	57	Lanthanum	13.5
14	Silicon	148	62	Samarium	13.3
77	Iridium	147	59	Praseodymium	12.5
11	Sodium	141	89	Actinium	12
42	Molybdenum	138	58	Cerium	11.4
44	Ruthenium	117	65	Terbium	11.1
30	Zinc	116	66	Dysprosium	10.7
27	Cobalt	100	64	Gadolinium	10.6
19	Potassium	102.4	95	Americium	10 (est.)
48	Cadmium	96.8	96	Curium	10 (est.)
24	Chromium	93.7	97	Berkelium	10 (est.)
28	Nickel	90.7	98	Californium	10 (est.)
76	Osmium	87.6	99	Einsteinium	10 (est.)
3	Lithium	84.7	100	Fermium	10 (est.)
49	Indium	81.6	101	Mendelevium	10 (est.)
26	Iron	80.2	102	Nobelium	10 (est.)
46	Palladium	71.8	103	Lawrencium	10 (est.)
78	Platinum	71.6	80	Mercury (liq.)	8.34
50	Tin (α)	66.6	83	Bismuth	7.87
32	Germanium	59.9	25	Manganese	7.82
37	Rubidium	58.2	94	Plutonium	6.74
73	Tantalum	57.5	93	Neptunium	6.3
90	Thorium	54.0	52	Tellurium	2.35
41	Niobium	53.7	34	Selenium	2.04
43	Technetium	50.6	85	Astatine	1.7
33	Arsenic (α)	50.0	53	Iodine	0.449
75	Rhenium	47.9	16	Sulfur (α)	0.269
91	Protactinium	47 (est.)	15	Phosphorus (P_4)	0.235
81	Thallium	46.1	1	Hydrogen (g)	0.1815
31	Gallium	40.6	2	Helium (g)	0.152
55	Caesium	35.9	35	Bromine (liq.)	0.122
82	Lead	35.3	10	Neon (g)	0.0493
38	Strontium	35.3	9	Fluorine (g)	0.0279
70	Ytterbium	34.9	8	Oxygen (g)	0.026 74
23	Vanadium	30.7	7	Nitrogen (g)	0.025 98
92	Uranium	27.6	18	Argon (g)	0.017 72
5	Boron	27.0	36	Krypton (g)	0.009 49
72	Hafnium	23.0	17	Chlorine (g)	0.0089
40	Zirconium	22.7	54	Xenon (g)	0.005 69
22	Titanium	21.9	86	Radon (g)	0.003 64 (est.)
84	Polonium	20			

Tables of Elements in alphabetical order by name and by formula, and table of relative atomic masses by atomic number

Table 1 The elements in alphabetical order with formulae and atomic numbers

Actinium	Ac	89	Mercury	Hg	80
Aluminium	Al	13	Molybdenum	Mo	42
Americium	Am	95	Neodymium	Nd	60
Antimony	Sb	51	Neon	Ne	10
Argon	Ar	18	Neptunium	Np	93
Arsenic	As	33	Nickel	Ni	28
Astatine	At	85	Niobium	Nb	41
Barium	Ba	56	Nitrogen	N	7
Berkelium	Bk	97	Nobelium	No	102
Beryllium	Be	4	Osmium	Os	76
Bismuth	Bi	83	Oxygen	O	8
Boron	B	5	Palladium	Pd	46
Bromine	Br	35	Phosphorus	P	15
Cadmium	Cd	48	Platinum	Pt	78
Caesium	Cs	55	Plutonium	Pu	94
Calcium	Ca	20	Polonium	Po	84
Californium	Cf	98	Potassium	K	19
Carbon	C	6	Praseodymium	Pr	59
Cerium	Ce	58	Promethium	Pm	61
Chlorine	Cl	17	Protactinium	Pa	91
Chromium	Cr	24	Radium	Ra	88
Cobalt	Co	27	Radon	Rn	86
Copper	Cu	29	Rhenium	Re	75
Curium	Cm	96	Rhodium	Rh	45
Dysprosium	Dy	66	Rubidium	Rb	37
Einsteinium	Es	99	Ruthenium	Ru	44
Erbium	Er	68	Samarium	Sm	62
Europium	Eu	63	Scandium	Sc	21
Fermium	Fm	100	Selenium	Se	34
Fluorine	F	9	Silicon	Si	14
Francium	Fr	87	Silver	Ag	47
Gadolinium	Gd	64	Sodium	Na	11
Gallium	Ga	31	Strontium	Sr	38
Germanium	Ge	32	Sulfur	S	16
Gold	Au	79	Tantalum	Ta	73
Hafnium	Hf	72	Technetium	Tc	43
Helium	He	2	Tellurium	Te	52
Holmium	Ho	67	Terbium	Tb	65
Hydrogen	H	1	Thallium	Tl	81
Indium	In	49	Thorium	Th	90
Iodine	I	53	Thulium	Tm	69
Iridium	Ir	77	Tin	Sn	50
Iron	Fe	26	Titanium	Ti	22
Krypton	Kr	36	Tungsten	W	74
Lanthanum	La	57	Uranium	U	92
Lawrencium	Lr	103	Vanadium	V	23
Lead	Pb	82	Xenon	Xe	54
Lithium	Li	3	Ytterbium	Yb	70
Lutetium	Lu	71	Yttrium	Y	39
Magnesium	Mg	12	Zinc	Zn	30
Manganese	Mn	25	Zirconium	Zr	40
Mendelevium	Md	101			

Table 2 The elements in alphabetical order of formula with names and atomic numbers

Ac	Actinium	89	Mn	Manganese	25
Ag	Silver	47	Mo	Molybdenum	42
Al	Aluminium	13	N	Nitrogen	7
Am	Americium	95	Na	Sodium	11
Ar	Argon	18	Nb	Niobium	41
As	Arsenic	33	Nd	Neodymium	60
At	Astatine	85	Ne	Neon	10
Au	Gold	79	Ni	Nickel	28
B	Boron	5	No	Nobelium	102
Ba	Barium	56	Np	Neptunium	93
Be	Beryllium	4	O	Oxygen	8
Bi	Bismuth	83	Os	Osmium	76
Bk	Berkelium	97	P	Phosphorus	15
Br	Bromine	35	Pa	Protactinium	91
C	Carbon	6	Pb	Lead	82
Ca	Calcium	20	Pd	Palladium	46
Cd	Cadmium	48	Pm	Promethium	61
Ce	Cerium	58	Po	Polonium	84
Cf	Californium	98	Pr	Praseodymium	59
Cl	Chlorine	17	Pt	Platinum	78
Cm	Curium	96	Pu	Plutonium	94
Co	Cobalt	27	Ra	Radium	88
Cr	Chromium	24	Rb	Rubidium	37
Cs	Caesium	55	Re	Rhenium	75
Cu	Copper	29	Rh	Rhodium	45
Dy	Dysprosium	66	Rn	Radon	86
Er	Erbium	68	Ru	Ruthenium	44
Es	Einsteinium	99	S	Sulfur	16
Eu	Europium	63	Sb	Antimony	51
F	Fluorine	9	Sc	Scandium	21
Fe	Iron	26	Se	Selenium	34
Fm	Fermium	100	Si	Silicon	14
Fr	Francium	87	Sm	Samarium	62
Ga	Gallium	31	Sn	Tin	50
Gd	Gadolinium	64	Sr	Strontium	38
Ge	Germanium	32	Ta	Tantalum	73
H	Hydrogen	1	Tb	Terbium	65
He	Helium	2	Tc	Technetium	43
Hf	Hafnium	72	Te	Tellurium	52
Hg	Mercury	80	Th	Thorium	90
Ho	Holmium	67	Ti	Titanium	22
I	Iodine	53	Tl	Thallium	81
In	Indium	49	Tm	Thulium	69
Ir	Iridium	77	U	Uranium	92
K	Potassium	19	V	Vanadium	23
Kr	Krypton	36	W	Tungsten	74
La	Lanthanum	57	Xe	Xenon	54
Li	Lithium	3	Y	Yttrium	39
Lr	Lawrencium	103	Yb	Ytterbium	70
Lu	Lutetium	71	Zn	Zinc	30
Md	Mendelevium	101	Zr	Zirconium	40
Mg	Magnesium	12			

Table 3 Relative atomic masses (atomic weights) of the elements

1	Hydrogen	1.0079	53	Iodine	126.9045
2	Helium	4.00260	54	Xenon	131.30
3	Lithium	6.941	55	Caesium	132.9054
4	Beryllium	9.01218	56	Barium	137.33
5	Boron	10.81	57	Lanthanum	138.9055
6	Carbon	12.011	58	Cerium	140.12
7	Nitrogen	14.0067	59	Praseodymium	140.9077
8	Oxygen	15.9994	60	Neodymium	144.24
9	Fluorine	18.998403	61	Promethium	(145)
10	Neon	20.179	62	Samarium	150.4
11	Sodium	22.98977	63	Europium	151.96
12	Magnesium	24.305	64	Gadolinium	157.25
13	Aluminium	26.98154	65	Terbium	158.9254
14	Silicon	28.0855	66	Dysprosium	162.50
15	Phosphorus	30.97376	67	Holmium	164.9304
16	Sulfur	32.06	68	Erbium	167.26
17	Chlorine	35.453	69	Thulium	168.9342
18	Argon	39.948	70	Ytterbium	173.04
19	Potassium	39.0983	71	Lutetium	174.967
20	Calcium	40.08	72	Hafnium	178.49
21	Scandium	44.9559	73	Tantalum	180.9479
22	Titanium	47.90	74	Tungsten	183.85
23	Vanadium	50.9415	75	Rhenium	186.207
24	Chromium	51.996	76	Osmium	190.2
25	Manganese	54.9380	77	Iridium	192.22
26	Iron	55.847	78	Platinum	195.09
27	Cobalt	58.9332	79	Gold	196.9665
28	Nickel	58.70	80	Mercury	200.59
29	Copper	63.546	81	Thallium	204.37
30	Zinc	65.38	82	Lead	207.2
31	Gallium	69.735	83	Bismuth	208.9804
32	Germanium	72.59	84	Polonium	(209)
33	Arsenic	74.9216	85	Astatine	(210)
34	Selenium	78.96	86	Radon	(222)
35	Bromine	79.904	87	Francium	(223)
36	Krypton	83.80	88	Radium	226.0254
37	Rubidium	85.4678	89	Actinium	(227)
38	Strontium	87.62	90	Thorium	232.0381
39	Yttrium	88.9059	91	Protactinium	231.0359
40	Zirconium	91.22	92	Uranium	238.029
41	Niobium	92.9064	93	Neptunium	237.0482
42	Molybdenum	95.94	94	Plutonium	(244)
43	Technetium	98.9062	95	Americium	(243)
44	Ruthenium	101.17	96	Curium	(247)
45	Rhodium	102.9055	97	Berkelium	(247)
46	Palladium	106.4	98	Californium	(251)
47	Silver	107.868	99	Einsteinium	(254)
48	Cadmium	112.41	100	Fermium	(257)
49	Indium	114.82	101	Mendelevium	(258)
50	Tin	118.69	102	Nobelium	(259)
51	Antimony	121.75	103	Lawrencium	(260)
52	Tellurium	127.60			

Index

The elements are listed alphabetically in the main part of the book. The following properties are given for each element where known. Some properties are listed in separate tables, pp. 217–49, for all the elements.